Plant Biology

Second Edition

BIOS INSTANT NOTES

Series Editor: B. D. Hames, School of Biochemistry and Molecular Biology, University of Leeds, Leeds, UK

Biology

Animal Biology, Second Edition	9781859963258
Biochemistry, Third Edition	9780415367783
Bioinformatics	9781859962725
Chemistry for Biologists, Second Edition	9781859963555
Developmental Biology	9781859961537
Ecology, Second Edition	9781859962572
Genetics, Third Edition	9780415376198
Human Physiology	9780415355469
Immunology, Second Edition	9781859960394
Mathematics & Statistics for Life Scientists	9781859962923
Medical Microbiology	9781859962541
Microbiology, Third Edition	9780415390880
Molecular Biology, Third Edition	9780415351676
Motor Control and Development	9780415391399
Neuroscience, Second Edition	9780415351881
Plant Biology, Second Edition	9780415356435
Sport & Exercise Biomechanics	9781859962848
Sport & Exercise Physiology	9781859962497

Chemistry

Analytical Chemistry	9781859961896
Inorganic Chemistry, Second Edition	9781859962893
Medicinal Chemistry	9781859962077
Organic Chemistry, Second Edition	9781859962640
Physical Chemistry	9781859961940

Psychology

Sub-series Editor: Hugh Wagner, Dept of Psychology, University of Central Lancashire, Preston, UK

Cognitive Psychology	9781859962237
Physiological Psychology	9781859962039
Psychology	9781859960974
Sport & Exercise Psychology	9781859962947

Plant Biology

Second Edition

Andrew Lack & David Evans

School of Biological & Molecular Sciences,
Oxford Brookes University, Oxford, UK

Taylor & Fran(
Taylor & Francis Group

Published by:

Taylor & Francis Group

In US: 270 Madison Avenue
          New York, NY 10016

In UK: 4 Park Square, Milton Park
          Abingdon, OX14 4RN

First published 2001; Second edition published 2005. Transfered to Digital Printing 2007.

ISBN: 0-4153-5643-1

Library of Congress Cataloging-in-Publication Data

Lack, Andrew (Andrew J.)
    Plant biology / Andrew Lack and David Evans.— 2nd ed.
        p. cm. — (BIOS instant notes)
    Includes index.
    ISBN 0-415-35643-1
1. Botany—Outlines, syllabi, etc. I. Evans, D. E. (David E.) II. Title. III. Series.
    QK45.2.L3 2005
    580—dc22   2005014654

Editor: Elizabeth Owen
Editorial Assistant: Chris Dixon
Production Editor: Karin Henderson
Typeset by: Phoenix Photosetting, Chatham, Kent, UK
Printed by: Biddles Ltd, Guildford, UK

Printed on acid-free paper

10 9 8 7 6 5 4 3 2 1

Taylor & Francis Group
is the Academic Division of T&F Informa plc.

Visit our web site at http://www.garlandscience.com

CONTENTS

Abbreviations ix
Preface to second edition xi

Section A **Introduction** **1**
    A1 Introduction 1

Section B **Understanding plants – methods in plant biology** **5**
    B1 Arabidopsis and other model plants 5
    B2 Methods in experimental plant science 8
    B3 Studying plant evolution and ecology 16

Section C **Plant cells** **21**
    C1 The plant cell 21
    C2 The cell wall 24
    C3 Plastids and mitochondria 27
    C4 Membranes 29
    C5 Nucleus and genome 33
    C6 Cell division 36

Section D **Vegetative anatomy** **41**
    D1 Meristems and primary tissue 41
    D2 Roots 45
    D3 Herbaceous stems and primary growth 50
    D4 Woody stems and secondary growth 53
    D5 Leaves 56

Section E **Plants, water and mineral nutrition** **59**
    E1 Plants and water 59
    E2 Water retention and stomata 65
    E3 Movement of nutrient ions across membranes 68
    E4 Uptake of mineral nutrients by plants 72
    E5 Functions of mineral nutrients 75

Section F **Metabolism** **79**
    F1 Photosynthetic pigments and the nature of light 79
    F2 Major reactions of photosynthesis 82
    F3 C3 and C4 plants and CAM 88
    F4 Respiration and carbohydrate metabolism 92
    F5 Amino acid, lipid, polysaccharide and
        secondary product metabolism 98

Section G **Reproductive biology** **105**
    G1 The flower 105
    G2 Pollen and ovules 109
    G3 Breeding systems 112
    G4 Self incompatibility 116
    G5 Ecology of flowering and pollination 121

Section H   **Seeds and fruits**   **125**
       H1   The seed   125
       H2   Fruits   129
       H3   Fruit and seed dispersal   131
       H4   Seed dormancy   135
       H5   Regeneration and establishment   138

Section I   **Sensing and responding to the environment**   **143**
        I1   Photoperiodism, photomorphogenesis and
              circadian rhythms   143
        I2   Tropisms   149
        I3   Nastic responses   153
        I4   Abscission   156
        I5   Stress avoidance and adaptation   158

Section J   **Growth and development**   **163**
        J1   Features of growth and development   163
        J2   Biochemistry of growth regulation   167
        J3   Molecular action of plant hormones and
              intracellular messengers   176
        J4   Physiology of floral initiation and development   183

Section K   **Plant genetic engineering and biotechnology**   **185**
       K1   Plant breeding   185
       K2   Plant cell and tissue culture   187
       K3   Plant genetic engineering   192

Section L   **Plant ecology**   **199**
        L1   Ecology of different growth forms   199
        L2   Physical factors and plant distribution   202
        L3   Plant communities   206
        L4   Populations   209
        L5   Polymorphisms and population genetics   215
        L6   Contribution to carbon balance and atmosphere   219

Section M   **Interactions between plants and other organisms**   **223**
       M1   Mycorrhiza   223
       M2   Nitrogen fixation   228
       M3   Interactions between plants and animals   231
       M4   Fungal pathogens and endophytes   235
       M5   Bacteria, mycoplasma, viruses and heterokonts   239
       M6   Parasites and saprophytes   241
       M7   Carnivorous plants   245

Section N   **Human uses of plants**   **249**
       N1   Plants as food   249
       N2   Plants for construction   255
       N3   Plants in medicine   259
       N4   Plants for other uses   261
       N5   Bioremediation   265

Section O   **Algae and bryophytes**   **269**
    O1   The algae   269
    O2   The bryophytes   273
    O3   Reproduction in bryophytes   278

Section P   **Spore-bearing vascular plants**   **283**
    P1   Early evolution of vascular plants   283
    P2   Clubmosses and quillworts   288
    P3   Horsetails   292
    P4   Ferns   295

Section Q   **Seed plants**   **303**
    Q1   Early seed plants   303
    Q2   Conifers   307
    Q3   Cycads, ginkgo and Gnetopsida   312
    Q4   Evolution of flowering plants   317
    Q5   General features of plant evolution   325

Further reading   **331**

Index   **341**

ABBREVIATIONS

2,4,5-T	2,4,5-trichlorophenoxyacetic acid	IAA	indole-3-acetic acid
2,4-D	2,4-dichlorophenoxyacetic acid	IP_3	inositol trisphosphate
ABA	abscisic acid	LDP	long-day plant
ABP	auxin-binding protein	LEA	late embryogenesis abundant
ACC	1-aminocyclopropane-1-carboxylic acid	LSD	lysergic acid diethylamine
AGP	Arabidopsis Genome Project	mRNA	messenger ribonucleic acid
AM	arbuscular mycorrhiza	NAA	naphthalene acetic acid
AMP	adenosine monophosphate	NADP	nicotinamide adenine dinucleotide phosphate
AS	asparagines synthase		
ATP	adenosine triphosphate	NADPH	nicotinamide adenine dinucleotide phosphate (reduced)
bp	base pair		
BP	before the present (era)	NE	nuclear envelope
CAM	crassulacean acid metabolism	NPA	1-N-naphthylphthalamic acid
CaMPK	calmodulin-dependent protein kinases	Pa	Pascals
		PCR	polymerase chain reaction
cDNA	complementary deoxyribonucleic acid	PEP	phosphoenol pyruvate
		PG	polygalacturonase
CDPK	calcium-dependent protein kinase	PGS	plant growth substance
		PIP_2	phosphatidyl inositol bisphosphate
CDPK	cyclin-dependent protein kinase		
		PLC	phospholipase C
CoA	coenzyme A	pm	plasma membrane
DAG	diacylglycerol	ppm	parts per million
DDT	1,1-bis(p-chlorophenyl)-2,2,2-trichloroethane	PS-I/II	photosystem I/II
		QTL	quantitative trait locus
DNA	deoxyribonucleic acid	RAPD	random amplified polymorphic DNA
EM	ectomycorrhiza		
ER	endoplasmic reticulum	RET	resonance energy transfer
EST	expressed sequence tag	RFLP	restriction fragment length polymorphism
FADH	flavin adenine dinucleotide (reduced)		
		RNA	ribonucleic acid
Fd	ferredoxin-dependent	RNase	ribonuclease
GA	Golgi apparatus	SAM	s-adenosyl methionine
GARC	gibberellic acid response complex	SDP	short-day plant
		SI	self incompatibility
GARE	gibberellic acid response element	SAUR	soybean auxin up-regulated gene
		T-DNA	transferred DNA
GM	genetically modified	TGN	*trans* Golgi network
GOGAT	glutamate synthase	UDP glucose	uridine diphosphoglucose
GS	glutamine synthase	UV	ultraviolet
GSH	glutathione	VIR	virulence region for infection

PREFACE TO SECOND EDITION

Plant science has always been a fundamental area of biology, but the emphasis in the subject has changed radically in the last two decades with a plethora of new information, much of it deriving from techniques of molecular biology. This has deepened our understanding of plant processes and has illuminated almost all aspects of plant biology. The ability to analyze genomes and to transfer genes has opened possibilities for plant biotechnology and genetic manipulation undreamed of in earlier decades. There have been advances in ecological knowledge that, with increased awareness of the richness of biodiversity, have shed new light on the relationships between plants, other organisms and their interdependence. Plant breeders, ecologists and many people outside plant biology have become acutely conscious of the aesthetic and economic value of the resources, so often dwindling, of the plant kingdom.

In this book we have covered all these aspects of modern plant biology. We have written it keeping in mind an undergraduate faced with a range of advanced courses, needing an affordable text that gives insight into the whole range of plant science. Its scope and depth are suitable for a first and second year undergraduate student of plant biology; specialism will need an advanced text. We have aimed it also at molecular biologists and biotechnologists needing an accessible route to understanding the basis of the systems on which they work. It is intended to provide the fundamental background required for true understanding. It should aid undergraduates in their learning and give insight for specialists into areas of plant science not their own. As in all Instant Notes books we have provided 'Key Notes' at the start of each section. These are intended solely as revision notes, e.g. before an exam, to prompt a reader's memory after reading the section fully. We have kept technical and jargon terms to a minimum needed for understanding and any such term is defined at first mention. We have assumed minimal previous knowledge of biology and hope that the book will prove useful to journalists, environmentalists and those with a genuine interest in the key issues of plant biology as they seek to be informed about the issues that they deal with.

For the second edition of this book we have made a number of structural changes and have updated the text throughout while keeping to the format and style of the first edition. A major change for this edition is adding Section B, Understanding Plants. This is designed to introduce readers to the major techniques in modern plant science and indicate the basis on which advances are being made. As such it is, inevitably, a 'technical' section but we hope one that the reader can refer back to. We feel it is needed to foster a critical understanding of the subject. In the rest of the book we have changed the order of the sections, introducing an outline classification and water relations and metabolism earlier and putting together the topics on flowering and seeds that had been separated in the first edition. This has allowed us to integrate structure, function and physiology more closely. Molecular techniques loomed large in the first edition and loom larger now. We have incorporated these and other advances throughout this new edition as they permeate all of plant science. The result is that some parts have been substantially rewritten and the emphasis has changed in many places. Our guiding principle remained the needs of an undergraduate wanting to understand all the latest advances in plant science, while remaining accessible to any with more than a passing interest in this diverse and rapidly moving subject.

Andrew J. Lack and David E. Evans

ACKNOWLEDGEMENTS

We would like to thank our families for their support throughout the writing, Margaret Evans for assistance with diagrams and our colleagues and associates who, sometimes unknowingly, have given advice, pointed us at relevant work, made comments and generally supported us in the task of authorship. We also wish to thank the editorial team at Taylor and Francis, especially Liz Owen, for their encouragement and persistence and their referees for valuable comments. Perhaps our main debt for the subject matter of this book is to those who introduced us, as students in Aberdeen and Aberystwyth, to the field of plant biology and to the students we have taught over many years at Oxford Brookes. Without their input, we would not have been able to come even close to communicating the fundamentals of plant science simply and understandably.

A1 INTRODUCTION

Key Notes

What is a plant?	Flowering plants are by far the most important plants and this book is primarily a study of them. The most fundamental dividing line between living organisms is that between prokaryote and eukaryote cells. Within the eukaryotes there are three main multicellular kingdoms, plants, animals and fungi and a heterogeneous group, the protists, belonging to several equivalent kingdoms. We include here only plants and some plant-like protists for comparison.
Unifying features of plants	Plants derive from a green algal ancestor. Plants are photosynthetic and autotrophic (with very few exceptions); have chlorophyll *a* and *b*; have a cellulose cell wall and a cell vacuole and have an alternation of diploid and haploid generations. Vegetative structure is similar across most vascular plants; reproductive structures differ.
Classification	There are four divisions of true plants, liverworts, hornworts, mosses and tracheophytes or vascular plants, the last including ferns, lycopsids, horsetails and the seed plants. Algae consist of several unicellular divisions and three main multicellular divisions: brown algae, red algae and green algae.
Life cycles	All plants have an alternation of diploid and haploid generations. Diploid sporophytes produce haploid spores. These germinate to produce gametophytes. Gametes from these fuse to form a diploid cell that can grow into a new sporophyte. In bryophytes the gametophyte is the main plant. In the tracheophytes the sporophyte is dominant, with gametophytes free-living in most ferns, horsetails and some clubmosses, but much reduced in other vascular plants, the female gametophytes being retained on the sporophyte. There is great variation among the algae with different groups showing reduction of either the sporophyte or the gametophyte.
Related topics	The algae (O1) Early evolution of vascular The bryophytes (O2) plants (P1)

What is a plant?　　The science of plant biology is primarily the study of **flowering plants** or **angiosperms**. Flowering plants are by far the most important group of plants in the world, providing the overwhelming majority of plant species, perhaps 400 000 in all, and most of the biomass on land. They are the basis for nearly all our food. This book is mainly about flowering plants.

Historically the science of **plant biology**, or **botany**, has included all living organisms except animals, but it is clear that there is a major division of life between cells with a simple level of organization, the **prokaryotes**, and those with much more complex cells, the **eukaryotes**. The prokaryotes include bacteria, or Eubacteria, and Archaea and will not be considered further in this book except in relation to plants, although some retain plant-like names, such as referring to the gut 'flora' for the bacteria in mammalian guts, and 'blue-green algae' for the **cyanobacteria**. Among eukaryotes three main multicellular kingdoms are recognized: **animals**, **plants** and **fungi**. The remaining eukaryotes are mainly unicellular but with a few multicellular groups such as **slime molds** and large **algae**. They are a heterogeneous group, forming several kingdoms of equivalent status to the three large ones and referred to for convenience, as the **protists**.

There is no clear boundary between protists and plants and authors differ in which organisms they include within the plants. Multicellular **green algae** have many features in common with land plants and are the modern group closest to the ancestors of plants. Along with the **brown algae** and **red algae**, they are the dominant photosynthetic organisms in shallow seas. These three algal groups form quite separate evolutionary lines. Unicellular planktonic groups, again from several different evolutionary origins among the protists, form the basis of the food chain in the deep sea. All these algae are photosynthetic and are considered in this book only for comparison with the true plants in Topic O1. Other protists, animals and fungi will not be considered further except in relation to plants. Plant groups other than flowering plants, such as mosses, ferns and conifers, differ in various ways and these are considered in sections O, P and Q.

Unifying features of plants

Evidence from morphology and from DNA suggests that all plants share an ancestor among the green algae dating from between 450 and 500 million years ago in the Silurian era, perhaps earlier. To characterize the features that define plants as different from other eukaryotes is almost impossible since every feature has exceptions, but usually these exceptions are among plants that have lost the feature or are shared with some algae.

- They are **photosynthetic** and obtain all their nutrients from inorganic sources, i.e. they are **autotrophic** and the start of a food chain. The large algae and many planktonic protists are also photosynthetic. A few plants derive all or part of their nutrients from other organisms (Topics M6, M7) but these are closely related to other, photosynthetic, flowering plants.
- The photosynthetic pigment is **chlorophyll**, and consists of two forms, *a* and *b*, contained within chloroplasts. The green algae and some unicellular protists share these pigments. Other large algae have chlorophyll *a*, some having another chlorophyll, *c*, as well.
- The cells have a **cell wall**, made predominantly of the polysaccharide **cellulose**, and a **vacuole** in addition to the **cytoplasm**.
- There is an **alternation of** diploid and haploid **generations**, another feature shared with some algae. Often one of these is much reduced and may not live independently.

Vegetative structure and physiology are similar throughout the seed plants (flowering plants, conifers and some smaller groups) and there are many similarities with other vascular plants as well, but the reproductive structures differ markedly. Bryophytes differ more fundamentally in vegetative and reproductive structure (Topics O2, O3).

Outline classification

A basic classification of plants and the larger photosynthetic algae is given in *Box 1*. Plants similar to present-day liverworts were probably the earliest land

---

Box 1. *Outline classification of multicellular algae and plants into their main divisions and subdivisions*

Algae
several unicellular groups
Division Phaeophyta Brown algae
Division Rhodophyta Red algae
Division Chlorophyta Green algae
Each of these groups should probably have 'kingdom' status equivalent to 'plants'
Plants
Division Marchantiophyta (Hepatophyta) Liverworts ⎫
Division Anthocerophyta Hornworts   ⎬ the bryophytes
Division Bryophyta Mosses   ⎭
Division Tracheophyta Tracheophytes or vascular plants
    + Rhyniopsida. Early land plants
Subdivision Lycophytina
    Lycopsida. Clubmosses, quillworts and fossil groups
    + Zosterophyllopsida. Early land plants
Subdivision Euphyllophytina
    + Trimerophytopsida. Early land plants
    Equisetopsida (Sphenopsida). Horsetails and fossil groups
    Filicopsida (Polypodiopsida). Ferns
    + Progymnospermopsida. Fossil group
    + some other fossil groups
Subdivision Spermatophytina Seed-bearing vascular plants (usually seen as part
of Euphyllophytina)
    + Lyginopteridopsida (Pteridospermopsida). Seed ferns
    + some other fossil groups of uncertain status
    Cycadopsida. Cycads
    Ginkgoopsida. Maidenhair tree
    Pinopsida (Coniferophyta). Conifers and + Cordaitales
    Gnetopsida. *Gnetum, Ephedra, Welwitschia*
    Angiospermopsida (Magnoliopsida). Flowering plants, divided into:
        Primitive dicotyledons
        Eudicotyledons
        Monocotyledons

Those marked + are known only as fossils

---

plants but separation of the other bryophyte groups, the mosses and hornworts, and the tracheophytes or vascular plants occurred early on and they diversified independently. It is possible that tracheophytes originated from a separate algal ancestor from bryophytes, and mosses separately from liverworts or hornworts, but DNA evidence suggests a single origin. Classifications differ in detail, especially in the endings of the words, e.g. some authors use -ales or -ophyta where -opsida is used here, but there is a broad consensus on the groups recognized and their relationships.

Life cycles   Fundamental to the life cycle of all plants and most multicellular algae is an **alternation** between a **sporophyte** generation and a **gametophyte** generation (*Fig. 1*). The sporophyte is **diploid**. To reproduce, cells of the sporophyte divide by meiosis to produce **haploid spores**. The spores germinate without any fertilization to form a haploid generation, the gametophyte. This produces gametes by mitosis. Male and female gametes fuse to form a diploid cell that

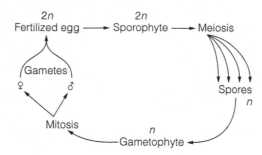

Fig. 1. *Life cycle of plants, showing the alternation between diploid sporophyte and haploid gametophyte generations.*

can germinate and grow into a new sporophyte. Divisions in plant classification are based on whether the sporophyte or gametophyte is the main plant, the degree of reduction of the subsidiary stage and the structures involved with reproduction.

In the three bryophyte groups the haploid gametophyte is the main vegetative plant, the sporophyte being a multicellular stalk and capsule that remains attached to the living gametophyte and dependent on it. In tracheophytes the diploid sporophyte is the main plant. The gametophyte in horsetails, most ferns and some lycopsids is up to 1 cm across, multicellular and free-living, but in a few ferns, some lycopsids and all seed plants it is much reduced and dependent on the sporophyte. The gametophytes can be **hermaphrodite**, producing gametes of both sexes, or **dioecious**, producing either only female or only male gametes, all seed plants having dioecious gametophytes.

In seed plants the male gametophyte is reduced to the three nuclei of the pollen grain and the female gametophyte to an embryo sac, commonly of eight cells, retained on the parent sporophyte (Topic G2).

Among multicellular algae, most green algae (Chlorophyta) have the gametophyte as the main plant, the sporophyte being represented by a single resting spore or zygote, although some have two multicellular generations looking alike. In the brown algae (Phaeophyta) the sporophyte is the dominant generation and the gametophyte much reduced in size. In a few, such as *Fucus* species (the wracks), the gametophyte is reduced to a gamete, so reproduction resembles that of a vertebrate. The red algae (Rhodophyta) are variable.

B1 ARABIDOPSIS AND OTHER MODEL PLANTS

Key Notes

Model plants
Model organisms are used because they are comparatively easy to study and sufficiently similar to other related organisms that findings can be broadly applied. *Arabidopsis thaliana* has recently been an important model plant; others include *Antirrhinum*, maize and rice.

Arabidopsis
Arabidopsis thaliana (arabidopsis) is a member of the cabbage family. It has become a preferred 'model organism' for the study of plants due to: its small genome size, short life cycle, small size, prolific production of small seeds and ease of mutagenesis and transformation.

The Arabidopsis Genome Project
The entire arabidopsis genome has now been sequenced and projects are in progress to sequence several other plants (maize, rice) and to identify every gene within them, together with understanding the regulation and function of the genes.

Rice
Rice has the smallest genome of the cereals, 430 000 kbp, only 3.5 times larger than arabidopsis. Cereal genomes show high levels of synteny so physical maps of the genome derived for rice are therefore useful for other cereals.

Antirrhinum
Antirrhinum majus is an important model of flowering mechanisms and to describe the genes controlling flower development. The large and elaborate flower is easy to dissect and is used in biochemical and proteomic studies of flower shape, organ identity, color and scent not possible with arabidopsis.

Related topics
Methods in experimental plant science (B2)

Model plants

A **model organism** is one studied by many scientists because it has characteristics which make it easy to study and because it is sufficiently similar to other organisms that conclusions from it can be widely applied. The insect *Drosophila melanogaster*, for instance, has been studied by many laboratories worldwide because of its simple genetics and short life cycle. The fact that many laboratories study the same organism has other advantages; knowledge can be shared, findings confirmed and progress is consequently more rapid. Several plants have been intensively studied and become models, the most important recently being *Arabidopsis thaliana* (see below). Corn (maize) and rice have also been studied widely as important crop species, and *Antirrhinum majus* has become a model for flowering. The mosses *Funaria hygrometrica* and *Physcomitrella patens* have also been used as models (as has the brown alga *Fucus serratus*). Subsequent topics illustrate findings from the use of model plants.

Arabidopsis

Arabidopsis thaliana (Fig. 1), now often called **arabidopsis** by plant biologists, is a small and rather insignificant agricultural weed. It is a member of the cabbage family (Brassicaceae). It has a number of key features that make it suitable as a model organism for the study of plants:

- **Small genome** size. 120000 kbp (kilobase pairs; compare with rice – 430000 kbp; haploid maize – 4500000 kbp). The entire arabidopsis genome has been sequenced. Pooling resources and working on one plant makes progress much more rapid.
- **Small size**. Arabidopsis can be grown easily in large numbers in plant growth chambers or cool greenhouses.
- **Short life cycle**. Seed germination to production of new seed in 6–8 weeks.
- **Small seed** – 20 µg each, 20000 per plant. Seed can be easily stored and large numbers of plants can be generated.
- Easy to induce mutations (**mutagenize)** by chemicals or radiation. Many have been recorded with a variety of morphological or physiological effects (Topic B2).
- Easy to carry out techniques of molecular biology (Topics B2 and K3).

Fig. 1.   Arabidopsis thaliana.

The *Arabidopsis* Genome Project

Each protein made by a cell is encoded by a section of the DNA of that cell, termed a gene (Topic C5). Various projects aim to obtain the entire DNA sequence of several model organisms and to identify every gene within that organism. For this information to be useful, it is important that the regulation and function of each of the genes is understood. The major plant **genome sequencing** project is the **Arabidopsis Genome Project (AGP)**, which was completed in December 2000. Details of the results of the project are available from The Arabidopsis Information Resource (TAIR; http://www.arabidopsis.org). Other genome sequencing projects include corn (maize) and rice.

Rice

Rice is a staple crop of global importance, but is also a very valuable model species. Rice has the smallest genome of the cereals, 430 000 kbp, only 3.5 times larger than arabidopsis and much smaller than wheat or maize (corn). It has a diploid chromosome number ($2n$) of 24. Cereal genomes show high levels of **synteny** – that is genes are arranged in the same order on the same chromosomes. Physical maps of the genome (Topic B2) derived for rice are therefore useful for other cereals. The International Rice Genome Project has undertaken sequencing the rice genome and a complete sequence for one of the major cultivars was completed in 2002.

Antirrhinum

Antirrhinum majus (the snapdragon) is a common ornamental species with attractive flowers. It has also been an important model for the understanding of flowering mechanisms since the early 1900s. Early studies used *Antirrhinum* to investigate classical genetics. *Antirrhinum* mutants were collected and used to study floral pigmentation and genetic instability and led to the discovery of transposons—mobile genetic elements (Topic B2). *Antirrhinum* was used to describe the genes controlling flower development (Topics G1, J4) and continues as a model system for investigations of perennial behavior or zygomorphy and to identify and isolate the genes that underlie species differences. The large and elaborate flower is easy to dissect and well suited to biochemical and proteomic studies of flower shape, organ identity, color, and scent not possible with arabidopsis.

B2 METHODS IN EXPERIMENTAL PLANT SCIENCE

Key Notes

Approaches to the study of plant function

Plant function and regulation were first studied using chemical, surgical and genetic approaches. Recently, research has made dramatic progress as a result of the techniques of molecular biology and the study of mutants.

Genetics and cytogenetics

The gene is the basic unit of inheritance. Classic genetics gives understanding of the processes of inheritance of genes and the characters they encode. Cytogenetics is the study of the behavior of the physical structure of genes and chromosomes.

Genetic maps

A genetic map shows the position (locus) of genes on chromosomes. Genetic maps are used as a framework for sequencing and gene cloning.

Gene cloning

Multiple copies of genes are made by cutting DNA with enzymes and inserting the fragment into a vector that is replicated in a bacterium. Genomic and cDNA libraries can be probed to identify specific genes or their product.

Genomics and proteomics

Genomics is the study of the structure and function of entire genomes. In structural genomics, the sequence is obtained and annotated – genes are identified and their location on the genetic map is recorded. In functional genomics, the activity and regulation of genes is investigated. Proteomics is the study of the properties and characteristics of the proteome, the proteins produced at any time in the life of an organism.

Mutants and transformed plants

Studying abnormal genes gives valuable information about gene function. Mutants are produced experimentally by chemical or X-ray treatment. Transposon and T-DNA tagging are forms of insertional mutagenesis, when foreign DNA inserted into the genome disrupts the function of a gene. The mutant is then analyzed to identify and characterize the gene.

Bioinformatics and systems biology

The information from genomics and proteomics is stored in computer databases. The information is then analyzed using a range of programs. Bioinformatics is the methodology for storing, retrieving, analyzing and comparing these data. Systems biology brings together the combined information from genomics, proteomics and bioinformatics to understand organisms and processes.

Related topics

Arabidopsis and other model plants (B1)
Studying plant evolution and ecology (B3)

Plant genetic engineering (K3)

Approaches to the study of plant function

Early studies of plant physiology and development involved growing plants in various environments, applying stimuli or carrying out surgery and observing changes in form and function. These were followed by biochemical analyses and the identification of compounds active in development. Similar methods were used to study photosynthesis, metabolism and transport, together with the use of radioactive tracer compounds. Studies like these, together with classical genetics, generated information on which much of our understanding of the function of plants is built. However, complete understanding requires information both on the cellular processes that occur and the genes that encode the components of those processes. Recently, plant scientists have combined the study of **mutants** with the techniques of **molecular biology** to gain detailed insights into genes and gene products in many plant processes. Model plants (Topic B1) have been central to this work.

Genetics and cytogenetics

The unit of inheritance, the gene, exists at a given position (**locus**) in macromolecular structures known as chromosomes (Topic C5) that replicate during cell division (Topic C6). Genes occur in pairs, **alleles**, which may exist in alternative forms, one form being **dominant** and the other **recessive**. When both types of allele are present, the plant is **heterozygous** and will show the characteristic (the phenotype) of the dominant allele. When both alleles are identical, it is **homozygous** for that allele. In some instances a condition of **incomplete dominance** exists, where the phenotype is intermediate between dominant and recessive. The heterozygote often has some effect and heterozygotes frequently have an advantage over both homozygotes.

The genetic basis of a characteristic (trait) is established experimentally by selectively pollinating plants. In a **monohybrid cross**, the offspring (F_1 generation) of a cross between homozygous parents of different characteristics is self-fertilized. The progeny in the second generation (F_2) are analyzed to determine which allele is dominant. In a **dihybrid cross**, the behavior of two genes is studied. **Back crossing**, in which a hybrid is crossed with one of the parents, is used to confirm the information gained. **Cross breeding** is also used in plant breeding (Topic K1) to generate plants with desirable characteristics.

Cytogenetics is the study of the behavior and structure of chromosomes. It includes mitosis and meiosis (Topic C6) and rarer events where chromosomes break and reorganize or change in number. **Inversion** (where a fragment is reinserted in the opposite orientation), **translocation** (where a fragment is inserted in a different place), **deletion** (where part of the chromosome is lost), and **duplication** (where copies of sections of DNA occur), may all have a large effect on the plant. Errors in pairing and separation of chromosomes in meiosis may result in **aneuploidy** – where a gamete is produced with several copies of a chromosome, or a missing chromosome. In **polyploidy**, meiosis fails to halve the chromosome number and progeny have multiple sets of chromosomes.

Genetic maps

Genes on separate chromosomes behave independently in genetic experiments, but ones on the same chromosome are linked in their behavior. This is known as **genetic linkage**. A **genetic map** showing the position of genes on a chromosome is produced by estimating the distance between genes, measured in **centimorgans**. Genetic maps are very important in genome sequencing. **Positional cloning** relates the location of a gene to identifiable DNA sequences in the genome and is used in sequencing newly identified genes. Genetic maps are used by plant breeders. Important characters, like disease resistance and yield (**quantitative traits**) are often controlled by several genes and the loci controlling these traits are known as **quantitative trait loci (QTLs).** QTLs can be used as part of the process to introduce advantageous characteristics into plants.

Gene cloning

A clone is an identical copy of a gene. Gene cloning is based on the fact that DNA can be cut at defined locations using enzymes called **restriction endonu-cleases.** These cut pieces of DNA are then inserted into a **vector** – commonly made from a plasmid – a circular piece of bacterial DNA that will replicate in a bacterium like *Eschericia coli (E. coli)* (*Fig. 1*). This system can be used to create a **genomic library**, where genomic DNA is cut and inserted into a vector. A **cDNA (complementary DNA) library** is generated by isolating messenger RNA (mRNA) and copying it into DNA using the enzyme **reverse transcriptase** (*Fig. 2*). The **polymerase chain reaction** (**PCR**) is also used in cDNA cloning (*Fig. 3*). PCR is used to make multiple copies of the DNA between two known regions (primers) using **DNA polymerase**. Copying is carried out using a **thermal cycler** which generates cycles of high and low temperature needed for the reactions to take place.

The clone of a gene can be used for a wide variety of purposes. It can be sequenced and regulatory regions identified; it can be modified, mutagenized or deleted; it can be over- or under-expressed and it can be labeled to create a

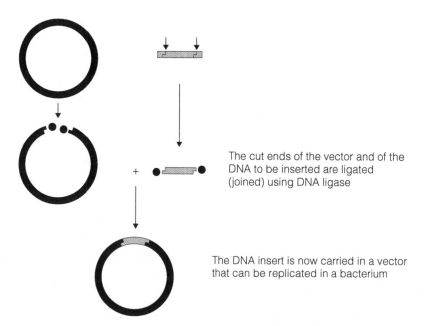

Both vector and insert DNA are cut with the same restriction endonuclease

The cut ends of the vector and of the DNA to be inserted are ligated (joined) using DNA ligase

The DNA insert is now carried in a vector that can be replicated in a bacterium

Fig. 1. Inserting a DNA fragment into a vector. Vectors are based on bacterial plasmids – circular DNA that can be easily handled and replicated in a bacterium. The DNA fragment is cut using the same restriction enzyme as the plasmid, leaving it with ends that will bind to the plasmid. At the end of the process, the vector now carries the DNA fragment and it is used to make multiple copies or to transform plants.

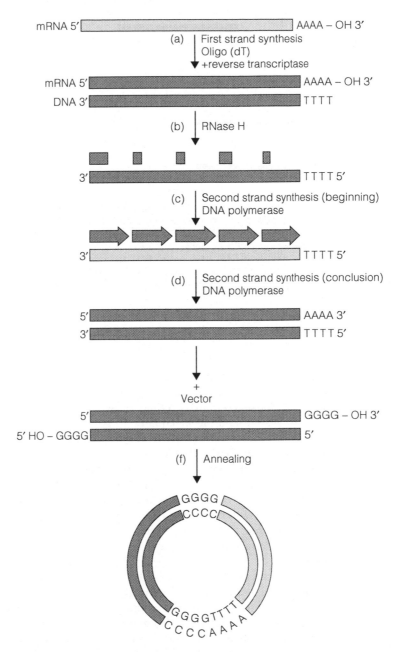

Fig. 2.   Making a cDNA library. Messenger RNA (either a single message or a collection of messages) is purified from a plant sample. Complementary DNA (cDNA) is then synthesized using the enzyme reverse transcriptase and the RNA removed. DNA polymerase is used to synthesize a second DNA strand and the cDNAs are inserted into a vector. The vector containing the cDNA is then placed in a bacterium in which it can be replicated. A typical library contains thousands of individual cDNA clones in a volume of less than 0.1 ml.

marker gene (for instance so that it produces a colored or fluorescent product). **Expressed sequence tags (ESTs)**, short regions of known cDNA sequence, are used to identify and locate genes that are active. An EST must originate from a

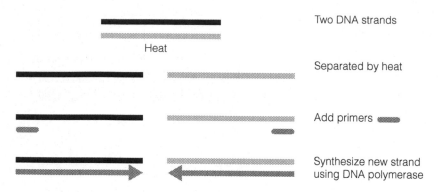

Two DNA strands

Heat

Separated by heat

Add primers

Synthesize new strand
using DNA polymerase

Fig. 3.   Polymerase chain reaction. The polymerase chain reaction (PCR) is a rapid means for making multiple copies of DNA strands. The DNA is first heated to separate the strands, then short lengths of primer DNA are hybridized to it. A new DNA strand is synthesized starting from the primer using a heat-stable form of DNA polymerase. At the end of this first cycle, two double strands of DNA have been obtained from one. The whole process is then repeated many times, copying the DNA each time. The process is carried out on a thermal cycler that heats and cools the DNA through cycles of chain separation and DNA synthesis.

gene that is expressed, as only sequences from active genes are present in a cDNA library.

Genomics and proteomics

Genomics is the study of the structure and function of entire genomes. The entire DNA sequence is known for the genome of several model plants, including arabidopsis (Topic B1). While knowing the genome sequence of an organism is very useful, it is equally important that the sequence is properly understood. In **structural genomics**, the sequence is obtained and **annotated** – genes are identified and their location on the genetic map is recorded. The process of identifying when genes are active, how they are regulated and their role in generating the phenotype of the organism is known as **functional genomics**. Gene expression is often studied using **microarrays** – microscope slides coated with thousands of tiny dots of DNA. Each dot is a short synthetic DNA sequence that will hybridize with DNA extracted from the cell (*Fig. 4*). The strength of signal obtained is proportional to the level of activity of a given gene.

The entire population of genes being transcribed at any time in the life of an organism is known as the **transcriptome**. The group of proteins produced during that lifetime is known as the **proteome** and the study of those proteins **proteomics**. Proteins are studied using a range of techniques. They may be sequenced or their interactions with other molecules studied. Their location in structures within the cell gives important information about function.

Mutants and transformed plants

A great deal can be learned about the function of genes by studying plants with one or more abnormal genes – a **mutant**. Mutants can be natural or generated by **chemical treatment**, commonly with ethylmethanesulfonate, by **irradiation** with X-rays or by inserting a segment of DNA into the genome (**insertional mutagenesis**). The seeds of mutagenized plants are known as M1 seeds. These are germinated, allowed to flower and self-pollinate producing the M2 generation. By the rules of Mendelian genetics, a quarter of these plants will have a mutant phenotype. Many possibilities follow when plants showing interesting

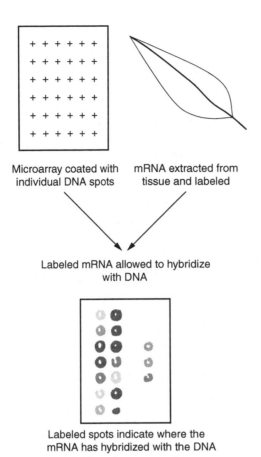

Microarray coated with   mRNA extracted from
individual DNA spots   tissue and labeled

Labeled mRNA allowed to hybridize
with DNA

Labeled spots indicate where the
mRNA has hybridized with the DNA

Fig. 4.   Microarrays. The image shows spots developed on part of a microarray – a glass slide coated with cDNAs representing more than 8000 genes. The microarray has been probed with a labeled cDNA made from mRNA extracted from the tissue of interest. Bright spots reveal genes where the fluorescent labeled cDNA has hybridized, meaning that the gene was being expressed and its mRNA was abundant. Using the technique, patterns of transcribed genes can be rapidly identified.

mutant phenotypes have been selected. The mutated gene is analyzed, giving information on the gene, its regulation and the protein it encodes. *Table 1* lists some of the many mutants that have been important in plant biology. **Deletion mutants** are commonly used to verify the function of a gene.

A variety of techniques has been developed using transformed plants where the genotype has been deliberately altered, often by insertion of foreign DNA. In **T-DNA tagging,** bacterial DNA from the organism *Agrobacterium tumefaciens* (Topic K3) is inserted into the plant genome. This T-DNA often inserts into a functioning gene and so acts as a mutagen (called **insertional** mutagenesis). The effect of the mutation is described and the gene identified by sequencing from the insertion. **Transposon tagging** is a similar process, but uses transposable

Table 1.   Some examples of arabidopsis mutants used in plant science research – mutants of other species are also being used to understand aspects of plant biology

Mutant	Phenotype	Related topic
aba	Plants wilty – mutant gene for ABA biosynthesis	Features of plant growth and development J1
abi	Plants insensitive to ABA	and
alf	**A**berrant **l**ateral root **f**ormation – used to understand the role of IAA in lateral root formation	Biochemistry of plant growth regulation J2
aux, axr	Do not show gravitropism; insensitive to auxin. AUX1 is an auxin influx carrier	
chl1	Insensitive to chlorate, an inhibitor of nitrate uptake – *CHL1* is a nitrate transporter	Movement of ions across membranes E3
clavata	Unusual meristem development – accumulate cells in the central zone of the shoot meristem	Features of plant growth and development J1
cop and *det*	Photomorphogenesis mutants – plant does not respond to darkness	Phytochrome, photoperiodism and photomorphogenesis I1
hy	Mutants with long hypocotyls in continuous white light – mutants of phytochrome synthesis	
lfy	Floral meristem mutant that forms shoots where flowers should be	Physiology of floral induction and development J4
rm1 (rootmeristemless)	Short roots – root meristem development altered	Features of plant growth and development J1
rsw1	Inhibited synthesis of cellulose at high temperatures	Cell walls C2
stm (shootmeristemless	Shoot meristem does not form	Features of plant growth and development J1
trp	Tryptophan biosynthesis mutants – IAA biosynthesis	Biochemistry of plant growth regulation J2

elements (**transposons**), natural DNA fragments that move and insert within the genome of some plants.

Bioinformatics and systems biology

Genomics and proteomics yield large volumes of complex information that can only be processed and explored using computer software. Major computer databases store the information obtained. Sequences are stored as nucleotide (DNA) sequences and as peptide (protein) sequences. Nucleotide and peptide sequences can be searched and investigated using a range of programs. The databases are annotated with information derived from analysis of the sequences and comparison with other sequences. **Systems biology** is an extension of the application of bioinformatics and molecular biology. It draws together information from genomics and proteomics with other techniques and uses powerful computational methods to obtain a holistic understanding of biological systems.

B3 STUDYING PLANT EVOLUTION AND ECOLOGY

Key Notes

Classification
Traditionally classification has been based on morphology, especially of reproductive characters.

Species level classification
Knowledge has been refined and the biological species concept introduced which has formalized some classification, although some plant groups do not fit into this system.

Molecular techniques
Variation in enzymes can be detected by electrophoresis and many hundreds of plants have been studied. A further range of variation is known through analysis of DNA using RFLP and RAPD techniques. Some proteins and DNA do not vary across a wide taxonomic range, but some are highly variable between individuals.

Plant phylogeny
Nuclear and organelle gene sequences have been used to classify plants at all levels from major divisions to the species level. Consistent classifications have been reached using many genes that have changed some traditional classifications. Some genes have a regular mutation rate and act as molecular clocks.

Studying plant ecology
Plant ecology requires minimal equipment, often just a simple frame or quadrat, but analysis often involves complex statistics. Models of ecological behavior have been developed with wide predictive implications.

Molecular studies in ecology
Increasingly molecular information is used in ecological study and the ecological significance of molecular variation is fundamental. This study is in its infancy.

Related topics   Sections L, O, P, Q

Classification   The study of plant classification has traditionally been based on **morphology**. This has proved effective at all levels of classification for dividing the major plant groups such as mosses, ferns and flowering plants to the species level. Reproduction and **reproductive characters** have been given greatest weight as these have always been seen as **conservative** characters. In reproduction many organs operate together and normally a change in one will be highly disadvantageous, so the form of these organs is strongly conserved. Some **vegetative characters** are used too, usually when they correlate with reproductive characters, but in general vegetative characters, such as leaf shape or plant form, are more responsive to immediate ecological conditions and are less use in classification. Using morphology alone has led to some anomalies, such as the position of *Psilotum* (Topic P4) or the Gnetopsida (Topic Q3) and there has been much dispute over species boundaries. Some authors have described different species on minor morphological differences, while others have lumped these together as one species, regarding such differences as varieties or subspecies.

Species level classification

Exploration and extensive collection over the centuries have indicated the extent of variation in many species leading to increasingly robust classifications. The **biological species concept** was introduced in the 1940s, basing the definition of a species on the potential or actual ability of its members to cross-fertilize and produce fertile offspring. This concept was developed for animals, mainly vertebrates, but has influenced the way all species are defined. It is not altogether appropriate for plants, since many species that normally remain distinct can **hybridize** to produce at least partially fertile offspring. This has happened frequently in nature and is the basis of plant breeding and much horticulture (Topics K1, N4, Q5). In addition some plants reproduce **asexually**, others are largely **self-fertilizing** and others have unusual **breeding systems** (Topic G3). Some plants have **diploid** and **polyploid** forms that may be morphologically identical or nearly so. Authors are divided as to how to classify these plants and where the species boundaries lie. Because of these differences there can be no fully common consensus on the definition of species boundaries across plant groups, but many **outbreeding** species are at least somewhat equivalent in status and, in any one geographical area, the great majority of species are distinct.

Molecular techniques

Differences in the detailed structure of proteins can be detected using a technique known as **electrophoresis** developed in the 1970s (*Fig. 1*). The most useful of these in studying plant evolution and, especially, genetics have been **isozymes**. Isozymes are genetic variants of enzymes that have differing electrophoretic **mobilities**, due to changes in overall charge following amino acid substitution or changes in size of the molecule. Measurement is usually made using enzyme-specific **colorimetric stains**. Reliance on measurable changes in mobility means that much variation must go undetected but the method is robust and fairly cheap. Many hundreds of plant species have been screened in this way and much variation has been detected within many sexually reproducing plants and between species. In a variant on this technique, **iso-electric focussing**, isozymes are focussed to a point of equal charge and migrate according to size.

Fig. 1.   *A representation of a gel showing isozyme variation detected by electrophoresis. Each column derives from leaf extract of a different plant and the isozymes have migrated across the gel (upwards in this picture) in an electric current. The gel is stained to detect their presence.*

Restriction fragment length polymorphism (RFLP)
From the 1990s DNA content has been analyzed directly. The RFLP method relies on the detection of sequence changes in DNA through the use of sequence-specific **restriction enzymes** isolated from certain bacteria. These cleave DNA at a target site and mutations in this site will prevent cleavage leading to changes in length of the target DNA. The sites are examined by a DNA hybridization technique known as **Southern blotting** or by using a **polymerase chain reaction** (PCR) to reproduce the DNA fragments that are then separated by electrophoresis and detected using probes of luminescent or radioactively marked DNA or RNA sequences that bind to the DNA.

Random amplified polymorphic DNA (RAPD)
In techniques involving RAPDs, copies of DNA are generated by synthesizing fragments of DNA between two identical short sequences known as **primers** (usually 10 base pairs) in a PCR. It is possible to sample many sites in the genome at once using this technique. Two sites for a single primer may exist in close proximity on a genome and the intervening fragment will be amplified, but for each primer a variable number of such sites will exist so a variable number of different-sized fragments will be produced. These can be analyzed using electrophoresis and the multiplicity of the sites means that different individuals may differ in the fragments amplified. The problems with this technique are that small changes in primer concentration or precise conditions in the experimental set-up can lead to different results and that certain DNA fragments may migrate at the same speed as other unrelated fragments so analysis of the bands can be problematic.

A combination of the RFLP and RAPD techniques may be used to give a **DNA 'fingerprint'**. This is widely used in forensic testing as well as study on classification and variation. Certain stretches of DNA do not vary at all within species or between species and sometimes across orders or even kingdoms, whereas others show so much variation that every individual is different. The most useful differences are those between these two extremes.

Plant phylogeny   Analysis of DNA sequences from certain genes has proved particularly useful in studying the relationships between plants and, from this, inferring a phylogeny, or tree of relationship. Some **nuclear genes**, such as those coding for **phytochromes**, and **chloroplast** and **mitochondrial** genes, known by code letters such as rbcL and atpB, vary little and have proved useful at higher level classification of the major plant groups. Others that vary a great deal are more useful at the species level. At any level a mutation such as an inversion or duplication (Topic B2) can be a defining feature of a plant group, be it an entire major group such as the flowering plants or one genus within a family.

Molecular analysis is always done using the principle of **parsimony**, i.e. using the smallest possible number of genetic changes required to reach the tree of relationship. If many samples are analyzed it can require considerable computing power to work out the most parsimonious phylogeny. Different genes can give apparently conflicting results, so the more gene sequences that are used, the more robust the classification will be. A sufficient number of such analyses have now been completed and combined together to suggest major changes in classification over those arrived at through traditional methods, e.g. incorporating the Psilotales among the ferns (Topic P4); establishing the liverworts as the earliest land plants (Topic O2); inferring which are the most primitive flowering plants (Topic Q4); and making many changes to genus and species definitions within families.

The studies are assuming that the gene in question is genuinely neutral in its effect. It is thought that many genes are, since there appears to be much **junk DNA** in many genomes, i.e. DNA that appears to have no function, either duplicating another functional gene or with no known effect. Rates of gene mutation vary, probably by several orders of magnitude, but although the mechanisms for this are poorly understood, it does appear that there can be constant rates of mutation, at least for some genes. By analyzing how many changes have occurred in these DNA sequences between different plant groups, we can infer how long it is since one group diverged from another. This idea is known as the **molecular clock** and has been used to verify and add a time-frame to the plant phylogenetic record.

Studying plant ecology

One of the advantages of studying plant ecology has been the lack of specialist equipment needed, making the subject accessible to amateurs as well as professionals. Often the only equipment is a **quadrat**, a metal or plastic frame, usually square, of a set size, standard sizes being with a side length of 10 cm, 25 cm, 50 cm or 1 m. This demarcates an area in which **numbers** can be counted, % **cover** estimated or, if set permanently at one spot, the fate of individuals can be followed over time. Larger squares can be marked with string and other simple marking tools can be used, often easily made.

Analyzing ecological information can be more problematic, as frequently there are many numbers in complex sequences or sets. **Parametric statistical tests**, such as *t*-tests, regressions and analysis of variance are widely used, often using transformations such as logarithms or arcsine to normalize the data. **Nonparametric ranking** tests, such as Spearman's or Mann-Whitney are used when the data cannot be normalized. More advanced techniques, using **multivariate** statistics, **principal components** analysis and related ideas are now commonplace.

Another way in which ecological information can be handled is developing mathematical **models** to fit sets of results. These are sometimes simple but, increasingly, simple ones are seen as giving too approximate a fit, and more complex quadratic or higher order models have been developed and models combined together to fit results more closely. These give rise to **predictions**, sometimes with far-reaching consequences, e.g. on pollen and seed flow in relation to genetic modification (Topics G5, H3, K3) or understanding plant population dynamics (Topic L4), vital in conservation.

Molecular studies in ecology

Molecular information is now used widely in ecological study as in most other aspects of biology. Indeed, the ecological significance of the wide-ranging variation in proteins and in DNA itself is fundamental to any interpretation. Much of this variation is assumed to be neutral and used as a genetic marker but, as more is studied, different isozymes or DNA variants are being shown to be associated with different ecological conditions, e.g. minor differences in soil conditions, temperature or water availability. These differences will be subject to **selection pressure** that will affect their distribution and abundance. They are widely studied in population ecology and genetics (Topics L4, L5) as ecological study focusses more on the ecological significance of variation within populations.

Although some variants are significant ecologically this does not stop them being useful in classification since many plant families or genera are associated with particular ecological conditions, such as halophytes belonging to a restricted group of plant families, each with many species involved. A few plant genomes may be fully sequenced (Topic B2) but which genes and which aspects of the study can be used most successfully for the various subjects within plant science is only beginning to be understood. The ecological importance of genetic variants is a study still in its infancy.

C1 THE PLANT CELL

Key Notes

Cell structure

The plant cell has a cell wall and plasma membrane enclosing the cytoplasm. Organelles, bounded by membranes, occur within the cytoplasm and are supported and moved by the cytoskeleton. The nucleus contains DNA and nucleoli. Many plant cells have a large vacuole.

Cell membranes

The endomembrane system of the cell is involved in synthesis and transport. The nucleus is surrounded by a nuclear envelope. The endoplasmic reticulum (ER) is divided into perinuclear ER and cortical ER, and may be smooth or rough (coated with ribosomes). Material from the ER is modified and sorted in the Golgi apparatus (GA) from which it travels in vesicles to the plasma membrane or the vacuole.

Organelles of metabolism

Mitochondria generate adenosine triphosphate (ATP) from stored food reserves. Chloroplasts carry out photosynthesis. Microbodies include peroxisomes containing catalase and glyoxysomes containing enzymes of lipid biosynthesis.

The cell wall

The cell wall is a dynamic, metabolic structure made up predominantly of carbohydrate. Adjoining cells are interconnected by plasmodesmata, in which membranes bridge the wall. Everything within the plasma membrane is the symplast; outside it is the apoplast, which is a water-permeated space, in which hydrophilic molecules are dissolved.

Related topics

The cell wall (C2)	Nucleus and genome (C5)
Cell division (C6)	Membranes (C4)
Plastids and mitochondria (C3)	

Cell structure

Plant cells show a wide range of shapes and internal structures, depending on their function. *Figure 1* illustrates the key features of a typical plant cell. Other cells, such as reproductive cells and conducting cells may be very different in appearance. It consists of a **cell wall** in close contact with a **plasma membrane** surrounding the **cytoplasm**, which is made up of aqueous fluid **cytosol** and many **organelles.** These organelles are supported and moved by a meshwork of fine protein filaments, the **cytoskeleton**, which includes **microtubules** made up of the protein **tubulin** and **microfibrils** made up of the protein **actin**. The nucleus contains genetic information in **chromosomes** and **nucleoli** that contain machinery for the production of **ribosomes**.

In most plant cells, there is a **vacuole** that may occupy up to 90% of the cell volume, surrounded by a membrane, the **tonoplast**. It contains solutes dissolved in water. It is important in storage and osmotic regulation (Topic C4).

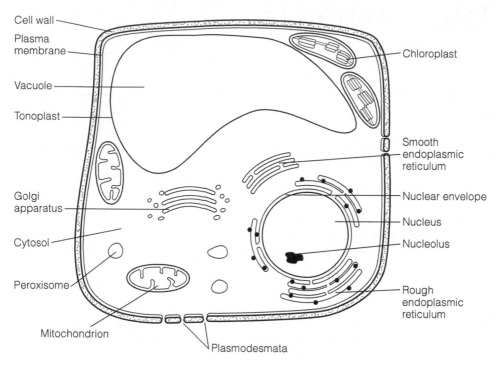

Fig. 1. Features of a typical plant cell.

Cell membranes

The cell contains a system of membranes termed the **endomembrane system** involved in the synthesis and transport of materials. The nucleus is bounded by two membranes, the **inner** and **outer nuclear envelope**, with **nuclear pores** to permit traffic of material. The outer nuclear envelope may be joined to a membrane system, the **endoplasmic reticulum (ER)** which may be **smooth** (site of lipid synthesis) or **rough** (coated with **ribosomes**; site of protein synthesis). In plant cells, the ER is often divided into **perinuclear ER** (ER around the nucleus) and **cortical ER** (ER at the cell periphery). Material from the ER is trafficked to the **Golgi apparatus (GA)**, a series of stacked membrane compartments (**cisternae**) in which modifications are carried out by enzymes. Material leaving the GA travels in **vesicles** to its destination, either the **plasma membrane** or the **tonoplast** (vacuolar membrane).

Organelles of metabolism

Plant cells have two major organelles of energy metabolism, one, the **chloroplast**, not being found in animal cells. **Mitochondria** (Topic C3) are bounded by a double membrane and generate adenosine triphosphate (ATP) from stored food reserves (carbohydrate, lipid; Topic F4). **Chloroplasts** (Topic C3) belong to a group of organelles known as **plastids**. Chloroplasts photosynthesize (Topics F1 and F2), using the energy of sunlight and carbon from carbon dioxide to produce carbohydrate. **Amyloplasts** are plastids modified to store starch. Plant cells also contain **microbodies** that are small membrane-bounded organelles. **Peroxisomes** contain catalase to remove toxic hydrogen peroxide produced in metabolism and **glyoxysomes** contain some enzymes of lipid biosynthesis.

The cell wall Surrounding the cell is a **cell wall** that is a metabolically active and constantly modified structure made up predominantly of structurally strong complex polysaccharides (Topic C2). Adjoining cells may be interconnected by **plasmodesmata**, pores through the wall where plasma membrane and ER connect. Everything within the plasma membrane is termed the **symplast**; everything outside it is the **apoplast**, which is a water-permeated space, in which hydrophilic molecules are present in solution. Cell walls function to adhere adjacent cells together in the formation of tissues and organs.

C2 THE CELL WALL

Key Notes

Cell wall structure
Primary cell walls are made up of cellulose microfibrils surrounded by a matrix of polysaccharides. Secondary cell walls contain cellulose microfibrils surrounded by polysaccharides and lignin.

Cell wall synthesis
Cellulose microfibrils are synthesized from uridine diphosphoglucose (UDP glucose) by cellulose synthase, an enzyme complex forming rosettes in the plasma membrane. Matrix materials are synthesized in the Golgi apparatus and deposited into the wall by secretory vesicles that fuse with the plasma membrane.

Cell wall function
Cell walls are essential for adhesion and the growth and formation of the plant body. Primary cell walls have high tensile strength and oppose turgor. Lignified secondary walls give greater strength. Cell walls act as a barrier to pathogens and deter herbivory. Primary cell walls are generally permeable to water and small molecules. The dynamic nature of primary cell walls permits cell expansion and plant growth.

Plasmodesmata
Plasmodesmata are structures in which membranes from adjacent cells connect through a pore in the cell wall. They link adjacent plasma membranes and cytoplasm. The desmotubule is a tube of endoplasmic reticulum in the center of the pore surrounded by globular proteins. The structure permits regulated transport between the cells.

Related topics
The plant cell (C1)
Roots (D2)
Herbaceous stems and primary growth (D3)

Woody stems and secondary growth (D4)

Cell wall structure Almost all plant cells have a **primary cell wall**. It is made of a long-chain poly-saccharide, **cellulose**, aggregated into bundles to form fibers, **microfibrils** 10–25 nm in diameter. The orientation of the microfibrils is governed by the **cytoskeleton** (see Cell wall synthesis, below) and the fibers are laid down in a coordinated fashion so that the plasma membrane is covered in layers (*Fig. 1*). The orientation of the fibers changes as the cell develops. Microfibrils have a great tensile strength; their strength is further enhanced by interlinking between the fibrils by a **matrix** composed of **matrix polysaccharides** and **pectins**. Between cells, there is a **middle lamella,** an adhesive region rich in **pectic poly-saccharides** where adjacent cell walls adhere to one another.

Some cells of strengthening and vascular tissues develop a **secondary cell wall**, between the primary wall and the plasma membrane. Secondary cell walls also contain cellulose **microfibrils**, infilled with polymerized phenolic compounds, **lignins**, that strengthen the wall. This is evident in **wood** (Topic D4). Lignin protects

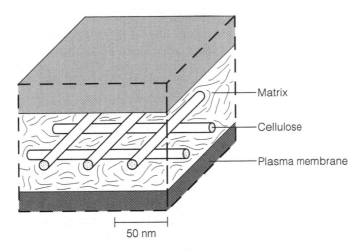

Fig. 1.   *The primary cell wall consists of cellulose microfibrils deposited in layers surrounded by a matrix of hemicelluloses and pectins.*

against digestion of the wall by fungal enzymes and against mechanical penetration by fungal hyphae and other pathogens (Topic M4). Secondary walls are produced in layers, with the cellulose fibrils orientated in different directions; this 'lamination' gives considerable strength to the structure.

Cell wall synthesis

The **primary cell wall** is deposited while the cell is increasing in size. Cellulose is deposited by an enzyme complex, **cellulose synthase** that appears as a rosette in the membrane (*Fig. 2*). Cellulose is synthesized from **uridine diphosphoglucose** (UDP glucose) which is added simultaneously to the end of several strands, forming a **cellulose fiber** (or **microfibril**) at the cytoplasmic face of the membrane. As the strand elongates, the rosette moves in the membrane, extruding the strand to lie along the outer face of the membrane. The rosettes

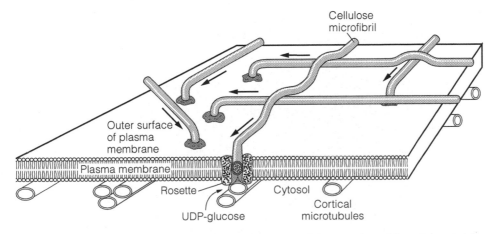

Fig. 2.   *The deposition of cellulose microfibrils by rosettes of cellulose synthase follows the underlying pattern of cortical microtubules, part of the cytoskeleton. (Redrawn from Raven, P.H. et al. (1992) Biology of Plants, 6th Edn, W.H. Freeman.)*

move parallel to fibers within the cell, the **cortical microtubules**. **Matrix materials** (matrix polysaccharides, lignin, pectic substances) are synthesized in the **Golgi apparatus** and transported to the plasma membrane in **secretory vesicles** that discharge their contents into the wall. New wall material is deposited at its inner face – that is the face adjacent to the plasma membrane.

Cell wall function

Primary cell walls have high **tensile strength**. They give strength to stems, leaves and roots, particularly by opposing turgor pressure from cell contents. They show **plasticity** and **elasticity**; as the cell grows, the dynamic structure of the wall means that it is able to adjust its structure to permit that growth. A new cell wall is rapidly formed after cell division. Cell walls prevent plasma membranes from contact except at pores, the plasmodesmata (see below). Primary cell walls are permeable to water and low molecular weight molecules (ions, organic compounds and small proteins) which reach the plasma membrane. This movement is restricted if the wall is lignified or contains suberin as at the endodermis (Topic D2). Adjacent cell walls adhere at the **middle lamella**, giving **cell adhesion** and allowing formation of tissues and organs. **Lignification** in secondary cell walls (Topic D4) greatly enhances **compressive strength**, permitting woody structures more than 100 m tall. Cell walls also provide resistance to pathogens and herbivory.

Plasmodesmata

Plasmodesmata (singular **plasmodesma**) are structures in which membranes from adjacent cells connect through a pore in the cell wall (*Fig. 3*). The plasma membranes of the two adjacent cells join around the pore and a **desmotubule**, a tube of endoplasmic reticulum (ER), is present in the center of the pore, surrounded by **globular proteins**. This means that the plasma membrane, ER and cytoplasm of two adjacent cells adjoin. The structure permits regulated transport between the cells. Careful study of plant tissues reveals that many cells connect through plasmodesmata forming what could be considered to be colonies of cells all interconnected with their neighbors.

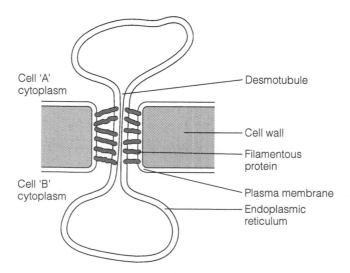

Fig. 3. The structure of a single plasmodesma in transverse section. Note the connecting membranes (endoplasmic reticulum via the desmotubule and plasma membrane) and cytoplasm penetrating the cell wall of two adjoining cells.

C3 PLASTIDS AND MITOCHONDRIA

Key Notes

Plastids	Plastids are a family of organelles bounded by two external membranes. Family members include: photosynthetic chloroplasts; chromoplasts containing pigments; leucoplasts involved in lipid biosynthesis; amyloplasts that store starch; and etioplasts, an intermediate stage in production of chloroplasts.
Chloroplast structure and origins	Plastids contain a small genome encoding some plastid proteins. Many other chloroplast proteins are encoded by nuclear genes. The endosymbiont theory suggests that they arose as primitive photosynthetic organisms that colonized cells. Stacks of thylakoid membranes (grana) containing chlorophyll are present in the matrix space (stroma).
Mitochondria	Mitochondria are the site of synthesis of adenosine triphosphate (ATP) using lipids, carbohydrates and other high energy compounds as fuel.
Mitochondrial structure	Mitochondria are bounded by an outer membrane with an inner fluid-filled stroma (matrix). An inner, selectively permeable membrane, folded into cristae, contains the components necessary for ATP synthesis. Mitochondria contain a small genome that encodes some, but not all, mitochondrial proteins.
Related topics	The plant cell (C1) Photosynthetic pigments and the nature of light (F1) Major reactions of photosynthesis (F2)

Plastids

Plastids are characteristic of plant cells and otherwise only occur in plant-like protists. They are organelles bounded by a double membrane. There are several types of plastid in plant cells. **Chloroplasts** are photosynthetic plastids found in the mesophyll cells of leaves, the cortex of herbaceous stems and in small numbers elsewhere in the plant. The green coloration is due to the presence of the pigment chlorophyll (Topic F1). **Chromoplasts** contain pigments other than chlorophyll and are associated with brightly colored structures like ripe fruit. **Leucoplasts** are colorless and are found in many cell types. They include **amyloplasts** that store starch and **elaioplasts** that synthesize lipid. **Etioplasts** are an intermediate stage in the production of photosynthetic chloroplasts in tissue exposed to light for the first time.

Chloroplast structure and origins

The two membranes of plastids surround a central fluid-filled **stroma** (*Fig. 1*). Plastids contain their own DNA in a small 'plastid genome' containing genes for some chloroplast proteins. This, together with the presence of the double outer membrane, has led to suggestions that they arose as **endosymbionts** – primitive photosynthetic organisms that colonized a nonphotosynthetic cell. However, while some chloroplast proteins are synthesized on **plastid ribosomes**, from genes in the

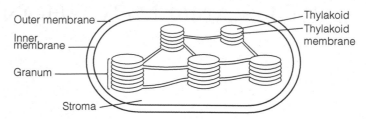

Fig. 1.   Chloroplast structure.

chloroplast genome, many others are encoded by nuclear genes and imported. Chloroplasts are highly organized for photosynthesis. Suspended in the stroma are **thylakoids**, membrane discs that form stacks or **grana**. Individual stacks are inter-connected by tubes of **thylakoid membrane**. The photosynthetic pigments are arranged in the stacks so that they can be orientated to capture as much light energy as possible. For details of the mechanism of photosynthesis see Topics F1 and F2.

Mitochondria

Mitochondria convert the energy in storage reserves, like lipid, starch and other carbohydrates, into the high energy compound **adenosine triphosphate** (**ATP**). The mitochondrion (*Fig. 2*) provides an isolated environment in which high energy intermediates can be formed without reactions with other cell constituents. Mitochondria contain a small **mitochondrial genome**, a circular piece of DNA encoding some (but not all) mitochondrial proteins, which are synthesized on mitochondrial ribosomes. Other proteins are encoded by nuclear DNA and synthesized on cytoplasmic ribosomes. Mitochondria can divide to give further mitochondria and can fuse to form an interconnected tubular mesh. Mitochondria are present in all cells and are abundant in those with high energy demands like phloem companion cells (Topic D3).

Mitochondrial structure

Mitochondria are bounded by two outer membranes. The inner of these membranes is invaginated (folded) to form **cristae** (singular **crista**) which project into the inner space, the **stroma** or matrix. This inner mitochondrial membrane therefore separates two compartments: the **intermembrane space** and the **matrix space**. It is selectively permeable and contains the transport proteins involved in ATP production (Topic F4).

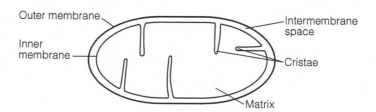

Fig. 2.   The structure of a mitochondrion.

C4 MEMBRANES

Key Notes

Membranes

Membranes are made up of a lipid bilayer with membrane proteins. The membrane is a hydrophobic region through which charged or large polar solutes cannot pass unless via a membrane protein. They separate compartments of different composition. Cells contain membrane-bounded organelles like mitochondria and chloroplasts and membrane systems like the endoplasmic reticulum and Golgi apparatus.

Endoplasmic reticulum

The endoplasmic reticulum (ER) is a system of flattened sacs or tubes of membrane. Smooth ER is the site of lipid synthesis and rough ER of protein synthesis. ER may be associated with the nuclear envelope, but may also be at the cell periphery. Proteins targeted for the ER lumen or membrane bear a signal sequence and are synthesized on ER ribosomes. Subsequently modification of the proteins may occur before they are exported to other locations by vesicles.

Golgi apparatus

The Golgi apparatus is a system of stacked membrane cisternae. It is a site of protein modification and polysaccharide biosynthesis. Proteins are delivered to the *cis* face by vesicles and move through the *medial* cisternae to the *trans* face for export by the vesicles of the *trans* Golgi network (TGN). Enzymes within the cisternae synthesize polysaccharides or modify proteins by glycosylation.

The nuclear envelope

The nuclear envelope is a double membrane separating the contents of the nucleus from the cytoplasm. Transport in and out of the nucleus occurs through nuclear pores. The nuclear envelope breaks down during cell division.

Plasma membrane

The plasma membrane (pm) is a single membrane bounding the cytosol. It generates a *trans*-membrane electrochemical gradient by pumping protons. It maintains ionic homeostasis of the cytoplasm and transports nutrients and other products. Other pm functions include: sensing and signaling (receptor proteins), secretion, turgor and communicating with adjacent cells through plasmodesmata.

The tonoplast

The tonoplast is a single membrane surrounding the vacuole. It maintains ionic homeostasis of the cytoplasm and transports nutrients and other products by maintaining an electrochemical gradient driven by proton pumping. Vacuoles form from pro-vacuoles formed in the Golgi apparatus. A mature vacuole contains inorganic ions, sugars, enzymes and organic acids. Some contain secondary products.

Related topics

The plant cell (C1)

Movement of nutrient ions across membranes (E3)

Membranes

Biological membranes are made up of a **lipid bilayer** in which membrane proteins are embedded, either deeply as integral membrane proteins or at the edge as peripheral membrane proteins (*Fig. 1*). Membrane lipids have

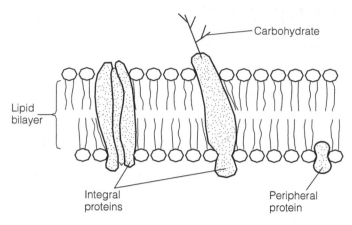

Fig. 1. *The membrane consists of proteins and a lipid bilayer. Proteins in a simple biological membrane exist suspended in a surrounding 'sea' of lipid within which they may move laterally and rotate.*

hydrophilic head groups and hydrophobic tails that are buried within the membrane, making a hydrophobic domain that is not readily crossed by charged hydrophilic (polar) substances. This barrier means that the membranes can separate two compartments of different composition. The proteins within the membranes govern the composition of those compartments by selective transport. This may be active (using ATP) or passive (allowing a substance to move from one compartment to another driven by an electrochemical gradient). As well as transport proteins, membranes also contain receptor proteins that transduce signals like phytohormones, gravity or light; anchor proteins that attach the membrane to the cytoskeleton and cell wall; and enzymes involved in the synthesis or degradation of cellular constituents.

Some organelles are membrane-bounded (e.g. mitochondria and chloroplasts). In other instances, the functional entity is the membrane itself, which may enclose a separate compartment or lumen (e.g. the ER and GA). The membranes of the cell may be considered as a more or less complete continuum from the sites of synthesis of new membrane to mature, functioning membranes. Old membrane components are constantly being recycled – removed from the membrane and reprocessed or hydrolyzed. Much of this flow of membrane through the cell occurs in vesicles, small sacs of membrane.

Endoplasmic reticulum

The endoplasmic reticulum (ER) is a network of membranes in the cytoplasm that form either tubules (tubular ER) or flattened sacs (cisternal ER). The ER may be divided into two types. Smooth ER is the site of membrane lipid synthesis and new membrane assembly. Rough ER appears rough due to the presence of ribosomes and is the site of the synthesis of membrane proteins and proteins destined for secretion or insertion into vacuoles. Free ribosomes, not bound to membrane, are the site of synthesis of proteins not associated with membranes. ER may be closely associated with the outer nuclear envelope (NE; Topic C5). The ER lumen and the lumen between inner and outer NE are a continuum. ER may also be separate from the NE; cortical ER, for instance, is located just beneath the plasma membrane. Proteins destined for insertion into the ER membrane or to the lumen bear a signal peptide, a short amino acid sequence targeting the protein synthetic apparatus to its destination. Some simple modifications of protein may occur in the ER – for instance simple **glycosylation** reactions in which carbohydrate residues are added to the protein. Further modifications require the protein to be transported to the Golgi apparatus (GA), the next stage in the pathway.

The Golgi apparatus

The **Golgi apparatus** (GA) is made up of a stack of flattened membrane sacs called **cisternae,** of which there may be only a few or many. Material enters the GA in vesicles, which fuse with the Golgi membranes. The stacks (**dictyosome**) are polar; new material is added at one face (the *cis* **face**) and transported through the *medial* cisternae to the *trans* **face**, from which a network of secretory vesicles, the *trans* **Golgi network** (TGN), exports the product (*Fig. 2*). Two models exist to explain how materials move through the GA; the **cisternal maturation** model suggests that new cisternae are constantly being added to the *cis* face, which then gradually mature until released as vesicles at the *trans* Golgi network. The **vesicle traffic model** suggests that the cisternae remain for the lifetime of a GA with material being moved from cisternum to cisternum by vesicles.

The GA prepares proteins and polysaccharides for secretion. Depending on cell type, it may be more or less dedicated to one or other activity. Enzymes in the GA either synthesize complex carbohydrates, or add carbohydrate groups to proteins (**glycosylation**). The *cis, medial* and *trans* GA contain different enzymes which progressively modify the material passing through. At the end of the

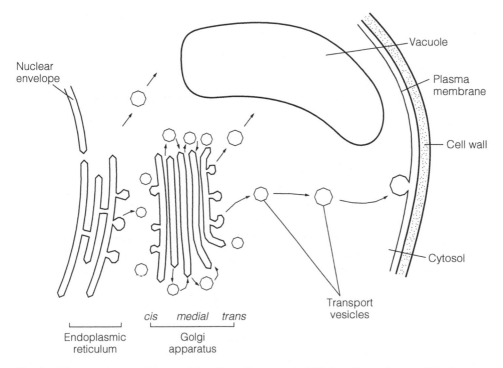

Fig. 2. The secretory pathway and insertion of new material into cell membranes. Membrane vesicles carry new membrane and material for export from the GA to the vacuole or plasma membrane. Membrane material is also returned by vesicles.

process, material for export is sorted and packaged in **secretory vesicles**, which move to their membrane destination before fusing with it and discharging their contents. Contents in the lumen of the vesicle are either secreted or enter the lumen of the membrane compartment they fuse to; proteins in the vesicle membrane are incorporated into the membrane itself.

The nuclear envelope

The nuclear envelope is a double membrane, surrounding the nucleus and separating it from the cytoplasm. The outer nuclear envelope is joined to the endoplasmic reticulum, while the inner membrane is attached to filamentous proteins that maintain the structure of the nucleus, the nuclear matrix. Nuclear pores perforate the two membranes and are complex structures through which all transport into and out of the nucleus occurs. The nuclear envelope breaks down in meiosis and mitosis and reforms around the daughter nuclei (Topic C6).

The plasma membrane

The plasma membrane is a single membrane bounding the cytoplasm. New material is added to it and old material is removed by vesicles. The plasma membrane has a number of functions.

- Maintaining ionic homeostasis of the cytoplasm and transporting nutrients and other products. To do this, a *trans*-**membrane electrochemical gradient** is maintained, driven by **a proton-pumping ATPase** (which uses the energy of ATP to transport protons across the membrane; Topic E3). Many other proteins located in the membrane transport ions and other molecules.
- Sensing and signalling the cells environment. **Receptor proteins** at the plasma membrane respond to the presence of signals (e.g. hormones) and cause changes in intracellular signalling molecules which result in altered cell functions.
- **Secreting materials** (e.g. the constituents of the cell wall; Topic C2).
- **Regulating turgidity** by osmotic effects resulting in the cytoplasm exerting force against the cell wall.
- **Communicating** with other adjacent cells through **plasmodesmata** (Topic C1).

The tonoplast

The **tonoplast** is a single membrane surrounding the vacuole (Topic C1). Like the plasma membrane, the tonoplast has a role in maintaining ionic homeostasis of the cytoplasm and in transporting nutrients and other products. A *trans*-membrane **electrochemical gradient** is maintained, driven by a **proton-pumping ATPase** (Topic E3). The tonoplast also contains other **transport proteins** which regulate the ionic homeostasis of the cytoplasm and vacuole. A mature vacuole contains inorganic ions, sugars, enzymes and organic acids. Specialized vacuoles may contain complex organic compounds (secondary products; Topic F5), long-term storage products like proteins or organic acids involved in carbohydrate metabolism.

C5 NUCLEUS AND GENOME

Key Notes

Structure of the nucleus	The nucleus is bounded by a double membrane, the nuclear envelope, which is perforated by nuclear pores that permit the movement of materials from the interior of the nucleus (nucleoplasm) to the cytoplasm. The genetic material (DNA) is present as a DNA–protein complex known as chromatin. The nucleolus is the site of synthesis of the ribosome subunits.
Structure and function of chromatin	The DNA double helix is entwined around histone proteins in units called nucleosomes, packed together to give a 30-nm fiber of chromatin. Genetic information is contained within the DNA as codons. Genes are made up of a structural region encoding the protein to be synthesized and promoter regions that control gene activity. In transcription, the DNA untwines and RNA polymerase II copies it as an RNA strand. Nontranscribed introns are removed giving messenger RNA (mRNA) which is translated into protein on ribosomes.
Plant chromosomes	Chromosomes are made up of tightly packed chromatin. Chromosomes in dividing cells have two chromatids joined at a centromere. Endoreduplication results in many copies of each chromosome, while gene amplification results in multiple copies of a few highly used genes. Polyploid cells have more than the usual number of chromosomes; normally only polyploids with an even number of chromosome copies are fertile as they can undergo chromosome pairing in meiosis.
Related topics	Membranes (C4) Features of growth and Cell division (C6) development (J1)

Structure of the nucleus

Each living plant cell contains a nucleus separated from the cytoplasm and other organelles by a double membrane, the **nuclear envelope** (Topic C4). The size and prominence of the nucleus varies depending on the major functions of the cell. Cells in meristems (Topic D1), which are to undergo cell division, have a large nucleus. Other cells, such as those of the parenchyma (Topic D3), have a smaller nucleus, with most of the cell occupied by the vacuole (Topic C4). The nuclear envelope surrounds the fluid **nucleoplasm**. Movement from nucleoplasm to cytoplasm is possible through complex protein structures, the **nuclear pores**, which span both membranes. Nuclear pores are complex multi-protein structures that regulate movement into and out of the nucleus; their most important function is to permit messenger RNA (mRNA) molecules and ribosomes to leave the nucleus. Within the nucleus, the DNA is entwined with proteins, the DNA–protein complex being termed **chromatin**. In interphase, the chromatin appears to be spread throughout the nucleoplasm; in mitosis, however, it condenses to form distinct chromosomes (Topic C6).

The **nucleolus** is often seen as a densely stained region in micrographs of the nucleus. It is the center for the synthesis and assembly of components of **ribo-**

somes, structures involved in protein synthesis. Ribosomes are exported from the nucleus where they function to synthesize proteins either free in the cytoplasm or attached to membranes like the outside of the nuclear envelope or the rough endoplasmic reticulum (Topic C4).

Structure and function of chromatin

DNA in the nucleus is associated with protein in a complex known as chromatin. The basis of chromatin is a double helix of DNA; this entwines around proteins called **histones**. Stretched out, the structure resembles beads on a necklace. Each 'bead' of DNA and histone is known as a **nucleosome**. An additional histone protein binds to the DNA, and causes an additional level of coiling. This results in the DNA–histone 'beads' packing closely together to give a 30-nm fiber of chromatin. This fiber then forms loops along a protein scaffolding. In metaphase (Topic C6), additional coiling and close-packing gives rise to highly condensed chromosomes.

The genetic information of the cell is contained within the DNA in the form of **codons**: triplets of nucleotides which encode an amino acid or indicate the start or finish point of each gene. Each gene is made up of two regions: a **structural region** which contains the information for the amino acid sequence of the protein and which is copied (**transcribed**) to mRNA when the gene is active and a **promoter region** which controls whether the gene is transcribed (*Fig. 1*).

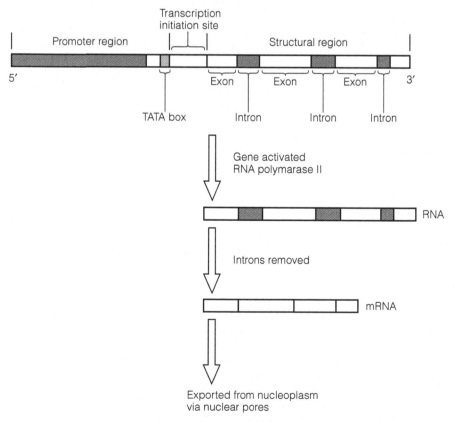

Fig. 1.   The structure of a plant gene. The promoter region is adjacent to the structural region that contains the code for the final protein, made up of exons and introns. The TATA box is important in RNA polymerase II binding; DNA copying begins 20–30 base pairs away at the transcription start site.

A single gene may be regulated by a number of factors, such as hormones and factors specific to where the cell is, for instance the root or shoot. Between the structural and promotor regions lies a sequence rich in adenosine and thymidine (A and T) known as the **TATA box** which is important in binding the enzyme that synthesizes mRNA. Genes contain regions which will be transcribed to mRNA (**exons**) and regions which will not (**introns**). Several (or many) introns may be found within one gene.

In order for a gene to be transcribed, the DNA double helix must unwind over a short region. **RNA polymerase II** commences copying a short distance from the promoter region, then moves along the gene and copies the DNA template as an RNA strand. The RNA strand contains a transcription of both introns and exons. Introns are then removed to form the mRNA which migrates out of the nucleus via the nuclear pores to be translated into protein at ribosomes.

Plant chromosomes

Plant **chromosomes**, which become clearly visible under the microscope in mitosis, are made up of tightly packed chromatin. When visible in this way, the chromosomes have passed through the S-phase of the cell cycle (Topic C6) and so a copy of the DNA has been synthesized by **DNA polymerase**. This remains attached so that the chromosomes have two **chromatids** joined at a **centromere** (*Fig. 2*).

Normally during cell division, each daughter cell receives a copy of the entire genetic information of its parent. At this point, the cell could cease synthesizing DNA and go on to differentiate; however, most plant cells go on to generate further copies of their genes in a process known as **endoreduplication** (Topic C6). In some instances, **gene amplification** results in multiple copies of a few highly used genes being made. If a gamete is formed by cells which have not undergone **meiosis** (Topic C6), the resultant cell has two sets of genes present (it is **diploid**). If it is then fertilized by a normal (**haploid**) gamete, the resulting offspring is **triploid** (has three sets of genes). In plants, this is frequently nonlethal and plants may show **polyploidy**, that is more than one copy of each chromosome. Plants with an even number of copies are fertile, such as **hexaploid** wheat (six copies), while those with an odd number are normally sterile as they cannot undergo pairing of homologous chromosomes during meiosis.

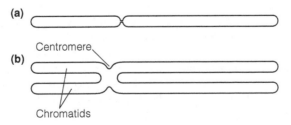

Fig. 2.   A chromosome before (a) and after (b) duplication of the single chromatid during S-phase. Note that the two chromatids are joined at the centromere. They later separate to donate one chromatid to each of the daughter cells formed.

C6 CELL DIVISION

Key Notes

The cell cycle	Cell division occurs in stages: interphase G_1 – cell enlarges, nucleus migrates to center, protein synthesis; interphase S – DNA replication occurs; interphase G_2 – preprophase band and structures of mitosis form, chromosomes condense. In M-phase mitosis, chromosomes replicate and divide to form two daughter nuclei. In M-phase cytokinesis, the cytoplasm divides and a cell plate and new cell walls form.
Cell cycle control	The cell cycle has two check-points: G_2/M and G_1/S. Progression through the cycle is controlled by cyclins that are synthesized and degraded through the cycle and activate cyclin-dependent protein kinases (CDPKs).
Meiosis	Meiosis occurs to produce haploid cells. It involves an extra round of cell division. The first phase, prophase to anaphase I, results in exchange of DNA between the pairs of chromosomes followed by their separation with both chromatids present. The second phase (metaphase II to telophase II) is a mitosis resulting in separation of the chromatids and the formation of four haploid cells.
Related topics	The plant cell (C1) Pollen and ovules (G2)
	Nucleus and genome (C5) Features of growth and
	Meristems and primary tissues (D1) development (J1)

The cell cycle

Cell division in plants occurs in meristems (Topic D1) and involves two parts: **mitosis** in which the chromosomes are replicated and sorted into two nuclei, and **cytokinesis** in which the cell wall, cytoplasm and organelles divide. In dormant meristems, the cells rest in G_0 **phase**. When conditions are correct, the cell begins the processes leading to division. The entire cycle may be considered as four phases, G_1, **S**, G_2 and **M** (*Fig. 1*).

In G_1 **phase** the cell doubles in size and new organelles and materials needed for two cells are formed. During this phase, the nucleus migrates to the center of the cell and is surrounded by a sheet of cytoplasmic strands called the **phragmosome** that bisects the center of the cell at the plane across which it will divide. The phase ends with the G_1/S **checkpoint**. The process can stop at this point (see Cell cycle control, below), or proceed to **S phase** in which DNA and associated nuclear proteins are replicated. At the end of S phase the cell contains two full copies of its genetic information. It proceeds to G_2 **phase** when the chromosomes begin to condense and structures required for division form. A distinct band of microtubules (Topic C1), the **pre-prophase band**, forms around the cytoplasm in a ring where the edge of the phragmosome lay, again predicting the plane of cell division. At the end of G_2, the cell has to pass another checkpoint (G_2/M) at which stage, if conditions are suitable, it enters **M phase** in which the cell divides.

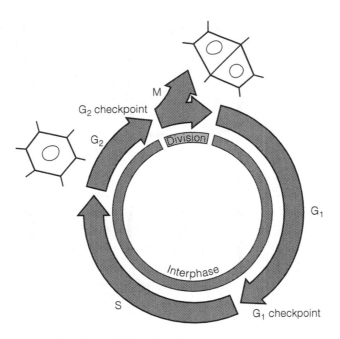

Fig. 1.   Key phases of the cell cycle.

Stages G_1 to G_2 are known as **interphase**. M phase, when division occurs, can be divided into a series of stages that can be recognized by microscopy (*Fig. 2* and *Table 1*).

Cell cycle control   In **meristems** (Topic D1), a population of cells characterized by thin cell walls and the lack of large vacuoles are constantly dividing. The daughter cells may undergo

Table 1.   Events in mitosis

Stage	Events
Early prophase	Chromosomes visible in nucleus
Mid prophase	Chromosomes shorten and thicken; the two parts (chromatids) making up each chromosome become visible with a join (centromere)
Late prophase	Kinetochores (specialized structures attached to microtubules) attach at the centromeres. Nuclear envelope breaks down
Metaphase	Chromosomes align at the center of the cell; chromosomes aligned by microtubules which run from the centromeres to the pole ends of the cell
Anaphase	Begins with coordinated movement of chromatids, drawn by the kinetochore microtubules. The two sets of chromatids (now called daughter chromosomes) are now separated to opposite ends of the cell
Telophase	Daughter chromosomes now visible at ends of cell; nuclear envelopes develop around chromosomes and a cell plate forms which will develop into the cell wall

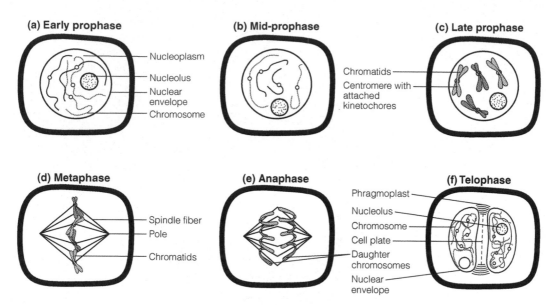

Fig. 2.   Stages of mitosis (see also Table 1).

a few further divisions, but then lose the capacity to divide and after a phase of cell enlargement generally develop large vacuoles. **Plant hormones**, auxin and cytokinin (Topic J2), are known to initiate the cell cycle. Auxin stimulates DNA replication, while cytokinin initiates the events of mitosis. The cell cycle is also controlled by the activity of cell proteins called **cyclins** and **cyclin-dependent protein kinases** (CDPKs; a kinase is an enzyme which will phosphorylate another protein). One group of cyclins, the G_1 **cyclins**, are manufactured by the cell in G_1 and activate CDPKs which stimulate DNA synthesis at the G_1/S control point. If sufficient G_1 cyclins are not formed, the cell will not progress to S. Having passed this point, the G_1 cyclins are degraded and a new family of cyclins, the mitotic or **M cyclins** are produced. These activate a second set of CDPKs which permit the cell to pass the G_2/M control point into mitosis (*Fig. 3*). Whereas animal cells which pass G_1/S are committed to undergo division, plant cells are not. This means that many plant

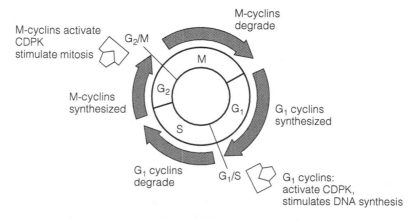

Fig. 3.   Cell cycle control by the synthesis and breakdown of cyclins and their binding to cyclin-dependent protein kinases (CDPK).

cells continue to replicate DNA without dividing. This is known as **endoredupli-cation**, which is shown by more than 80% of all plant cells and particularly cells with a high metabolic activity and requirement for protein synthesis.

Meiosis

Meiosis occurs in the reproductive tissues of the plant. To do so, it must result in a halving of the number of chromosomes, so that each cell has only one set (**haploid**, rather than the usual two sets, **diploid**) of chromosomes. The full complement of chromosomes is restored after fertilization, when the two sets (one from each gamete) combine. In **interphase**, DNA synthesis occurs and each chromosome exists as a pair of sister chromatids joined by a centromere. In **prophase I**, the homologous chromosomes (originally from the maternal and paternal generative cells) pair up to give a **synaptonemal complex**. Each chromosome can be seen to be composed of two **chromatids**. The chromatids join at points called **chiasmata**, at which genetic material can cross over from one chromatid to another. This can be between homologous chromatids or between sister chromatids. In **metaphase I**, the paired chromosomes move together at the **metaphase plate**. In **anaphase I**, homologous chromosomes, each with its two chromatids, separate to the spindle poles, drawn by microtubules (Topic C1). The daughter nuclei now have a haploid set of chromosomes. Each chromosome has two chromatids (compare mitosis, where at this stage the chromatids separate so each chromosome has only one). Instead of forming new nuclei and stopping division, the cells go on to another phase of division.

In **metaphase II**, a new metaphase plate forms in daughter cells and the chromosomes line up at the equator of the cell. In **anaphase II**, chromatids separate and move to the poles. By **telophase II**, the chromosomes have completed movement and four new nuclei, each having half the original number of single chromosomes, have been formed.

As only one period of chromosome duplication has occurred, the result is four haploid cells. In pollen formation, all four cells survive; in ovule formation, three normally abort leaving one to form the ovule. The stages are shown diagramatically in *Fig. 4*.

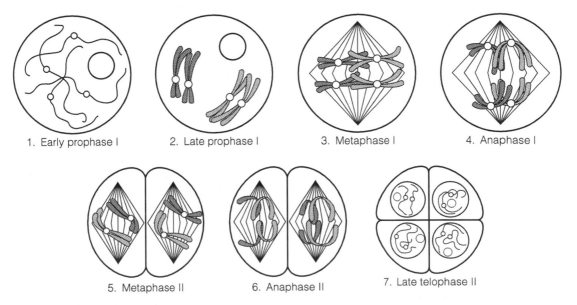

1. Early prophase I 2. Late prophase I 3. Metaphase I 4. Anaphase I

5. Metaphase II 6. Anaphase II 7. Late telophase II

Fig. 4.   Key stages of meiosis.

D1 MERISTEMS AND PRIMARY TISSUES

Key Notes

Meristems	Meristems form the new cells of a plant. Apical meristems occur near the tips of roots and shoots and give primary growth. They give rise to the epidermal tissue, vascular tissue and ground meristems that form the parenchyma. Intercalary meristems are formed at the nodes of grass stems. Lateral meristems generate secondary growth. The vascular cambium is a cylinder that forms new phloem and xylem. The cork cambium is also a cylinder, that forms the bark.
Tissues and organs	A tissue is a collection of cells with common function. It may be simple or complex. An organ is a multicellular functional unit. Plant organs are made of three types of tissue: dermal, vascular and ground.
Ground tissues	Ground tissue is made up of parenchyma, collenchyma and sclerenchyma. Mature parenchyma cells have large vacuoles and intercellular spaces. Leaf parenchyma cells with chloroplasts are chlorenchyma. Collenchyma cells have strong cell walls providing support. Sclerenchyma also provides support. It is made up of dead cells with a thickened, lignified secondary wall, in sclereids (small groups of cells) or fibers (long strands of elongate cells).
Dermal tissue	The outermost cell layer is known as the epidermis. It is formed of parenchyma-like cells and specialized cells for transport of gases, water and nutrients.
Vascular tissue	Vascular tissue is made up of xylem and phloem. Xylem carries water and dissolved minerals, and phloem the solutes from sites of synthesis to sites of storage or use. Xylem cells have thickened secondary cell walls and no cell contents. Xylem vessel elements join to form vessels. Tracheids are tapered cells that interconnect via pits. Phloem has two cell types: sieve tube elements and companion cells. Sieve tube elements have no nucleus, but contain organelles and link through sieve plates. Companion cells are narrower than sieve tubes and function with them.
Related topics	The cell wall (C2) Plants and water (E1) Features of growth and development (J1)

Meristems

Meristems are the site of formation of new cells within the plant (*Fig. 1*). **Apical meristems** occur near the tips of roots and shoots and give length increase (**primary growth**). They contain **initials**, the equivalent of stem cells in animals, which retain the ability to divide and generate new cells throughout the life of

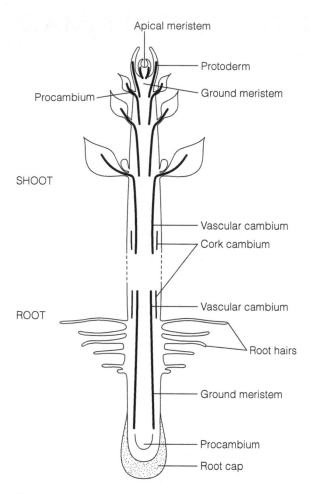

Fig. 1. The location of meristems in a dicot.

the meristem. They form **primary meristems** that produce the tissues of the stem and root. These primary meristems are the **protoderm** (which forms the **epidermis** or outer cell layer of the shoot), the **procambium** (which forms **phloem** and **xylem**, see Vascular tissue, below), and the **ground meristems** (which form **parenchyma**). Grasses have **intercalary meristems** close to the nodes of the stem (the point where leaves attach) which are also responsible for length increase.

   Lateral meristems give increase in girth (**secondary growth**). The **vascular cambium** (sometimes known just as **cambium**) is a cylinder of cells that forms new phloem and xylem (see Vascular tissue, below). New cells formed by the cambium are termed **derivatives**, while those that remain to divide again are **initials**. The **cork cambium** is a meristematic tissue found in woody plants. It is a cylinder of cells, located beneath the bark, which it forms (*Fig. 1*).

Tissues and organs

A **tissue** is a collection of cells with common function. It may be **simple**, made of one cell type or **complex**, made of several cell types. An organ is a collection of tissues with a specific function such as flowers, roots and leaves. Plant organs are made of three types of tissue: **dermal** (the epidermis or outer cell layer), **vascular** (transport tissue) and **ground** (all the remaining cells). The type and quantity of each tissue varies for each organ.

Ground tissues

Ground tissues lie under the epidermis and contribute to structural strength and function. **Parenchyma** cells are the most abundant cell types found throughout the plant and form the bulk of organs such as leaves, roots and stems. They are formed with thin flexible cell walls and are initially cuboid, later becoming nearly spherical or cylindrical. As they form mature tissue, their shape is constrained by surrounding cells. In a typical parenchyma tissue, cells are joined at the points of contact, but there are frequent large **intercellular spaces**. This means they take a variety of shapes. Parenchyma cells have large vacuoles, and parenchyma are frequently storage tissues for starch and other food reserves and for water. Leaf parenchyma cells are photosynthetic, have chloroplasts and are called **chlorenchyma**. **Aerenchyma** is a parenchyma tissue found in some species in which some cells are lost to form large gas spaces. These promote buoyancy in some aquatic species or supply oxygen to the growing tip of roots in waterlogged soil. Parenchyma cells often retain the ability to divide when the plant is wounded, forming a **callus** (an amorphous mass of cells; see Topic J1) which fills the wound site and may eventually form new tissues to repair the wound.

Collenchyma is made up of living cells which are similar to parenchyma, but with stronger, unevenly thickened cell walls. Collenchyma cells are often found as flexible support beneath the epidermis of growing tissues and in many other locations, including stems, leaves and fruits. Cotton is a fabric made of plant collenchyma tissue.

Sclerenchyma is a supporting tissue found in organs which have completed lateral growth. It is made up of dead cells with even, lignified secondary cell walls either in **sclereids** (small groups of cells) or **fibers** (long strands of elongate cells). Plant fibers have been widely used in textiles and rope making.

Dermal tissue

All organs are surrounded by a layer of cells, the **dermal tissue** that forms a protective covering, known as the **epidermis**. It consists of parenchyma or parenchyma-like cells and usually forms a complete covering, except where specialized pores (e.g. for gas exchange; see Topic E2) are present. The epidermis protects the tissues beneath from mechanical damage and pathogens (Topic M4). The root epidermis (or **rhizodermis**) is specialized for the absorption of mineral nutrients and water (Topics E1 and E4).

Vascular tissue

Xylem and **phloem** together make up the **vascular tissue**, the conducting tissues for fluids through the plant (*Fig. 2*). Xylem carries predominantly water and dissolved minerals from root to shoot (Topics E1 and E4); phloem carries solutes like sugars and amino acids from sites of synthesis or storage (sources) to sites of storage or use (Topic F4).

Xylem cells have lost their cellular contents, and have unevenly thickened secondary cell walls that are strengthened with lignin. There are two main types of xylem cells. **Vessel elements** are elongate, open-ended cells which lose their end walls to form continuous pipes (**vessels**). **Tracheids** are tapered cells, with overlapping ends and often spiral thickening of the cell wall. Rather than being open-ended, perforations or pits allow water to move from one cell to another. The first formed xylem cells are known as **protoxylem**. They have

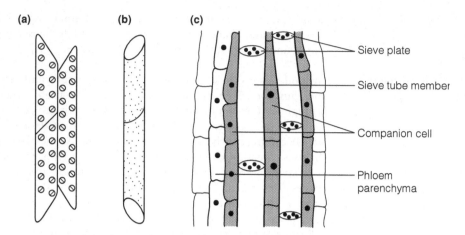

Fig. 2.   Structure of xylem and phloem (a) xylem tracheids showing pits (b) a xylem vessel (c) phloem.

helical or annular secondary thickening, so the cells can elongate when stretched by the surrounding tissue. Metaxylem cells form after elongation has ceased and have more complex secondary thickening. **Ray cells** radiate from the vascular cambium of roots or stems of woody plants and permit lateral movement of water.

Phloem tissue has two cell types. **Sieve tube** members form an elongated pipe, the sieve tube. The cells are living and are linked through perforations in adjoining end cell walls (the **sieve plate**). The cytoplasm of these adjoining cells also connects, and transport of solutes takes place through it (Topic H4). Sieve tube members have no nuclei, but retain other functional organelles. The other cell type is the **companion cell**, which has many mitochondria and provides energy required for phloem transport (Topic F4).

D2 ROOTS

Key Notes

Primary structure	The tip of a root is protected by a root cap behind which is a meristem and elongation zone. Nutrient and water uptake occurs through root hairs, formed as protrusions of epidermal cells in the maturation zone. The vascular cylinder becomes mature in this zone, with a water-impermeable endodermis and functional phloem and xylem.
Root apical meristems	Closed roots have three layers of cells (initials) in the apical meristem which form all the tissues present. Open roots have three meristems close to the apical meristem: the procambium which forms the vascular cylinder, the ground meristem which forms the cortex and the protoderm which forms the epidermis.
Root cap	The root cap is a cone of cells which protects the meristem and secretes mucilage. Root cap cells are constantly sloughed off. The mucilage provides an environment for bacteria and fungi which live in the rhizosphere, the immediate environment of the root.
Root hairs	Root hairs are outgrowths of epidermal cells giving a high surface area of contact with the soil for water and mineral nutrient uptake. They are short-lived and the mature regions of the root are hairless.
Root architecture	The primary root persists as a taproot and lateral roots arise from it in many plants, though in monocots it dies and adventitious roots grow from the stem. These give rise to a mass of fine, fibrous roots, all of similar size. Root architecture is responsive to soil depth, water availability and nutrients.
Related topics	Meristems and primary tissues (D1) Uptake of mineral nutrients by Plants and water (E1) plants (E4)

Primary structure   **Roots** extract water and minerals from soil. They penetrate the soil, growing by elongation near the tip. As soil is a resistant medium, the growing tip has a near conical protective **root cap** (*Fig. 1*) which lubricates the root surface as it is pushed by cell expansion between soil particles. Growth occurs when new cells, formed in meristems in the **zone of cell division** elongate in the **elongation zone** (*Fig. 1*). This elongation is driven by **hydrostatic pressure** within the cell (Topic E1) that propels the root tip deeper into the soil. The elongation zone is effectively cylindrical, allowing growth in the soil. Behind it, in the **maturation zone**, root hairs develop which provide large areas of contact with the water and nutrient films surrounding soil particles and anchor the older parts of the root.

Internally, the ground tissue of the root is the **cortex**, made up of large, vacuolate cells which fill the space between the epidermis and **vascular cylinder** (**stele**).

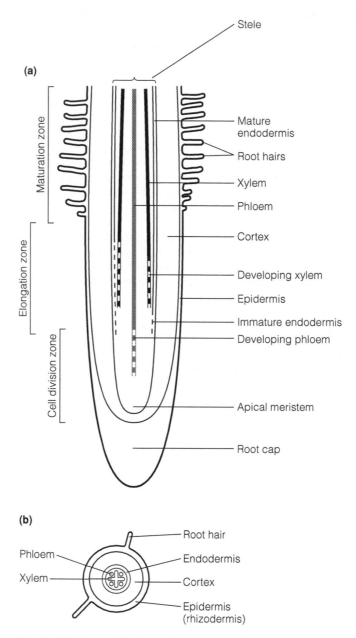

Fig. 1. (a) A generalized representation of a longitudinal section of a typical root. (b) The transverse section is from the maturation zone and is of a root with four groups of xylem cells forming a cross-like appearance. This is known as a tetrarch root; other species are diarch (two), triarch (three) or more.

While all the cells in the root can be traced back to their origins in the root meristems (see below), many of the key tissues of the root cannot be observed until they form in the maturation zone (*Fig. 1*). The vascular bundle (stele) contains phloem (which develops first) and xylem surrounded by a cylinder of cells with water-resistant cell walls, the **endodermis** (Topic E4). Endodermal cells possess a characteristic thickening of the **anticlinal cell walls** (radial and transverse walls, which are perpendicular to the root surface). This wall thickening, the **Casparian strip**, is impregnated with a water-impermeable substance, **suberin**.

Immature xylem and endodermis can be observed below the area where root hairs form (*Fig. 1*). Later, above the root hair zone, **lateral roots** may form. These originate in the cells near the vascular tissue and force their way out through the cortex and epidermis. Each lateral root has a meristem and its vascular tissue interconnects with that of the primary root.

Root apical meristems

The meristem lies behind the root cap and comprises small, thin-walled cells with prominent nuclei and no large vacuole (Topic D1). Two types of root have been suggested. In **closed roots** (for instance in corn [maize]; *Fig. 2a*), three layers of cells (**initials**) can be identified in a single region in the **apical meristem**, which form all the tissues of the root. In the second type, **open roots** (for instance onion; *Fig. 2b*), there appear to be three meristems which separately form the tissues of the root.

Root cap

The **root cap** is a conical collection of parenchyma cells, which in some plants is large and conspicuous and in others small. It protects the root tip as it is forced through the soil and secretes **mucilage** as lubricant made in the Golgi apparatus of **slime-secreting cells**. Root cap cells have a short life (up to 1 week, depending on species) and are constantly sloughed off and may remain alive in the mucilage for some time. The mucilage provides an environment for bacteria and fungi which live in the **rhizosphere**, the immediate environment of the root.

Root hairs

Root hairs are single cell structures formed as outgrowths of epidermal cells (*Fig. 3*). They form a **root hair zone** that gives a very high surface area of contact with the soil. While root hairs are normally only 0.1–1.0 cm long, they can be extremely numerous, with more than 20 000 hairs cm^{-2} of root and billions per plant, giving a vast surface area for absorption. Root hairs are also short-lived and the mature regions of the root are hairless. The root hair zone is

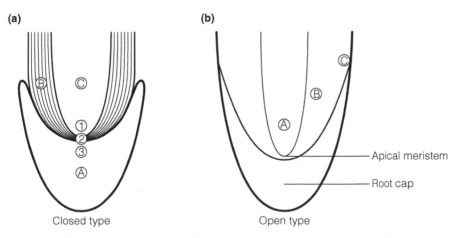

Fig. 2. (a) Closed and (b) open roots. Closed roots have a single region of cell division where three layers of cells (initials; 1, 2, 3) give rise to the epidermis (A), the cortex (B) and the vascular cylinder (C). In an open root, there are three separate meristems near the apical meristem. The procambium (A) gives rise to the vascular cylinder; the ground meristem (B) gives rise to the cortex; and the protoderm (C) gives rise to the epidermis.

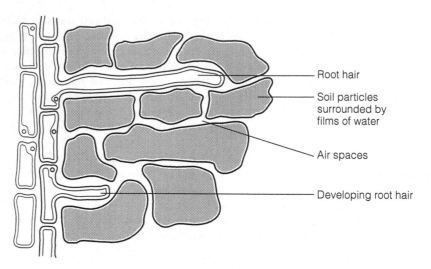

Root hair

Soil particles surrounded by films of water

Air spaces

Developing root hair

Fig. 3.   Root hairs. A root hair forms as an outgrowth of an epidermal cell and is a unicellular structure.

therefore extremely important for water and mineral nutrient uptake, though in plants with mycorrhizal fungi (Topic M1) associated with the root, root hairs may be much less important or absent.

Root architecture   All plants first form a **primary root** and in most plants, including the gymnosperms, primitive dicotyledons and eudicotyledons this persists as a **taproot**. The **lateral roots** arise from it at various points (*Fig. 4a*). In the monocotyledons the primary

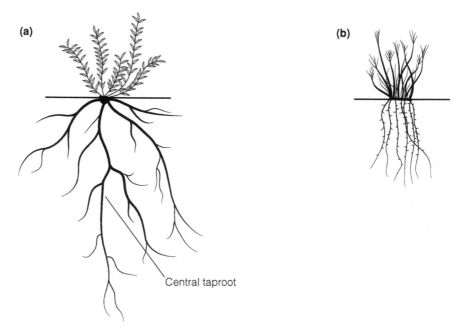

(a)

(b)

Central taproot

Fig. 4.   Root architecture. (a) Taproot systems result when lateral roots branch out from a single central taproot. (b) Fibrous roots form from adventitious roots, branching out from the stem base.

root dies and **adventitious roots** are formed which grow out of the stem. These give rise to a mass of fine, **fibrous roots**, all of similar size and prominence (*Fig. 4b*). Root architecture always follows one of these basic plans, but is also responsive to soil depth, water availability and nutrients.

D3 HERBACEOUS STEMS AND PRIMARY GROWTH

Key Notes

Shoot structure

Herbaceous stems are green and nonwoody. They have a range of growth forms generated by an apical meristem that produces leaf and bud primordia. The shoot is surrounded by an epidermis, within which lies the cortex. Vascular tissues occur as a ring of separate bundles towards the outside of the stem in dicotyledons and in a scattered pattern in monocotyledons. The center of the stem is the pith parenchyma.

Meristems

The shoot meristem has a tunica producing the epidermis and the layers of cells beneath it while the corpus produces the cortex, pith and vascular tissues. The meristem can also be divided into the central zone where cell division occurs, the peripheral zone where leaves, shoots and new meristems form, and the rib zone where stem growth occurs.

Vascular tissue

Vascular bundles containing phloem and xylem form a ring or a complex array throughout the inner part of the stem. In some dicotyledons the ring is almost continuous.

Architecture

The shape of the stems is governed by the controlled positioning of leaves and buds by the meristem. Their positioning may be spiral, distichous, opposite, decussate or whorled.

Related topics

Meristems and primary tissues (D1)
Woody stems and secondary
   growth (D4)

Features of growth and
   development (J1)

Shoot structure

Herbaceous stems are green throughout without deposition of lignin that would make them woody. Shoots are more complex than roots and take on a wide range of growth forms reflecting function. Like the root, the shoot has an **apical meristem.** Unlike the root (where lateral branching only occurs in mature tissue), the shoot meristem produces lateral shoots and leaves (*Fig. 1*).

The apical meristem produces **leaf primordia** (which will form leaves) and **bud primordia** (which will form shoots). These are produced in position and order which gives rise to the characteristic form of the shoot which is recognizable for each species (see Architecture, below).

The shoot is surrounded by an **epidermis.** This outer layer provides the protective barrier between the stem and its environment and is covered in a lipid-based protective substance, cutin (Topics E1 and M4). Within the epidermis, cells of the **ground tissues**, the **cortex**, may be photosynthetic and occupy the space surrounding the **vascular bundles.** In some species, the cortex is only

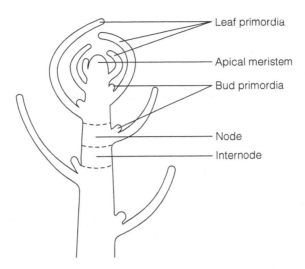

Fig. 1.   A longitudinal section of a simplified dicot stem.

a few cell layers, in others many more. Within the cortex, bounded by the vascular bundles, the center of the stem is occupied by **pith** (*Fig. 2*).

Meristems   The shoot meristem produces the cells that will form the major tissues of new stems and leaves. It is organized in layers of cells each producing a different tissue and in zones with different functions (*Fig. 3*). **Layer 1** produces the

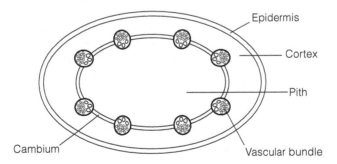

Fig. 2.   A transverse section of a simplified dicot stem, showing the vascular bundles containing phloem and xylem arranged in a cylinder and separated by interfascicular cambium.

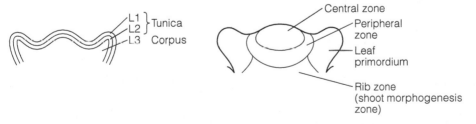

Fig. 3.   A schematic representation of a longitudinal section through a shoot apical meristem. Layer L1 forms the epidermis, L2 the cell layers beneath the epidermis and L3 the pith, cortex and vascular tissue. (Redrawn from R. Twyman (2001), Instant Notes in Developmental Biology, BIOS Scientific Publishers Ltd.)

epidermis that surrounds the shoot. Here cells divide to generate only a single layer. Beneath it, **layer 2** divides to generate the layers of cells beneath the epidermis. Layer 1 and layer 2 together are known as the **tunica**. **Layer 3** produces all the other tissues of the shoot: the cortex, pith and vascular tissues, and is known as the **corpus**. The **central zone** contains actively dividing cells. Leaves and shoots (including new meristems) form in the **peripheral zone**. Stem growth occurs in the **rib zone** and cells mature to form the tissues of the shoot.

Vascular tissue

The vascular tissue of the shoot commonly occurs as a ring of separate vascular bundles, each containing **phloem** and **xylem**. They are located between the cortex and pith (*Fig. 2*) in dicotyledons, while in monocotyledons they are found in a scattered array throughout the inner part of the stem. All vascular bundles contain phloem and xylem strands in parallel to one another. Associated with the xylem strands are **xylem parenchyma** cells and fibrous **sclerenchyma** tissue may be associated with the phloem. In some dicotyledons, the vascular bundles form an almost continuous ring; however, individual vascular bundles remain, separated by narrow bands of **interfascicular parenchyma** (the site of origin of the cambium for secondary growth) which separate the cortex from the pith (*Fig. 2*).

Architecture

The shapes of the stems of different species are clearly recognizable from one another (*Fig. 4*). For instance, before flowering, *Arabidopsis thaliana* (Topic B1) forms a **rosette** of leaves, each initiated 137.5° from the preceding one formed. This process, termed **spiral** (or helical) **phyllotaxy**, is common in many species and results in the formation of a spiral pattern of growth. In other species, new leaves may be formed singly at each node in two opposite rows (**distichous phyllotaxy**), in pairs opposite one another (**opposite phyllotaxy**) or in whorls of five or more at one node (**whorled phyllotaxy**). Where pairs of leaves form at right angles to a previous pair of leaves, the plant has **decussate phyllotaxy**.

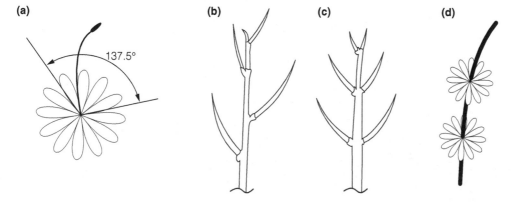

(a)   **(b)**   **(c)**   **(d)**

137.5°

Fig. 4.   Forms of phyllotaxy: (a) spiral phyllotaxy, (b) distichous phyllotaxy, (c) opposite phyllotaxy, (d) whorled phyllotaxy.

D4 WOODY STEMS AND SECONDARY GROWTH

Key Notes

Secondary growth
Secondary tissues are formed by secondary meristems and give increases in diameter and strength. Secondary tissues are frequently woody, strengthened by deposition of lignin, a polymeric phenolic compound, mostly in secondary cell walls.

Vascular cambium and cork cambium
Increase in diameter results from the action of the vascular cambium and the cork cambium. The vascular cambium lies between the primary phloem and xylem and divides to form phloem on the outside and xylem on the inside. The cork cambium lies within the stem cortex and generates cells filled with suberin. Gas exchange occurs through lenticels where stomata once were.

Wood anatomy
The center of a tree trunk contains heartwood, where the vascular system no longer functions. Around it lies the sapwood that contains functional phloem and xylem. Vascular rays of long-lived parenchyma cells radiate across the trunk, conducting nutrients and water. Bark is made up of the outermost layers of the trunk, including cork and outer phloem cells.

Related topics
Meristems and primary tissues (D1)
Herbaceous stems and primary growth (D3)

Features of growth and development (J1)

Secondary growth
Many species of plant complete their life cycle using only the primary tissues generated by primary meristems. These are the herbaceous species described in Topic D3. In many other species, production of primary tissue and elongation growth are followed by the deposition of **secondary tissue**. These tissues give increases in diameter and the tissues formed are strengthened in comparison with primary tissues by deposition of extensive secondary walls and **lignin**, a polymeric phenolic compound, in the cell wall making them **woody**. Plants showing secondary growth usually live for many years, the wood making them resistant to damage by herbivores and weather. Monocots do not normally generate secondary tissues, but some, like palm trees, undergo additional primary growth to form thick stems. Some palms also continue cell division in older parenchyma tissue to give what is known as diffuse secondary growth.

Vascular cambium and cork cambium
Increase in diameter results from the action of two secondary meristems, the **vascular cambium** and the **cork cambium**. The vascular cambium is a narrow band of cells between the primary phloem and xylem (*Fig. 1*), which remains a meristem. This tissue goes on dividing indefinitely with active growth in spring and early summer in temperate trees, with the new cells being formed to the

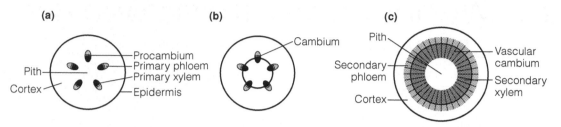

Fig. 1. *Development of a woody angiosperm stem. When primary growth is complete, the stem has a number of vascular bundles with procambium separating primary phloem and xylem (a). At the onset of secondary growth, this cambium forms a ring (b) which develops into the vascular cambium which generates secondary xylem on its inner, and secondary phloem on its outer face (c).*

outside of the cambium to form phloem and the inside to form xylem (*Fig. 2*). The newly formed **secondary phloem** and **xylem** function like the primary vascular tissue in transport. As xylem cells are deposited on the inner face, the vascular cambium moves outwards. In temperate regions early in the growing season, the cambium produces xylem cells of large diameter and these get progressively smaller as the season progresses. In consequence, growth appears as more- and less-dense bands of cells, the familiar **annual rings** observed in cross section of a tree trunk. Some tropical species on the other hand, such as ebony, produce xylem consistently throughout the growing season and their wood is of a consistent appearance without annual rings, though many species of the seasonal tropics have annual cycles similar to those of temperate trees.

The cork cambium arises within the stem cortex and generates cuboid cells that quickly become filled with the waxy substance **suberin** (also found as water-proofing in the root endodermis; Topic E1). Suberinized cells die, but the dead cells remain as a protective layer (*Fig. 3*), required because the original epidermis can no longer function. **Cork tissue** forms part of the **bark** of the tree (the remainder being secondary phloem) and replaces the epidermis. Its character and thickness vary from species to species. Some gas exchange to the stem occurs through **lenticels**, pores remaining through the bark.

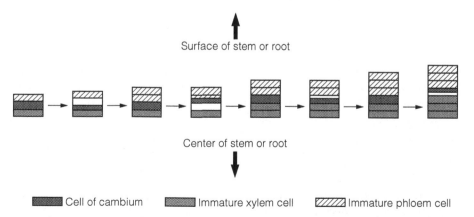

Fig. 2. *The cells of the vascular cambium divide to form secondary phloem to the outside and secondary xylem to the inside. (Redrawn from Stern, K.R. (2000)* Introductory Plant Biology, *8th Edn, McGraw-Hill Publishing Company.)*

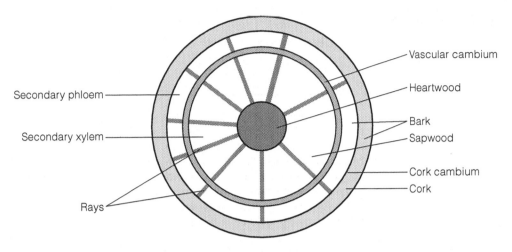

Fig. 3. A transverse section of a woody stem, showing the major layers of heartwood, sapwood and bark.

Wood anatomy   If a transverse section of a tree trunk is studied (*Fig. 3*), a number of features appear. Concentric circles of annual growth rings are evident in most trees from seasonally varying environments. The center of the trunk of many species is the **heartwood**, where the vascular system no longer functions. It provides structural support and is often darker and impregnated with tannins (Topic F5), but may be completely removed without killing the tree. Around the heartwood lies the **sapwood** that contains functional phloem and xylem. **Vascular rays** radiate from the center of the trunk. These are formed of long-lived parenchyma cells which conduct nutrients and water across the trunk, crossing both phloem and xylem, and are responsible for secretions into the heartwood.

The outermost layers of the trunk are known as **bark**. Bark includes the **corky tissues** produced by the cork cambium and the underlying layers outside the vascular cambium, including the secondary phloem.

D5 LEAVES

Key Notes

Leaf types

Leaves are the main photosynthetic organs of plants. The leaf blade lamina generally has a large surface area to maximize light capture and may have a petiole. Compound leaves have a lamina divided into leaflets and a central rachis. The lamina may be much reduced (e.g. in conifer needles). The vascular tissue of the leaf occurs as veins, which are branched from a central midrib forming a network, or are parallel to one another.

General leaf structure

A typical dicot leaf has upper and lower surfaces protected by the epidermis with a waxy cuticle and may have leaf hairs. The photosynthetic cells are the palisade and spongy mesophyll. Both cell types are surrounded by gas spaces. Carbon dioxide enters these by way of stomatal pores in the epidermis connecting with the gas spaces of the mesophyll. Transport to and from the leaf occurs in veins which contain xylem and phloem.

Other types of leaf

Some species show Krantz anatomy in which the mesophyll surrounds bundle sheath cells around the vascular bundle in a ring. Leaves adapted to drought may be thickened and fleshy with a low surface area. Conifer needles have a small surface area, sunken stomata, a hypodermis beneath the epidermis and an endodermis surrounding the vascular tissue.

Related topics

Plastids and mitochondria (C3)
Herbaceous stems and primary
  growth (D3)
Features of growth and
  development (J1)

Water retention and stomata (E2)
C3 and C4 plants and CAM (F3)

Leaf types

The major photosynthetic organ of most plants is the **foliage leaf** (*Fig. 1*). This is mainly made up of cells containing chloroplasts. They are generally flat and thin with a lamina of large surface area attached to the plant by a stalk or **petiole**. The structure is arranged so that both the leaf lamina and the chloropasts can be orientated to the sunlight. The lamina is commonly thin so that light does not have to penetrate far (losing energy) before it impinges on a chloroplast. However, such large, flat surfaces are easily desiccated or damaged and a variety of leaf adaptations exists.

Leaves with a stalk or **petiole** are **petiolate**. A lamina without a petiole is a **sessile** leaf. Some monocotyledons, such as the grasses, have a long narrow leaf blade without a petiole, the base of which forms a sheath around the stem. In some species, the lamina is divided into smaller leaflets, forming a **compound leaf**. In these, the petiole is extended to form a **rachis**, with **leaflets** (*Fig. 2*). The vascular tissue of the leaf appears as **veins**, which give strength and rigidity. Most dicots and some monocots have a prominent central midrib and a network of veins around this; others have several parallel veins. In conifers, the

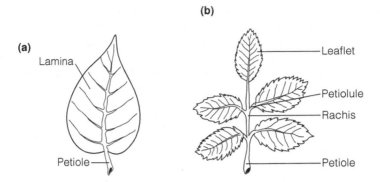

Fig. 1.   (a) A simple and (b) compound leaf showing the names of the major structures of
the leaf.

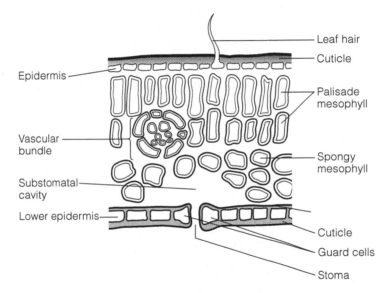

Fig. 2.   A transverse section of a leaf, showing the major tissues and cell types present. Note
the photosynthetic tissues of the palisade and spongy mesophyll.

leaf is reduced to a **needle** or scale with a single vascular strand and much
reduced surface area.

General leaf structure

A dicotyledonous leaf, like that of the bean or pea, is shown in *Fig.* 2. The upper
and lower surfaces are protected by an **epidermis** of closely packed cells, coated
with a **waxy cuticle** to minimize desiccation. **Leaf hairs** or **trichomes** may also
project from the epidermis, which lessen air movement across the leaf (and therefore water loss). Beneath the epidermis lies the **palisade mesophyll**, normally one
or two layers of elongated cells orientated perpendicular to the epidermis with
many chloroplasts. Beneath this lies more photosynthetic tissue, the **spongy
mesophyll**, surrounded by many air spaces, beneath which lies the **lower
epidermis**. Carbon dioxide required by the leaf enters the leaf gas spaces by way
of **stomata** (singular **stoma**), pores in the epidermis which are regulated by specialist
epidermal cells called **guard cells** (Topic E2). The stomata lead to a cavity, the
substomatal cavity, which connects with the gas spaces of the mesophyll.

Other types of leaf

Some species such as corn (maize; *Zea mays* L.) that show **C4 photosynthesis** (Topic F3), have a leaf structure in which a ring of **bundle sheath cells** surrounds the vascular bundle with mesophyll cells radiating from it in a **ring** (**Krantz [wreath] anatomy**; *Fig. 3*). These leaves have one type of mesophyll cell surrounding the bundle sheath with gas spaces beneath the stomata. Some **succulent** species that are adapted to drought have highly thickened, fleshy leaves that store water, a low surface area and thick cuticle that minimizes water loss. Frequently they only open their stomata at night (Topic F3). The needles of pines and other conifers have a small surface area, sunken stomata and a thick layer of protective tissue, the **hypodermis** just beneath the epidermis (*Fig. 4*), together with a protective layer (the **endodermis**) surrounding the vascular tissue. Other leaf adaptations to drought include having more stomata on the surface less exposed to desiccation and folds and tube-like structures that minimize air movement across leaf surfaces.

Leaves may be reduced to spines, as in cacti, or lost altogether in some plants in which the stem is the most important photosynthetic organ. A few other plants have outgrowths from the petiole or main stem that resemble leaves and act like them. The acacias of Australia, for instance, have flattened petioles but no lamina, while the butchers' broom of Europe has no leaves, but flattened stem outgrowths that bear flowers. **Modified leaves** also have a number of important roles in the plant, for instance as storage tissue in the **fleshy scales** of bulbs, as the **protective scales** of buds or as **tendrils**, which give support to climbing plants.

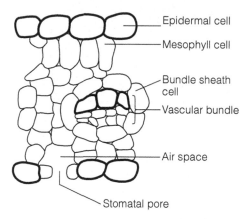

Fig. 3. *A transverse section of the leaf of a species showing Krantz anatomy. Note the absence of spongy mesophyll and the arrangement of mesophyll around bundle sheath cells forming a wreath-like structure.*

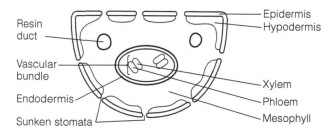

Fig. 4. *A transverse section of a pine needle, which shows adaptation to drought. Note the low surface area, the protective hypodermis below the epidermis, sunken stomata and the endodermis surrounding the central vascular bundle.*

E1 Plants and water

Key Notes

Properties of water	Water is polar and a good solvent for charged or polar solutes. Hydrogen bonding means that water is a liquid at temperatures common for plant growth and has high cohesive (tensile) strength.
Water movement	Water moves from high water potential to low water potential. Water movement occurs by diffusion and by mass flow, where flow rate depends on cross-sectional area of the tube, the pressure gradient and the viscosity of the liquid. Embolism may block flow. Water potential is the sum of the pressure potential and the osmotic potential. A cell generating a positive hydrostatic pressure is turgid; one in which it is negative is flaccid.
Transpiration	Evaporation of water from the leaves through stomata generates a low water potential and results in the movement of water from the soil through the root system and into the xylem.
Xylem water flow	The cohesion-tension theory suggests that water is drawn upwards through the xylem by tension created by transpiration at the leaves. Other forces, including root pressure and capillary action contribute.
Water transport in roots	Water predominantly enters roots via root hairs and mycorrhizae. Water transport may be apoplastic or symplastic (transcellular via plasmodesmata or transmembrane via membrane transporters). At the endodermis, water movement must be cytoplasmic. Water enters the xylem as a result of the low water potential generated by the transpiration stream.
Related topics	Roots (D2)
	Water retention and stomata (E2)
	Movement of nutrient ions across
	membranes (E3)

Uptake of mineral nutrients by
  plants (E4)

Properties of water

Water is essential for plant growth. All cells contain water as the solvent in which biochemical reactions take place and in which cell structures are maintained. Water shows **hydrogen bonding** (between electronegative oxygen and electropositive hydrogen), which means that it is a liquid at temperatures common for plant growth. As it is polar, it is a good solvent for polar molecules like ions and charged organic molecules. The strong forces between water molecules (**cohesion**) give it several other key properties; it has a very high **surface tension** and adheres strongly to surfaces. It also has a high tensile strength, and is able to form columns under high tensions without breaking, for example in xylem.

Water movement

Water molecules are in constant random motion. **Diffusion** occurs when molecules migrate as a result of this motion. Water will move progressively from regions of high free energy (high water concentration) to regions of low free energy (low water concentration) down a concentration gradient. Diffusion is important over short distances, for instance within a plant cell, but not over long distances, like from soil to leaf.

Mass or bulk flow

Movement of water through xylem is largely by **bulk flow** that occurs as a response to a pressure gradient. Fluid flow through a pipe depends on the pressure gradient between the ends of the pipe, the radius of the pipe and the viscosity of the fluid. As the radius doubles, the flow rate increases by a factor of 2^4 (16); therefore, flow in larger pipes can be much faster than in small ones. Larger pipes are much more susceptible to **embolism** and **cavitation**, the formation of air bubbles and the break up of the water column. This occurs when the pressure gradient is provided by a tension, a force drawing from above, rather than a pressure below. Vessels are broader in general than tracheids and by increasing water flow may allow the flowering plants to grow more quickly than most other seed plants.

Water potential

The **chemical potential** of water is the amount of free energy associated with it. **Water potential** is defined as the chemical potential of water divided by the volume of a mole of water. It is measured in $J \, m^{-3}$ or Pascals (Pa; $N \, m^{-2}$). The symbol used for water potential is ψ_ω; it has two major components, **solute or osmotic potential**, ψ_s, and **pressure potential**, ψ_p, such that $\psi_\omega = \psi_s + \psi_p$. The solute or osmotic potential, ψ_s, is dependent on the solute concentration and the temperature. The **pressure potential**, ψ_p, is the hydrostatic pressure in excess of atmospheric pressure developed by the cell or tissue. Water moves from areas of high water potential to areas of low water potential. Water entering a cell will result in an increase in volume. If the cell wall stops that volume increase, the **hydrostatic pressure** will increase. Eventually, the positive hydrostatic pressure equals the negative osmotic potential and the water potential of the cell reduces to zero (i.e. $\psi_\omega = 0$ on both sides of the membrane). At this point there is no net movement of water into or out of the cell.

Turgidity and plasmolysis

A cell will take up water when it is immersed in a solution of lower solute concentration (a **hypotonic** solution). This will generate a **hydrostatic pressure (turgor pressure)** in the cell. In such a cell, the cell contents exert a pressure on the cell wall and the cell is turgid. A cell will lose water in a **hypertonic** solution (i.e. a solution with a higher solute concentration – and therefore a more negative osmotic potential than the cell cytoplasm), until the hydrostatic potential becomes negative. At this point, the plasma membrane will pull away from the cell wall and the cell will be **plasmolysed** (flaccid). The **point of incipient plasmolysis** occurs when the plasma membrane is in contact with the cell wall, but no hydrostatic (turgor) pressure is generated; at this point, $\psi_\omega = \psi_s$ as $\psi_p = 0$ (*Fig. 1*).

Water channels (aquaporins)

The movement of water across cell membranes is limited by its low solubility in the lipid bilayer. Aquaporins are proteins that permit water to cross a membrane. Their regulation is important in regulating water potential.

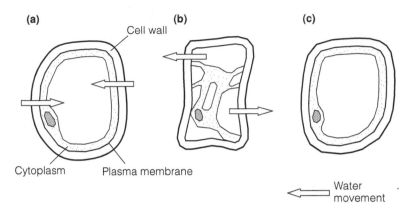

Fig. 1. *A cell in a hypotonic medium is fully turgid; (a) ψ_ω and ψ_p are positive. Cells in a hypertonic solution will lose water by osmosis and both ψ_ω and ψ_p are negative (b). At the point of incipient plasmolysis $\psi_p = 0$ and $\psi_p = \psi_s$ (c).*

Transpiration

Transpiration is the process by which water is drawn from the soil through the plant as a result of evaporation from the leaves. Surfaces exposed to the air are generally covered with a layer that resists water loss (Topic D5). **Stomata** in the leaf surface permit water loss by evaporation from the leaf (Topic D5). Most transpiration (90–95%) occurs through these pores. The rate of transpiration increases with temperature and with wind speed. Changing stomatal aperture (Topic E2) varies the rate of water loss in changing environmental conditions.

Xylem water flow

The water conducting tissue of the plant is the **xylem**. It is made up of elongated **cells** with walls thickened and strengthened by secondary wall deposits. Its structure is given in detail in Topic C1. Three possible driving forces exist for water flow in the xylem: **root pressure, capillary action** and **cohesion-tension** (in which a column of water is drawn up from the soil by forces generated by evaporation at the leaf surface).

Root pressure
When the stem of a plant is cut, xylem fluid often exudes from the cut. This exudation is driven by root pressure. It occurs where accumulated solutes in the xylem cause the influx of water into the xylem by osmosis. The suberinized endodermal layer prevents back-flow of water and a hydrostatic pressure is generated, causing water movement. Root pressure is insufficient to explain water movement to the upper leaves of a tall tree and is not observed in all plants. It is therefore unlikely to be the major cause of xylem water flow.

Capillary action
Capillary action is generated by the adhesive forces between the surface tension in the meniscus of water and the wall of a tube. While capillary effects occur, the total elevation of water achieved by capillaries of a diameter typical of xylem elements is less than a meter – insufficient to explain water transport to the top of a tall tree.

Cohesion-tension
Cohesion-tension is a theory commonly used to explain water flow through plants. The driving force is provided at the leaf, where evaporation generates

tension within the aqueous parts of the tissue. This is particularly strong where the water surface bridges microscopic gaps, for instance among the microfibrils of the cell walls mesophyll cells (*Fig. 2*). Evaporation from the surface of a leaf causes the water to retreat to these microscopic spaces, where it adheres to hydrophilic wall components. Cohesion of the water molecules (surface tension) results in the formation of a concave **meniscus**. This is pulled by adhesion and cohesion of water molecules to the walls and of water molecules to each other generating a negative pressure. A meniscus drawn into a pore of radius 0.01 μm has a tension of 15 mPa, more than sufficient force to drive water movement from the soil; the spaces present in the cell walls are even smaller than this, so the forces may be even greater. The theory then suggests that leaf water is part of a continuous column running through the xylem to the root. The whole water column is under tension and water is drawn upwards from the soil. If cohesion-tension is the sole mechanism drawing water up the plant, the column of water must be under considerable tension; in a 100-m tree, a force of up to 3.0 MPa is present. There are problems with cohesion-tension as an explanation for water movement, however. Water under extreme tension is very unstable and may vaporize. Dissolved gas results in bubbles (**embolisms**) that block xylem flow. Damage due to embolisms is minimized as the xylem is divided

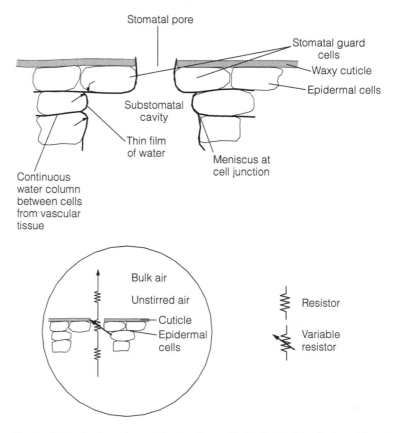

Fig. 2.   Water is drawn by evaporation from cells bordering the substomatal cavity. As water retreats to the small spaces between the cells, a meniscus with a high surface tension develops and a negative pressure results.

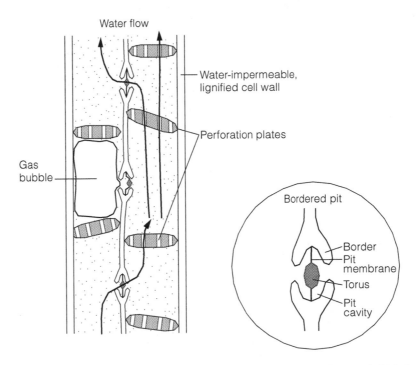

Fig. 3.   The action of bordered pits and perforation plates to contain a gas bubble (embolism) formed in either a vessel element or tracheid. Note that the central torus acts as a valve to seal the pit in the region of the bubble.

into many small pipes, with interconnections via pits (see above). *Figure 3* illustrates the action of pits and perforation plates in containing an embolism. Recent re-analysis of the evidence has suggested that a variety of forces must be involved in water transport through the plant, which is best considered as occurring in a series of stages, rather than a single continuous column.

Water transport in roots

In most plants, water uptake occurs predominantly through **root hairs**, fine, extensions of single epidermal cells that enter the water film on soil particles (*Fig. 4*). They provide a very large surface area for absorption. Mycorrhizal fungi may also fulfill this function (Topic M1). The internal anatomy of the root is also important. The xylem and endodermis (a water-impermeable cell layer with suberinized cell walls, through which water movement must be symplastic; Topic D2) develops some distance behind the root tip. Water flows either through the cell wall spaces (the **apoplast**) or through the cell contents (the **symplast**). Water may be taken up from, or lost to, the apoplast by any cell in the pathway. Movement of a water molecule across the root involving both pathways is termed **transcellular transport**.

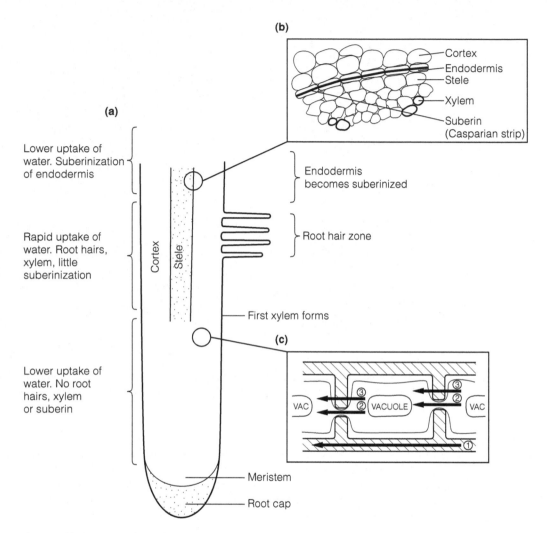

Fig. 4. (a) Key features of water transport in a primary root. Root hairs increase root surface area in contact with films of water around soil particles. The suberinized endodermis of older tissue restricts water uptake. Inset (b) illustrates the position of the endodermis between the cortex and vascular tissue. Inset (c) shows the possible pathways of transport through root cells: (1) apoplastic; (2) symplastic through plasmodesmata; and (3) symplastic (transmembrane).

E2 WATER RETENTION AND STOMATA

Key Notes

Plant types	Plants are categorized as mesophytes adapted to an environment in which water is available, xerophytes adapted for areas of low water availability and hydrophytes adapted to the presence of large amounts of water or to growth entirely in water.
Plant structure and surfaces	Leaf structure is adapted to regulate and minimize water loss in unfavorable conditions. The close-packed, flattened epidermal cells of leaf surfaces are covered in hydrophobic cuticles made of cutin and wax, which minimize evaporation. Stomata are most abundant on the lower side of the leaf and may be sunken into the leaf surface.
Stomata: structure	Diffusion of gas is essential for photosynthesis. Stomata (singular, stoma) permit gas diffusion into the leaf while minimizing and regulating water loss. The stomatal pore penetrates the epidermis into the substomatal cavity in the mesophyll. The pore is bounded by two guard cells, which are associated with subsidiary cells in the epidermis. Stomata are open when the guard cells are turgid and closed when they lose turgor.
Stomata: action and regulation	Stomatal opening responds to a number of environmental factors, in addition to drought, including: CO_2 concentration, light and temperature. Stomata are held open by turgor from high solute concentrations in guard cells. Abscisic acid causes release of solutes via regulated anion channels in the plasma membrane. This reduces the osmotic potential of the guard cell and water flows from the cells. Opening results when the anion channels close and solutes re-enter the cell.
Stomatal adaptations of xerophytes	Xerophytes are adapted to low water availability. Typical characteristics include: thick fleshy leaves or no leaves (succulence), protected stomata and thick cuticles. Plants showing crassulacean acid metabolism (CAM) open their stomata at night to fix CO_2 as malate which is stored in the vacuole, while they remain closed in daytime when evaporative losses are greatest.
Related topics	Roots (D2) Plants and water (E1)
	Stress avoidance and adaptation (I5) C3 and C4 plants and CAM (F3)

Plant types

Mesophytes are adapted to an environment in which water is generally available in the soil and can be extracted by transpiration to supply the needs of the plant. Mesophytes can restrict water loss by the closure of stomata, but desiccation to below 30% total water content normally results in death. Mesophytes may show adaptations to either drought or waterlogging to different degrees depending on species. Some, like corn, form **aerenchyma** in response to water-logging, while in others (e.g. rice) it is always present (Topic D1). Presence of thicker **cuticles**, fleshier leaves, sunken stomata and photosynthetic modification indicate increasing adaptation to drought conditions. Other adaptations to drought include **drought avoidance**, for instance over-wintering as seeds and loss of leaves or other aerial parts.

Xerophytes are adapted to environments in which water is generally very scarce. They display a range of adaptations, including: **sunken stomata** (that entrap a layer of unstirred air); stomata open at night; modified photosynthetic mechanisms (Topic F3); thick cuticles and **succulence** (thick, fleshy leaves or the absence of leaves and presence of fleshy, modified stems). **Cacti** and some **euphorbias** are examples of xerophytes.

Hydrophytes are adapted to live submerged or partially submerged in water. They display modified leaves and stems, and frequently contain air spaces (**aerenchyma**) to supply oxygen to underwater organs.

Plant structure and surfaces

Land plants face a dilemma; how to acquire sufficient carbon dioxide and light for photosynthesis without losing large quantities of water by exposing large evaporative surfaces to the atmosphere. In xerophytes, the area exposed is reduced and lower growth rates occur. Mesophytes with a large leaf area also show adaptations to minimize water loss. Surfaces are coated with a **cuticle** of lipids, cutin, suberin and waxes which reduces both water loss and the access of pathogens to the cells of the leaf.

Cutin is a polymer of **long-chain fatty acids** forming a rigid mesh through ester linkages. **Suberin** contains long-chain fatty acids joined through ester linkages with dicarboxylic acids and phenolics. It is the predominant coating of underground parts of the plant and is also found in the Casparian strip of the endodermis (Topic D2). Associated with cutin and suberin are waxes, long-chain acyl lipids which are solid at ambient temperatures. The cuticle is deposited in layers (*Fig. 1*): first a cuticular layer of cutin, wax and carbohydrates at the epidermal cell wall, then a layer of cutin and wax and finally a waxy surface layer exposed to the atmosphere.

Stomata: structure

Stomata are pores through which gas exchange to the leaf takes place. The pore is formed by two specialized cells, the **guard cells**, that open and close the pore, frequently associated with **subsidiary cells** (*Fig. 2*). The pore structure leads to the **substomatal cavity** surrounded by cells of the **spongy mesophyll**. The stomata may themselves be sunken in order to minimize evaporative air movement.

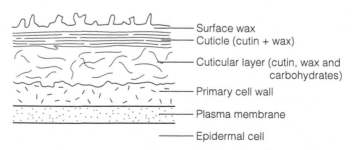

Fig. 1.   The structure of the cuticle of a higher plant leaf.

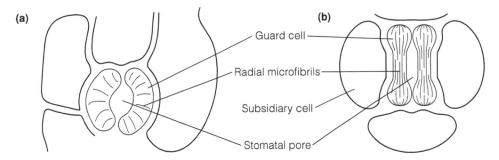

Fig. 2. *Stomata of (a) a dicot and (b) a grass.*

Stomata of dicots are typically kidney-shaped, while those of grasses are dumb-bell-shaped. In both types, **microfibrils** of the cell wall (Topic C2) are arranged radially, causing enlargement of the pore when the guard cells swell (*Fig. 2*).

Stomata: action and regulation

Stomatal aperture is tightly regulated by the plant, in a system, which integrates **carbon dioxide** requirement with **light** and **water stress**. Stomatal aperture often varies according to a **circadian** (day/night) **rhythm**. Low CO_2 concentrations in the guard cells result in opening, while high CO_2 concentrations result in closure. Stomata in most species open at dawn and remain open in daylight, given adequate water availability; this may in part be due to changing CO_2 concentrations as a result of photosynthesis and respiration.

Stomatal guard cells are also very sensitive to **water stress**. Localized **turgor** loss results in wilting of the guard cells (**hydro-passive closure**) and stomatal closure. Water stress elsewhere in the plant results in the production of **abscisic acid** (**ABA**; Topic J2), which results in stomatal closure. The mechanism of ABA sensing probably involves **ABA receptor proteins** (Topic J3) and may involve the action of calcium as an intracellular messenger that alters membrane ion channel activity, giving altered turgor and stomatal closure.

Guard cells take up K^+ and increase in **turgor** in favorable conditions, resulting in swelling of the cells and opening of the pore. Potassium is rapidly lost from the guard cells during stomatal closure. The driving force for potassium uptake is provided by the **plasma membrane proton pump** (Topic E3). This is stimulated during stomatal opening and the resulting **membrane hyperpolarization** is believed to open K^+ channels in the plasma membrane which permit a passive K^+ influx. This, together with influx of Cl^- and organic anions such as malate provides the increase in turgor. When stomata close, other plasma membrane ion channels open – resulting in a rapid efflux of anions and a drop in turgor.

Stomatal adaptations of xerophytes

Xerophytes show a number of adaptations to water stress, including sunken stomata, thickened cuticles and succulence. A key adaptation is the presence of crassulacean acid metabolism (CAM; Topic F3). CAM plants show a specialized rhythm of stomatal action that minimizes water loss. Coupled with their unique metabolism their stomata only open at night, when they fix CO_2 as malate in the vacuole. The stomata then remain closed during the day, when evaporative losses would be greatest.

E3 MOVEMENT OF NUTRIENT IONS ACROSS MEMBRANES

Key Notes

Transport of nutrients into cells

Mineral nutrients are transported as ions. They are soluble in water but cannot cross membranes without the presence of transport proteins. This transport is coupled to the transport of protons (H^+), which are pumped actively across the membrane by the activity of primary pumps using adenosine triphosphate (ATP) as energy source.

Studying membrane transport

Advances in understanding membrane transport have been made by use of membrane vesicles and radioactive isotopes, by electrophysiology and by molecular techniques. The identity of many nutrient transport proteins is now known.

Primary pumps

There are two major primary ion pumps in plant cells: the plasma membrane proton pump, which uses ATP as energy source and pumps H^+ out of the cell; and the vacuolar (or tonoplast) proton pump which pumps H^+ into the vacuole. Both pumps are electrogenic (generate a membrane potential). The apoplast and vacuole are therefore more acidic than the cytoplasm. Other primary pumps exist for calcium and the vacuole contains a second proton pump using pyrophosphate as energy source.

Secondary coupled transporters

Primary pumps generate a proton electrochemical gradient across the plasma membrane and tonoplast. Secondary coupled transporters couple the energy in this gradient to move other ions against their own electrochemical gradients. Antiporters transport an ion in the opposite direction to the transported ion; symporters transport two ions in the same direction.

Ion channels

Ion channels permit nutrient ions to move across membranes driven by the electrochemical gradient. Channels are gated (they open and close) in response to changes in membrane potential (voltage-gated channel), binding signal or other molecules (ligand-gated channel) or tension at the membrane (stretch-activated channels).

Related topics

Membranes (C4)
Plants and water (E1)

Uptake of mineral nutrients by plants (E4)

Transport of nutrients into cells

Membranes are impermeable to charged molecules and large molecules that are polar (have a charge imbalance). As mineral nutrient ions are charged, they cannot cross cell membranes without the action of specific membrane proteins. These proteins fall into three categories: primary pumps, which use the energy of adenosine triphosphate (ATP) to transport ions actively; secondarily coupled transporters, which couple the movement of one ion to that of another; and channels that permit the passive movement of ions. Transport of most ions is coupled to the transport of protons (H⁺), which are pumped actively across the membrane by the activity of primary pumps using ATP as energy source. The action of the transport proteins is tightly regulated and they are specific for given nutrients. They provide selectivity and specificity of uptake and, together with the properties of membranes, result in the concentration of nutrients within a cell or tissue being different from that outside it.

Movement of nutrient ions into and out of a cell, or subcellular compartment, is driven by a combination of concentration and electrical effects at any membrane (the electrochemical gradient). The force driving an ion across a membrane is made up of two parts, an electrical driving force, and a chemical driving force. The two forces are balanced (equal and opposite) at equilibrium, when no net movement of the ion occurs across the membrane. Plant cells generally maintain a membrane potential of −120 to −200 mV at the plasma membrane as a result of the action of primary proton pumps. This electrical driving force is used to maintain the required concentration of anions and cations on each side of the membrane. *Figure 1* illustrates the key transporters for ions in a typical plant cell.

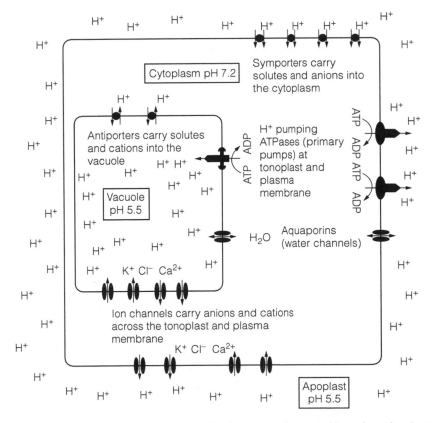

Fig. 1.   *Transporters for nutrient ions at the plasma membrane and tonoplast of a plant cell. The primary driving force for ion movement is created by electrogenic H⁺ pumps at the plasma membrane and tonoplast which create a trans-membrane electrical potential ($\Delta\psi = -120\,mV$ at the plasma membrane and $-90\,mV$ at the tonoplast) as well as steep transmembrane proton gradients (pH 5.5 in apoplast and vacuole; pH 7.2 in the cytoplasm).*

Studying membrane transport

The transport of an ion across plant membranes was first investigated using **radioactive isotopes** and **membrane vesicles**, sealed spheres of membrane which are formed when membranes are purified. The properties of many transporters have been determined by these methods.

The technique of **patch clamping** (*Fig. 2*) permits the transport properties of single proteins in the membrane to be assayed. A small piece of membrane (patch) is attached to the tip of a fine pipette and sealed to it by gentle suction. By measuring currents flowing through the membrane, the activity and properties of ion channels can be described.

Advances in describing nutrient transporters have resulted from cloning transport proteins. The overall structures of membrane proteins have been established and the role of regions of the protein investigated by altering key amino acids and expressing them in foreign host cells, like yeast cells, or giant egg cells of the African frog *Xenopus laevis*. Use of mutants has also been important; for instance, the nitrate transporter CHL1 was identified using an arabidopsis mutant *chl1* that is insensitive to chlorate, an inhibitor of the transport of nitrate. Finally, new membrane transport proteins have been detected by studies based on their homology with known ion transporters from other organisms.

Primary pumps

The primary pumps of plant cell membranes are proton (H^+) pumps, located at the plasma membrane and vacuolar membrane (tonoplast). The plasma membrane proton pump uses ATP as substrate and expels H^+ from the cell. The process generates an electrical gradient of about -120 mV across the membrane. The tonoplast has two primary pumps: an adenosine triphosphatase (ATPase) and a pyrophosphatase (which uses pyrophosphate rather than ATP as substrate) which pump H^+ into the vacuole. These pumps generate a membrane potential across the tonoplast of about -90 mV. The transport of all other nutrient ions except Ca^{2+} depends on these electrochemical gradients. Ca^{2+} concentrations are regulated by active Ca^{2+} pumps.

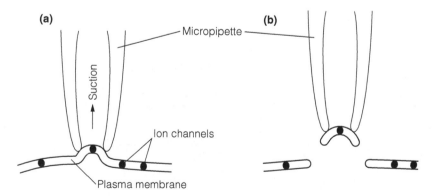

Fig. 2. *The patch clamp technique. A fine glass micropipette is pushed against the plasma membrane. Gentle suction is applied (a) and the pipette withdrawn (b) with a small patch of membrane adhering to the tip. The patch is sufficiently small to contain only a few membrane proteins. Their transport properties can then be measured by measuring the electrical current carried by ions as they cross the membrane.*

Secondary coupled transporters

Transport of many other ions is coupled to the electrochemical gradient developed across the plasma membrane and tonoplast by the primary proton pumps. **Antiporters** are membrane proteins which couple the flow of H^+ down its electrochemical gradient to the movement of another ion against its electrochemical gradient. The tonoplast **calcium proton antiporter**, for instance, accumulates calcium in the vacuole by coupling its transport to the outflow of H^+ from the vacuole. **Symporters** similarly couple transport of an ion to the proton electrochemical gradient, but with the movement of the ion occurring in the same direction as the proton movement.

Ion channels

Some transport proteins permit the passive movement of nutrients across membranes driven by the electrochemical gradient for that nutrient. These passive transporters are also membrane proteins and are known as **uniporters**. **Ion channels** are a specialized type of uniporter; they generally show a high degree of **selectivity** for individual ions and are **gated** (can either open to permit ions to cross the membrane or close to prevent them from crossing). Channels may be gated by the membrane potential (**voltage-gated**), by signal molecules within the cell (**ligand-gated**) and by tension in the membrane (**stretch-activated**). The combination of these properties permits very close regulation of ionic concentration within the cytoplasm or vacuole.

E4 UPTAKE OF MINERAL NUTRIENTS BY PLANTS

Key Notes

Key properties of nutrients

The majority of plant nutrients are taken up by the plant in ionized form from films of water surrounding soil particles. Nutrients move in aqueous solution. They cannot cross lipid membranes unless transport proteins are present.

The soil–root interface

The root surface (the rhizodermis) makes limited contact with the nutrient film surrounding soil particles. Root hairs and, in many species, mycorrhizal fungi greatly increase the surface area in contact.

Symplastic, apoplastic or cellular?

Transport within the root may be across cell membranes (transmembrane) then through the cell cytoplasm (cellular transport) or between cells (apoplastic). Plasmodesmata provide continuous contact between the cytoplasm of adjacent cells giving direct cell-to-cell (symplastic) transport without contact with the apoplast.

The endodermis

Cells of the endodermis have suberinized cell walls which form a water-impermeable barrier surrounding the vascular tissue of the root. It prevents apoplastic movement of nutrients which must therefore either travel symplastically through the endodermis or enter the vascular system from below the endodermis.

Transport into the xylem

Water and nutrients leaving the endodermis enter xylem parenchyma cells that surround xylem vessels. These cells actively accumulate nutrients to a high concentration before they are loaded into the xylem for transport to the rest of the plant.

Distribution in the plant

Xylem extends throughout the plant and water flows to wherever transpiration is taking place. The apoplast of all tissues is in close contact with xylem fluid and nutrients are taken up from this space by cells. Rapidly growing tissues (e.g. fruits, tubers) may have low transpiration and in these tissues redistribution of ions in the phloem may be important.

Related topics

Membranes (C4)
Roots (D2)
Plants and water (E1)

Movement of nutrient ions across
 membranes (E3)
Mycorrhiza (M1)

Key properties of nutrients

Nutrients are taken up as ions dissolved in water. Bulk movement occurs in the **transpiration stream** (via the xylem) to shoots and leaves. Movement depends on the unique size, charge and solubility of each nutrient ion. As ions cannot cross membranes without a specific transport protein being present, their uptake into cells and into the root can be regulated. This means that roots accumulate some ions against a concentration gradient, while others are excluded and some ions move through cells while others move in intercellular spaces.

The soil–root interface

Roots elongate into the soil by growth near the tip, pushing the root cap between soil particles (Topic D2). They deplete their immediate soil environment of nutrients creating a depletion zone. Good root-to-soil contact is provided by the secretions of the root cap (Topic D2) and by root hairs. **Fibrous root systems,** made up of fine roots with many root hairs, maximize the area available for uptake. In addition, many species have **mycorrhizal fungi** in symbiotic association with the root which greatly enlarge the available soil area from which nutrients are extracted (Topic M1). Nutrients become available at the root surface as a result of three processes: **interception**, growth of roots into new nutrient-rich area; **mass flow**, movement of ions in the water flow driven by transpiration; and **diffusion**, passive movement of ions to regions depleted in nutrients (*Fig. 1*).

Symplastic, apoplastic or cellular?

Nutrient ions entering the root at the root hairs may travel directly through the cell cytoplasm until they reach the vascular tissue (*Fig. 2*). Cells of the rhizodermis (root epidermis) are linked by plasmodesmata to adjacent cells and solutes can move from cell to cell directly. Transport through the cell via plasmodesmata is termed **symplastic transport**. The second form of transport, **apoplastic transport**, is through the walls of root cells. Water and nutrients can travel through this region (the apoplast) until the endodermis, where water and nutrients must cross a plasma membrane and enter a cell symplasm (**cellular transport**). Water and nutrients may also enter cells via the plasma membrane (**transmembrane transport**), then leave the cell to enter the apoplast before being taken up across the plasma membrane of an adjacent cell.

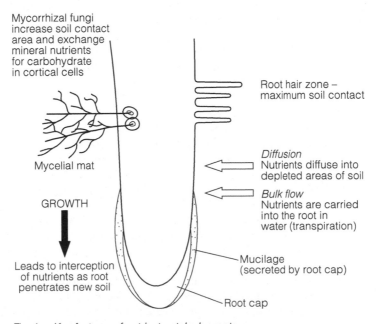

Fig. 1.   Key features of nutrient uptake by roots.

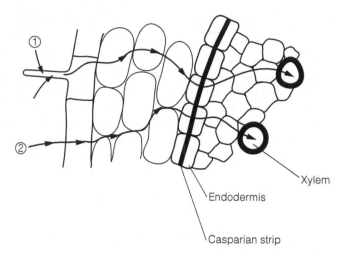

Fig. 2. Pathways of nutrient transport in roots. (1) Symplastic throughout; (2) apoplastic until endodermis, then symplastic.

The endodermis

The **endodermis** develops near the point of development of the root vascular tissue (Topic D2). Cells of the endodermis have suberinized cell walls which form a water-impermeable barrier surrounding the vascular tissue of the root. This prevents water and nutrients moving into the vascular tissue through the apoplast and the only transport possible at this point is **symplastic**. Most nutrient ions entering the vascular tissue will therefore have passed through living root cells at some point. This permits **selectivity** and filtration of the transported ions. Older endodermal cells become entirely enclosed in a suberin layer. These cells provide a barrier to the back-flow of water from the xylem.

Transport into the xylem

After crossing the endodermis, ions enter the xylem from the cells surrounding it, the **xylem parenchyma**. These cells actively accumulate nutrients to a high concentration before they are loaded into the xylem for transport to the rest of the plant. Influx into the xylem occurs via specific membrane proteins down the concentration gradient. Nutrient flow in the xylem to leaves and shoots occurs in the transpiration stream.

Distribution in the plant

Mineral nutrient ions and water move in the xylem and reach all parts of the plant. Xylem tubules branch out from the main **vascular bundles** of the stem to reach leaves and buds and branch again to form finer tubules in the **leaf veins.** Water and nutrients leave these tubules as water evaporates from the leaf surface; the cells of the leaf are surrounded by water and dissolved nutrients which permeate the leaf wall spaces (leaf apoplast). Cells extract water and nutrients from this. Some nutrient ions may be redistributed through the plant in the **phloem**, while others only move in the xylem. Some areas of the plant undergoing rapid growth, e.g. fruits and tubers, do not have high transpiration rates and the xylem flow is low. This may lead to nutrient deficiency if nutrient transport in the phloem does not occur. Blossom end rot of tomato occurs in these circumstances, due to lack of calcium. Ions may be moved from xylem to phloem by **transfer cells** that lie between the two pathways or ions may leave source tissue as it loads organic nutrients into the phloem. Not all ions are **phloem mobile** and there are marked differences between species.

E5 FUNCTIONS OF MINERAL NUTRIENTS

Key Notes

Essential macro- and micronutrients

Plants depend on a range of essential mineral nutrients, which are extracted from the soil by the roots. These are categorized as macro- and micronutrients depending on the quantity required. Examples of macronutrients are sulfur, phosphorus, nitrogen, magnesium, potassium and calcium.

Essential, beneficial and toxic elements

Minerals may be categorized according to their effects on the plant. Essential elements are those without which a plant cannot reproduce; beneficial elements have beneficial effects on plant growth, but the plant can complete its life cycle without them. Toxic elements are deleterious to growth. Some elements are essential at low concentrations, but toxic at higher concentrations.

Key macronutrients

Nitrogen is a constituent of amino acids and proteins and is taken up as either nitrate or ammonium from soils or by nitrogen-fixing organisms. Nitrogen is transported as reduced nitrogen compounds. Sulfur is required in sulfur-containing amino acids to maintain protein structure. Phosphorus is required for membranes, nucleic acids and ATP. It is transported either as inorganic phosphate or as sugar phosphates. Its uptake is enhanced by mycorrhizal fungi. K^+, Mg^{2+} and Ca^{2+} are all water-soluble cations. K^+ is required for enzyme activity and osmoregulation, Ca^{2+} for membrane stability and as an intracellular regulator and Mg^{2+} for chlorophyll and enzyme activity.

Nutrient deficiency and toxicity

Plants growing without enough of a nutrient show deficiency symptoms related to the function of the nutrient in the plant. Toxic ions such as aluminum may result in deficiency symptoms as they act by restricting the availability or uptake of nutrients. Good agricultural practice seeks to maximize the availability of nutrients and minimize toxic ions.

Related topics

Plants and water (E1)
Movement of nutrient ions across
  membranes (E3)

Uptake of mineral nutrients by
  plants (E4)

Essential macro- and micronutrients

A nutrient is **essential** if: (i) it is required for the plant to complete a normal life cycle; (ii) it can be shown to be a component of the plant, either as part of structure or metabolism; and (iii) its function cannot be substituted for another element. Nutrients are either **macronutrients** (required in large amounts) or **micronutrients** (**trace elements**) required in much smaller amounts. *Table 1* lists a range of both macro- and micronutrients. The **available form** (i.e. the form in which the nutrient is transported into the plant) is usually either an **anion** (negative) or a **cation** (positive). The distinction between macro- and micronutrients is to some extent arbitrary as some species contain more and others

Table 1. Nutrient elements and their functions

Element	Available form	Typical concentration (mmol kg^{-1} dry weight of plant)	Typical use
Macronutrients			
Hydrogen	H_2O	60 000	Turgor; photosynthesis; carbohydrates
Carbon	CO_2	40 000	Carbohydrate; protein; metabolism
Oxygen	O_2, CO_2, H_2O	30 000	Carbohydrate; metabolism
Nitrogen	NO_3^- (nitrate) NH_4^+ (ammonium)	1000	Amino acids; proteins; nucleic acids
Potassium	K^+	250	Remains a free ion for turgor regulation; cofactor for many enzymes
Calcium	Ca^{2+}	125	Cell signaling; cell wall linkages
Magnesium	Mg^{2+}	80	Chlorophyll (photosynthesis)
Phosphorus	HPO_4^- (phosphate) HPO_4^{2-}	60	Phospholipids; nucleic acids; ATP metabolism
Sulfur	SO_4^{2-} (sulfate)	30	Amino acids and proteins
Micronutrients			
Chlorine	Cl^- (chloride)	3	Turgor regulation; photosynthesis
Iron	Fe^{2+} (ferric) Fe^{3+} (ferrous)	2	Photosynthesis, respiration and nitrogen fixation in cytochromes and nonheme proteins
Boron	BO_3^{3-}	2	Complexed in cell walls
Manganese	Mn^{2+}	1	Cofactor for various enzymes
Zinc	Zn^{2+}	0.3	Cofactor for various enzymes
Copper	Cu^{2+}	0.1	Cofactor for enzymes and electron carrier proteins
Nickel	Ni^{2+}	0.05	Constituent of urease
Molybdenum	MoO_4^{2-} (molybdate)	0.001	Constituent of enzymes in nitrogen metabolism

less of a particular nutrient; macronutrients are conveniently defined as those present at >10 mmol kg^{-1} dry matter.

Essential, beneficial and toxic elements

Some minerals are important in plant nutrition, but not essential. Elements which result in improved growth or reduced disease susceptibility, but without which the plant can still complete its life cycle are known as beneficial. An example of such an element is silicon (Si), which causes increased structural strength of cell walls, increased tolerance of toxic elements in soils and improved resistance to fungal pathogens. Some other elements are usually toxic, or may be toxic at high concentrations. Aluminum is almost always toxic when available (at acidic soil pH) as it complexes phosphate. Sodium, zinc, copper, manganese, boron, molybdenum and iron can all be toxic if present at high concentrations in the soil. Species differ in their ability to tolerate toxic ions and in some instances ions that are normally toxic are known to be beneficial. The growth of the tea plant, for instance, is enhanced by the presence of available soil aluminum, which it takes up as an organic acid complex, and the growth of many halophytes by sodium.

Key macronutrients	*Nitrogen*

Nitrogen

Nitrogen gas is abundant in the aerial and soil environment, but unlike oxygen cannot be used directly. Nitrogen is **fixed** from the atmosphere by a number of microorganisms, which may be free-living, or in symbiotic association with some species of plants, mainly legumes (Topic M2). Other sources of available soil nitrogen may be decaying organic material, animal excreta and chemical fertilizers, frequently added to agricultural land.

Nitrogen assimilation

Nitrogen assimilation occurs by a number of pathways depending on species and form of available N. Plants take up either nitrate (NO_3^-) or ammonium (NH_4^+) from the soil, depending on availability and species. Nitrate will be more abundant in well-oxygenated, nonacidic soils, whilst ammonium will predominate in acidic or waterlogged soils. After uptake, nitrate is reduced to ammonium in two stages by **nitrate reductase** (NR) and **nitrite reductase** before assimilation into amino acids.

$$2H^+ + NO_3^- + 2e^- \xrightarrow{\text{Nitrate reductase}} NO_2^- + H_2O$$

$$8H^+ + NO_2^- + 6e^- \xrightarrow{\text{Nitrite reductase}} NH_4^+ + 2H_2O$$

NR is a dimer of two identical subunits. Expression of the gene for NR requires light and nitrate. The gene is expressed in both shoots and roots, but in low nitrate conditions, almost all **nitrate assimilation** will occur in the roots. Crops grown at high temperatures but only moderate light intensity may accumulate nitrate in the vacuole as a result of low induction of the NR gene.

Ammonium is toxic to plants, and is therefore rapidly incorporated into amino acids via the **glutamine synthase-glutamate synthase (GS-GOGAT)** pathway:

$$\text{Step 1: glutamate} + NH_4^+ + ATP \xrightarrow{\text{Glutamine synthase}} \text{glutamine} + ADP + P_i$$

$$\text{Step 2: glutamine} + \text{2-oxoglutarate} + NADH \xrightarrow{\text{Glutamate synthase}} 2\,\text{glutamate} + NAD^+ \begin{array}{l}\nearrow \textit{one returned to GS} \\ \searrow \textit{one exported for use}\end{array}$$

The process runs as a cycle, one of the glutamates produced being used as substrate by GS while the other is exported to the plant. Inputs to the cycle are ammonium and **2-oxoglutarate** (from photosynthate imported via the phloem).

In plants with nitrogen-fixing nodules, nitrogen is exported to the plant from the nodule in the xylem flow as high nitrogen-containing compounds such as **amino acids** or **ureides**.

Sulfur

Sulfur is taken up in the form of **sulfate** (SO_4^{2-}) by high-affinity transport proteins at the plasma membrane. The expression of these proteins varies with sulfate availability, the genes being repressed by high sulfate and activated by low sulfate. Once in the plant, sulfate is reduced to the sulfur-containing amino acid **cysteine**. The entire process involves the donation of 10 electrons and a variety of **electron donors** is required. It is more active in photosynthesizing leaves as the chloroplast provides a supply of electron donors. Sulfur is predominantly transported around the plant in the **phloem**, as **glutathione**. Glutathione (GSH) is a tripeptide formed of three amino acids (γ-glutamyl-cysteinyl-glycine) and acts as a storage form of sulfur in the plant.

Phosphate

Phosphate is not reduced in the plant. Root cells contain **phosphate transporters** at the plasma membrane and after uptake it either travels in xylem as inorganic phosphate (P_i) or is **esterified** through a hydroxyl group on a sugar or other carbon compound. Phosphate taken up by the root is rapidly incorporated into **sugar phosphates**, but is released as P_i into the xylem. Roots of plants are frequently in symbiotic association with **mycorrhizal fungi** that extract phosphate efficiently from the soil (Topic M1). Phosphate is an essential constituent of nucleic acids and of many of the compounds of energy metabolism and it is utilized throughout the plant.

Nutrient cations

The **nutrient cations** (e.g. **potassium**, **magnesium** and **calcium**) are water-soluble and transported as cations in the xylem. Potassium is very soluble and highly mobile, calcium being the least mobile of the three. Potassium is required for enzyme activity and osmotic regulation, calcium maintains cell membrane stability, is an intracellular regulator and forms **calcium pectate** links in cell walls at middle lamellae (Topic C2). Magnesium is required as a central component of chlorophyll (Topic F1) and for the activity of some enzymes.

Nutrient deficiency and toxicity

Plants grown in the absence of sufficient quantities of a particular nutrient show visible **deficiency symptoms** that relate to the function of that nutrient in the plant. Key terms are **chlorosis**, a lack of chlorophyll and yellowing of the leaves, and **necrosis**, the death of cells, often the growing tip or in lesions in the leaf surface. The extent and nature of the symptoms observed depend on where the nutrient is required and whether it can be redistributed in the plant. Potassium deficiency, for instance, causes necrosis of leaf margins and tips, whereas deficiency of manganese (a micronutrient) causes necrosis of tissue between the leaf veins (interveinal necrosis).

Development of symptoms such as those for deficiency can result from the presence of **toxic elements**, which interfere with the availability or transport of nutrients. **Aluminum** is more available in acidic soils and complexes with phosphate, creating phosphate deficiency. Plants show adaptations for the nutritional characteristics of the soils in which they grow. **Calcicoles** are adapted for growth in an alkaline, high calcium environment, where other nutrients are of low availability, while **calcifuges** are adapted to acidic soils, with high levels of aluminum and low levels of phosphate. Sodium competes with potassium for uptake resulting in potassium deficiency and failure to osmoregulate. **Halophytes** show a range of adaptations for growth in saline conditions, including salt-extruding glands, high levels of discrimination between sodium and potassium for transport and **xerophytic characteristics** to minimize transpiration and salt uptake.

Many agricultural crops make severe demands on the nutrients in the soils in which they are grown; growth of cereals, for instance, results in the removal of large quantities of soil nitrogen and sulfur in seed protein. Good agricultural practice aims to maintain soil nutrient levels and minimize the availability of toxic compounds. For instance **liming**, the application of calcium or magnesium oxides, hydroxides or carbonates, is used to neutralize acidic soils and reduce aluminum toxicity, thereby increasing phosphate availability.

F1 PHOTOSYNTHETIC PIGMENTS AND THE NATURE OF LIGHT

Key Notes

The nature of light	Sunlight is made up of high-energy photons that can energize molecules capable of absorbing their energy. Light can also be described as a wave motion, with the longest wavelengths having the lowest energy.
Photosynthetic pigments	Chlorophyll *a* is the major photosynthetic pigment. It has a porphyrin ring structure containing a central magnesium atom and a long hydrocarbon tail. Accessory pigments pass their harvested energy to chlorophyll *a*. Blue and red light are absorbed most. Absorption of light energizes an electron in the chlorophyll, the energy being transferred to an electron acceptor.
The reaction center	The pigments are grouped into photosystems in the thylakoid membrane. Two reaction center chlorophylls are surrounded by pigment molecules of the antenna complex. Energy harvested by this complex is passed by resonance energy transfer to the reaction center chlorophylls, which pass on the electron to an electron acceptor.
Related topics	Plastids and mitochondria (C3) C3 and C4 plants and CAM (F3) Major reactions of photosynthesis (F2)

The nature of light

The energy for **photosynthesis** is derived from sunlight, which originates from the exothermic reactions taking place in the sun. Light has the properties of both a wave motion and of particles (**photons** or **quanta**). The amount of energy contained in a photon is inversely proportional to the wavelength, short wavelengths having higher energy than long wavelengths. In order to photosynthesize, plants must convert the energy in light to a form in which it can be used to synthesize carbohydrate.

Photosynthetic pigments

Plants appear green because **chlorophyll** absorbs blue and red light and reflects green. This can be shown as an **absorption spectrum**, in which absorbance is plotted against wavelength (*Fig. 1*). The **action spectrum** of photosynthesis, i.e. the photosynthetic activity at different wavelengths, is also shown and corresponds to the absorption spectrum.

When a chlorophyll molecule is struck by a photon, it absorbs energy. This energizes an electron within the chlorophyll that may be used in photosynthesis or return to its original energy state with a loss of heat and transmission of light of lower wavelength.

The major photosynthetic pigments are the chlorophylls (*Fig. 2*). A chlorophyll molecule is made up of a 'head' group, a nitrogen-containing **porphyrin**

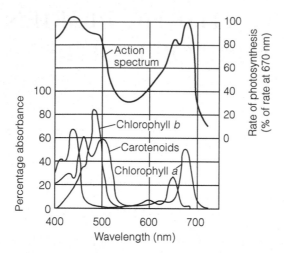

Fig. 1. Absorption spectra of three major pigments of photosynthesis: chlorophyll a, chlorophyll b and carotenoids, together with the action spectrum of photosynthesis. The absorption spectrum represents the degree to which the pigment is able to absorb energy at each wavelength of light; the action spectrum represents rate of photosynthesis at each wavelength (Govindjee, unpublished data, 1961; redrawn with permission of Govindjee, University of Illinois at Urbana, Illinois, USA).

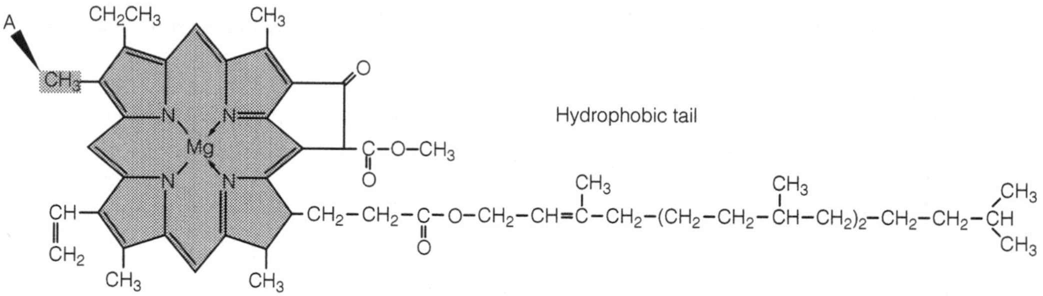

Fig. 2.   The structure of chlorophyll. Chlorophyll b differs from chlorophyll a by the substitution of a CHO group for the

ring structure, with a **magnesium atom** at its center, and a **tail of hydrocarbons**, which anchors the molecule to a membrane. Chlorophyll *a* is the main pigment of photosynthesis. Most chloroplasts also contain **accessory pigments**, pigments that broaden the range of wavelengths at which light can be absorbed and that pass the energy obtained to chlorophyll *a* or protect against damage. **Chlorophyll *b*, carotenoids** and **xanthophylls** are examples of these pigments.

The reaction center

To harvest the energy of light, the pigments of photosynthesis must be arranged so that the energy is focused to a point from which it can be used. Energy absorbed by the pigment molecules is transferred to one of a central pair of chlorophyll *a* molecules (termed the **reaction center chlorophylls**) by **resonance energy transfer** (*Fig. 3*). From here, the high energy electron is passed on to an **electron acceptor**. The pigments are arrayed in flat sheets in the

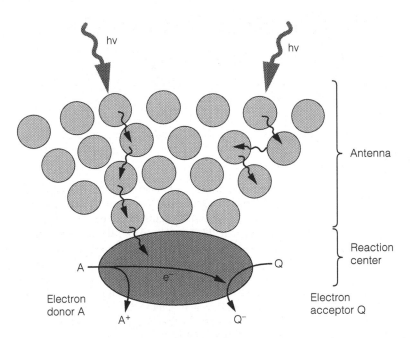

Fig. 3.   The reaction center. Chlorophyll and accessory pigments are grouped around a central pair of reaction center chlorophylls. The absorbed energy is collected and conveyed to the central pair of reaction center chlorophylls by resonance energy transfer (RET; arrows). The energized electron from an electron donor, A, in the reaction center is then passed on to an electron acceptor, Q. (Redrawn from Hopkins (1998) Introduction to Plant Physiology, *2nd Edn, John Wiley & Sons.)*

thylakoid membrane, orientated to capture the incoming radiation. Each reaction center is surrounded by 200–400 pigment molecules, the **antenna complex**, and the whole structure constitutes a **photosystem**. The thylakoids are stacked in **grana** held within the **stroma** of the chloroplast (Topic C3).

There are two types of photosystem, known as **photosystem I (PS-I)** and **photosystem II (PS-II)**. PS-I was the first to be discovered, and the reaction center chlorophylls in it are known as P_{700} because they absorb maximally at 700 nm. The reaction center chlorophylls of PS-II absorb maximally at 680 nm (P_{680}). The functions of PS-I and PS-II are explained in Topic F2.

F2 MAJOR REACTIONS OF PHOTOSYNTHESIS

Key Notes

The light reactions

The energy of light is used to energize electrons in the reaction center chlorophyll molecules of a photosystem. The electrons are then passed on to an electron transport chain. The missing electrons in the reaction center are replenished when $2H_2O$ is split to O_2, $4H^+$ and $4e^-$. The electron moves through the electron transport chain until it energizes the cytochrome b/f complex to pump protons into the thylakoid lumen and is then passed to photosystem I (PS-I) where it is re-energized and passed to NADP reductase, generating NADPH. PS-I can also function alone to transport H^+. Adenosine triphosphate (ATP) is generated by H^+ passing through the enzyme ATP synthase located in the thylakoid membrane. The two products of the light reactions are therefore NADPH and ATP.

The carbon fixation reactions

The carbon fixation reactions (Calvin cycle) in the stroma of the chloroplast use ATP and NADPH to fix CO_2 as carbohydrate in a cyclic process which does not directly require light. The cycle involves (i) carboxylation, in which ribulose bisphosphate carboxylase/oxygenase (Rubisco) carboxylates 5-carbon ribulose bisphosphate using CO_2 to give a transient 6-carbon compound, which forms two molecules of 3-carbon 3-phosphoglycerate; (ii) reduction, where NADPH and ATP are used to form two molecules of glyceraldehyde 3-phosphate; and (iii) regeneration, where one molecule of 3-carbon glyceraldehyde 3-phosphate is converted to 5-carbon ribulose bisphosphate using ATP. Overall, fixation of three molecules of CO_2 requires 6NADPH and 9ATP and leads to the net synthesis for export of one glyceraldehyde 3-phosphate.

Photorespiration

Ribulose bisphosphate carboxylase/oxygenase also has oxygenase activity which generates 3-phosphoglycerate and 2-phosphoglycolate. The overall efficiency of photosynthesis is decreased by about 25% as a consequence. The photorespiratory cycle partially recovers the fixed carbon and involves the peroxisomes and mitochondria. It results in the loss of CO_2 and the use of ATP to convert the 2-phosphoglycolate to 3-phosphoglycerate which is converted to ribulose 1,5-bisphosphate by the Calvin cycle.

Related topics

Plastids and mitochondria (C3)
Photosynthetic pigments and the
　nature of light (F1)

C3 and C4 plants and CAM (F3)

The light reactions

In photosystem II (PS-II) (Topic F1) a pair of electrons energized by light are passed from the reaction center chlorophyll to an electron transport chain. These electrons are replaced by a unique process called **photolysis** in which water is oxidized to yield molecular oxygen. The overall reaction, termed the **S state mechanism**, is as follows:

$$2H_2O \rightarrow O_2 + 4H^+ + 4\ e^-$$

It requires manganese and is energized by PS-II. The protons generated are released into the lumen of the **thylakoid**. The high energy electrons are captured by an electron acceptor and passed on to the P_{680} **chlorophyll** (the reaction center chlorophyll in PS-II, having an absorption maximum at 680 nm; Topic F1).

The electrons then move through an **electron transport chain** located in the thylakoid membrane. The first stage is **plastoquinone**, a quinone molecule which is able to move within the membrane. Plastoquinone accepts two electrons and two protons to form PQH_2. The electrons are then passed to the **cytochrome *b/f*** complex. This is a proton pump and pumps H^+ into the **thylakoid lumen** (*Fig. 1*). The electron is then transferred to **plastocyanin**, a copper-containing protein that accepts electrons, by the copper cycling between Cu^{2+} and Cu^+ that then supplies it to **PS-I**. It is energized again by light and transported by another electron acceptor, **ferredoxin**, a protein. The electron is then passed on to the enzyme nicotinamide adenine dinucleotide phosphate (**NADP**) **reductase**, that reduces $NADP^+$ to nicotinamide adenine dinucleotide phosphate (reduced form; NADPH). Overall, two photons absorbed by PS-II result in: the oxidation of a water molecule to give O_2 and the release of H^+ into the lumen of the thylakoid; the formation of NADPH by the reduction of $NADP^+$; and the transport of H^+ into the lumen of the thylakoid via the cytochrome *b/f* complex. The process, known as **noncyclic electron flow**, produces NADPH and a proton gradient across the thylakoid membrane. NADPH

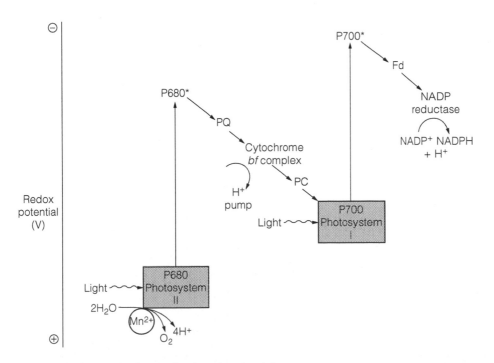

Fig. 1. The Z-scheme of noncyclic photophosphorylation.

is used directly in the **Calvin cycle** (below). The proton gradient is used to drive **adenosine triphosphate (ATP) synthase**, the enzyme that makes ATP.

The second photosystem, PS-I, can work independently of PS-II. This occurs when electrons are transferred from cytochrome b/f back to P_{700} via plastocyanin and re-energized by light. This process, termed **cyclic electron flow** is illustrated in *Fig. 2*. As the electrons never reach NADP reductase, the system only generates a proton gradient.

ATP is produced by **ATP synthase** (*Fig. 3*). This is a large protein complex located in the thylakoid membrane. As the protons flow down the electrochemical gradient from the thylakoid lumen into the stroma, the energy is used to synthesize ATP by the phosphorylation of adenosine diphosphate (ADP) by the ATP synthase complex. ATP synthesis driven by cyclic electron flow (PS-I only) is termed **cyclic photophosphorylation**, while ATP synthesis driven by noncyclic electron flow (PS-I and PS-II) is **noncyclic photophosphorylation**.

The carbon fixation reactions

The next stages of photosynthesis do not require light and are termed the **carbon fixation reactions**. They do require the ATP and NADPH generated by the light reactions and result in the incorporation (fixation) of carbon into carbohydrates. The carbon fixation reactions in many plants occur in the stroma of the chloroplast by the Calvin cycle.

The Calvin cycle

The **Calvin cycle** has three stages: **carboxylation**, the incorporation of CO_2; **reduction**, utilizing ATP and NADPH; and **regeneration** where the CO_2 accep-

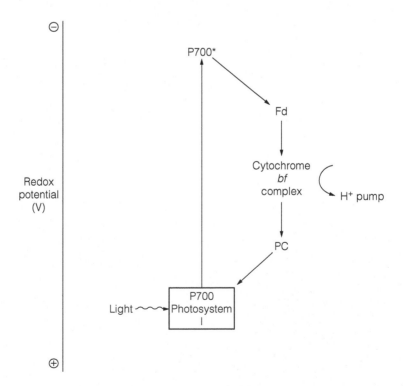

Fig. 2.   *Cyclic phosphorylation in green plants.*

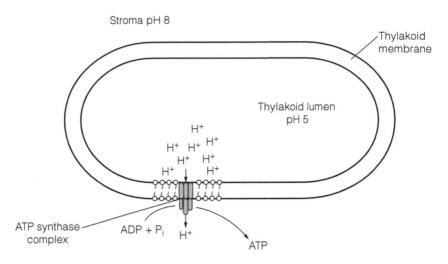

Stroma pH 8

Thylakoid membrane

Thylakoid lumen
pH 5

H⁺
H⁺ H⁺ H⁺
H⁺ H⁺
H⁺ H⁺

ATP synthase complex

ADP + P_i H⁺

ATP

Fig. 3. ATP synthase.

tor is formed again. The entire cycle, together with the number of moles of each molecule produced and the number of moles of ATP and NADPH used is shown in *Fig. 4* .

Stage 1: Carboxylation

$$\text{Ribulose bisphosphate (5C)} + CO_2 \rightarrow 2 \times \text{3-phosphoglycerate (3C)}$$

In this stage, a carbon atom from CO_2 is added to one molecule of **5-C ribulose bisphosphate** by the enzyme **ribulose bisphosphate carboxylase/oxygenase (Rubisco)** to yield two molecules of **3-C 3-phosphoglycerate**. The enzyme is the world's most abundant protein, often constituting around 40% of the soluble protein in a leaf. Rubisco is a CO_2 **acceptor**, which binds with sufficient affinity to ensure carboxylation of ribulose 1,5-bisphosphate. The reaction is energetically favorable, so the cycle runs in favor of 3-phosphoglycerate without additional energy input.

Stage 2: Reduction

$$\text{3-phosphoglycerate (3C)} \rightarrow \text{1,3-bisphosphoglycerate (3C)} \rightarrow \text{glyceraldehyde 3-phosphate (3C)}$$

The process involves two enzymes, **3-phosphoglycerate kinase** and **NADP glyceraldehyde-3-phosphate dehydrogenase**. In the first enzymatic reaction, one mole of ATP is used and in the second, one mole of NADPH is used for each mole of 3-phosphoglycerate. **Glyceraldehyde-3 phosphate** is a 3-carbon sugar, some of which is used in the next stages of the cycle and some removed as the product of the cycle (*Fig. 4*).

Stage 3: Regeneration

Regeneration involves the steps from glyceraldehyde-3-phosphate to ribulose 1,5-bisphosphate. Most of the stages are energetically favorable and do not consume further ATP or NADPH. The pathway involves the activity of seven enzymes in total. Finally, ribulose 5-phosphate (5C) is produced, which is phosphorylated by the enzyme **ribulose 5-phosphate kinase** using ATP to generate ribulose 1,5-bisphosphate (5C) ready for carboxylation.

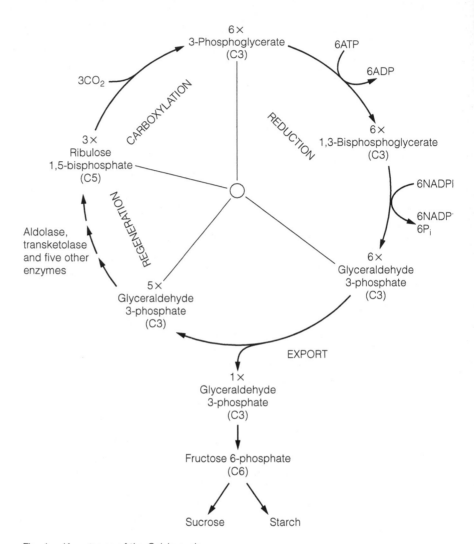

Fig. 4.   Key stages of the Calvin cycle.

Photorespiration   Rubisco can either carboxylate ribulose bisphosphate, giving 3-phosphoglycer-ate or can oxygenate it to give a 2-carbon sugar 2-phosphoglycolate and a 3-carbon sugar 3-phosphoglycerate (*Fig. 5*). In equal concentrations of CO_2 and O_2, carboxylation is favored over oxygenation by about 80:1; however, at ambient CO_2 concentrations, the ratio falls to 3:1. The rate of assimilation of carbon by a leaf is therefore the result of two opposing pathways: the Calvin cycle and photorespiration. The overall effect of photorespiration is a reduction of about 25% in carbon assimilation.

3-Phosphoglycerate is salvaged for carboxylation by the Calvin cycle (see above), but recovery of the phosphoglycolate requires the enzymes of the **photorespiratory cycle** which are located in peroxisomes (Topic C1) and mito-chondria (Topic C3). The stages are shown in *Fig. 5*. Overall, the pathway consumes oxygen and ATP; equal amounts of NADH are used up in the peroxi-some and produced in the mitochondrion.

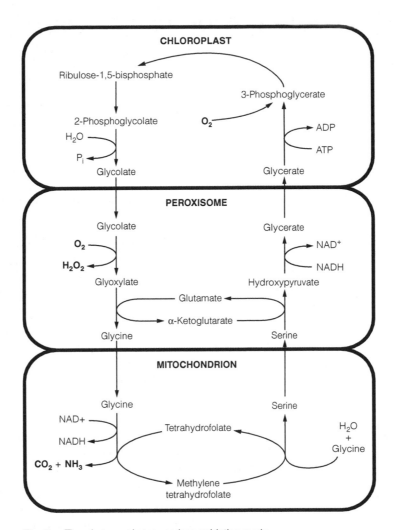

Fig. 5. The photorespiratory carbon oxidation cycle.

The function of photorespiration is unknown; it may be a consequence of the structure and reaction mechanism of Rubisco; alternatively, it may be a safety mechanism to protect the photosystems in conditions where light intensity is high and carbon dioxide low. Some plants show a modified form of photosynthesis termed C4 in which photorespiration does not occur (Topic F3).

F3 C3 AND C4 PLANTS AND CAM

Key Notes

CO₂ concentration and water conservation	Land plants have to open their stomata to admit CO_2. This means water loss is inevitable while photosynthesis occurs. Atmospheric concentrations of CO_2 do not saturate Rubisco and it therefore functions at less than its maximal rate. C4 photosynthesis and crassulacean acid metabolism (CAM) are two adaptations which decrease water loss.
C4 anatomy	C4 plants show Krantz anatomy, a ring of bundle sheath cells containing chloroplasts surrounding the leaf veins. Surrounding mesophyll cells also contain chloroplasts and are in close contact with the bundle sheath cells by plasmodesmata.
C4 biochemistry	C4 plants carry out the functions of the Calvin cycle in the bundle sheath cells. Rubisco in these cells is supplied with CO_2 released from the C4 compounds malic acid or aspartic acid generated in the mesophyll cells and transported through the plasmodesmata. CO_2 is first fixed by phosphoenolpyruvate (PEP)-carboxylase, which has a high affinity for HCO_3^-. The effect of the system is to increase the efficiency of CO_2 fixation and reduce the requirement for stomatal opening.
CAM anatomy	CAM plants show adaptations to drought and are succulents, having fleshy leaves with a minimum surface area to volume ratio. The photosynthetic cells have a substantial vacuole in which C4 acids are stored.
CAM biochemistry	In CAM plants, the stages of CO_2 fixation by PEP-carboxylase and of the Calvin cycle occur in the same cells but at different times. CO_2 is fixed at night and the C4 compounds formed are stored in the leaf cell vacuole. The C4 compounds break down to release CO_2 during the day, permitting the Calvin cycle to function when stomata are closed.
Distribution of CAM and C4 photosynthesis	Both C4 and CAM are adaptations to drought occurring in a wide range of species and families, generally from arid zones. Some families of plants contain C3, C4 and CAM members.
Related topics	Plants and water (E1) Major reactions of photosynthesis Water retention and stomata (E2) (F2)

CO₂ concentration and water conservation Land plants have a dilemma: allowing a free diffusion of CO_2 into leaf cells without excessive water loss. The plant's environment is often most desiccating when light intensities are highest, meaning that stomata close during times when CO_2 could be used most effectively. In addition, the photosynthetic apparatus of many plants showing **C3 photosynthesis**, the type of photosynthesis described in Topic G2 (so called because its first stage involves formation of two C3 sugars) operates maximally at CO_2 concentrations above that found in the atmosphere. Two adaptations have been described which minimize water loss and maximize CO_2 usage; they are known as the **C4-syndrome (C4 photo-synthesis)**, so called because CO_2 is fixed to give 4-carbon compounds, and **crassulacean acid metabolism (CAM)**.

C4 anatomy

In a typical C3 leaf, the majority of chloroplasts are distributed throughout the palisade mesophyll (below the epidermis) and the spongy mesophyll. These cells are somewhat randomly arranged between large gas spaces that connect to the stomatal pores (Topic D4). In the leaf of a C4 plant, the vascular bundles are surrounded by a ring of **bundle sheath cells** containing chloroplasts. These are surrounded by loosely packed mesophyll cells and air-spaces (Topic D5). This is known as **'Krantz' anatomy** (German for 'a wreath'). The bundle sheath and mesophyll cells are interconnected by many plasmodesmata (Topic C1) and no bundle sheath cell is separated by more than a few cells from a mesophyll cell.

C4 biochemistry

In a C4 leaf, the first product of CO_2 fixation is a **4-carbon acid**. This is produced by the action of an additional cycle, the **Hatch/Slack pathway**, found in the mesophyll cells. The first stage of this pathway is catalyzed by the enzyme **phosphoenolpyruvate (PEP)-carboxylase**. PEP-carboxylase uses HCO_3^- (formed when CO_2 dissolves in the cytoplasm) and **PEP** as substrates, yielding the C4 acid **oxaloacetate** which is converted to either **malate** or **aspartate** (both C4 acids) and they are transported directly to the bundle sheath cells, where the C4 acid is decarboxylated to yield a C3 acid and CO_2. At this stage, the CO_2 released is fixed by the Calvin cycle and the C3 acid transported back to the mesophyll cell (*Fig. 1*). The system therefore functions as a C3 pathway to which a 'CO_2 concentrator' has been added, transporting CO_2 into the bundle sheath cells, and generating a high CO_2 environment for Rubisco. The system has two major effects. First, it reduces photorespiration to undetectable levels, as the Rubisco is saturated with CO_2 giving conditions in which its oxygenase function is inhibited (Topic F2). This improves photosynthetic efficiency by up to 30%. Second, Rubisco is able to function at optimal CO_2 concentrations, well above atmospheric levels. PEP-carboxylase is also saturated at the concentration of HCO_3^- in the cytosol which results from atmospheric CO_2.

There is an energy cost to C4, however. The regeneration of the C3 acid PEP from the C3 acids transported requires the consumption of adenosine triphosphate (ATP) with the cleavage of both phosphates to yield adenosine monophosphate (AMP).

CAM anatomy

CAM plants are generally adapted to survive drought; they are succulents with fleshy leaves and few air spaces. Unlike C4 plants, they lack the physical compartmentation of photosynthesis into two cell types, but instead fix CO_2 in the night as C4 acids that are stored in a vacuole which occupies much of their photosynthetic cells' volume. The stomata of CAM plants are closed during the most desiccating periods of the day and are open at night.

CAM biochemistry

In common with C4 plants, the first stage of CO_2 fixation in a CAM plant involves PEP-carboxylase, which incorporates CO_2 into a C4 compound, oxaloacetate, in the dark. NADPH is used in the conversion of oxaloacetate to malate; malate is then stored in the vacuole (*Fig. 2*). In the light, malate is released from the vacuole and broken down to yield CO_2, which is then used in

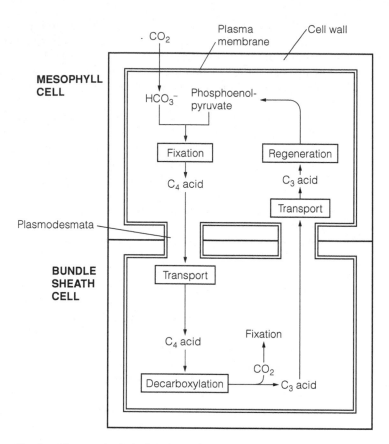

Fig. 1. Photosynthesis in C4 plants. Carbon is fixed from HCO_3^- in the mesophyll cells by PEP-carboxylase to produce C4 acids. These are transported to the bundle sheath cells, where they are broken down to release CO_2 which is fixed by the Calvin cycle. C3 acids produced are returned to the mesophyll cells, where they are converted to phosphoenolpyruvate with the consumption of ATP.

the Calvin cycle. CAM photosynthesis eliminates photorespiration and increases efficiency, though losses result from the energy expended in transporting C4 acids to the vacuole and in the formation of malate. CAM plants may have a major advantage in drought conditions, with the ability to photosynthesize in daylight with closed stomata.

Distribution of CAM and C4 photosynthesis

The C4 syndrome occurs in many plant families. The syndrome is most common in arid environments. C4 plants are distributed throughout the world and C4 crops such as maize have been successfully introduced into temperate climates. C4 plants are found in a number of families (*Table 1*), but by no means all the members of those families are C4 species. CAM plants are similarly distributed in many families, mostly succulents, and are distributed in arid zones throughout the world. Some families contain C3, CAM and C4 species. Although there appears to be an advantage to C4 and CAM, many arid zone plants are C3 and use other adaptations to survive.

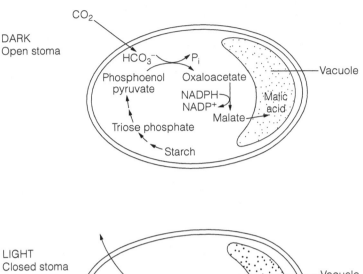

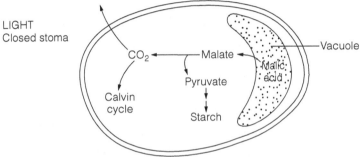

Fig. 2.   CAM plants use PEP-carboxylase to fix carbon as a C4 compound. However, carbon fixation and the operation of the Calvin cycle are separated temporally, rather than spatially. C4 acids are synthesized in the dark and deposited in the vacuole. In the day, the C4 compounds are released to supply CO_2 to the Calvin cycle that can function with the stomata fully closed.

Table 1.   Examples of C4 and CAM plants

Family	Common examples
C4 plants	
Amaranthaceae	Amaranths
Asteraceae	Asters and daisies
Euphorbiaceae	Euphorbias
Poaceae	Grasses including corn (maize), sugarcane and sorghum
Nyctaginaceae	Bougainvillea
CAM plants	
Agavaceae	Agaves
Asteraceae	Asters and daisies
Cactaceae	Cacti
Crassulaceae	Crassulas
Euphorbiaceae	Euphorbias
Liliaceae	Lilies
Orchidaceae	Orchids
Vitaceae	Grape vines

F4 RESPIRATION AND CARBOHYDRATE METABOLISM

Key Notes

Starch degradation	Starch is the major carbohydrate storage product of most plants. It is broken down to 6-carbon sugars by either hydrolytic or phosphorolytic enzymes. Products are exported from the chloroplast to the cytosol for glycolysis.
Glycolysis	Glycolysis breaks down glucose (6C) to two molecules of pyruvate (3C). Two molecules of ATP are used and four formed, giving a net yield of 2ATP and 2NADH. In anaerobic conditions, pyruvate is fermented to ethanol and carbon dioxide, with no further ATP production and use of the NADH.
The citric acid cycle	In aerobic conditions, pyruvate binds to coenzyme A (CoA) and enters the citric acid cycle as acetyl CoA. In total, the 3C sugar yields three CO_2, one ATP, three NADH and one $FADH_2$. As each glucose molecule supplies two 3C sugars, it takes two turns of the cycle to fully oxidize glucose.
The electron transport chain	NADH and $FADH_2$ supply energized electrons to the electron transport chain in the mitochondrial inner membrane. The electrons are carried through the chain and result in the pumping of protons across the membrane, giving a proton gradient, which is used to drive the enzyme ATP synthase, which generates ATP by oxidative phosphorylation. In all, about 12 ATP molecules are formed per glucose.
Sources and sinks	Tissues which are net suppliers of carbohydrate to the plant are known as sources; those which are net consumers are sinks. Young leaves are sinks, but become sources when mature. Storage organs are major sinks in a mature plant.
Phloem transport	Phloem transport is driven by a pressure gradient. Active loading of assimilates at the source creates a high solute concentration in the phloem, which results in water influx, creating a high hydrostatic pressure. Assimilate unloading at the sink tissue is accompanied by efflux of water, creating a low hydrostatic pressure. Loading is an active process. Assimilates unload down a concentration gradient maintained by the constant metabolism and incorporation of assimilates into storage reserves at the sinks.
Related topics	Plastids and mitochondria (C3) Amino acid, lipid, polysaccharide and secondary product metabolism (F5)

Starch degradation

The chief carbohydrate storage products from photosynthesis are **sugars** and **starch**. These are respired by plant cells to provide ATP. **Respiration** occurs in all cells, whether photosynthetic or not. Starch is made up of glucose polymers, either as long, straight chain molecules, **amylose**, or in highly branched form, **amylopectin**. Starch is mostly deposited in plastids. *Figure 1* shows the two major pathways for starch breakdown, either involving hydrolytic enzymes, α-amylase, dextrinase, α-glucosidase or phosphorolytic enzymes, starch phosphorylase. Starch breakdown

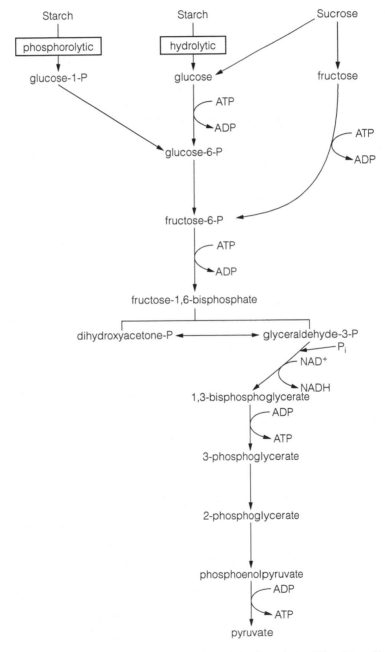

Fig. 1.   *The major pathways of starch degradation in a plant cell, involving either hydrolytic or phosphorylytic enzymes. All pathways lead into glycolysis.*

occurs initially in the plastid. The products are exported to the cytoplasm by a glucose transporter and a triose-P transporter in the chloroplast envelope.

Glycolysis

Glycolysis occurs in the cytoplasm without the consumption of oxygen. The products of starch breakdown, or sucrose, must be converted to a 6-carbon sugar phosphorylated at both ends for glycolysis. This 6-carbon sugar is **fructose 1,6-bisphosphate**; it is produced by the enzyme **aldolase**, which adds a phosphate group taken from ATP to fructose-6-P (*Fig. 1*). Production of fructose 1,6-bisphosphate consumes two molecules of ATP.

Fructose 1,6-bisphosphate is cleaved into two **3-carbon sugars**, each having one phosphate: **dihydroxyacetone-P** and **glyceraldehyde-3-P**. They are readily interconvertible and glyceraldehyde-3-P is then phosphorylated again to generate **1,3-bisphosphoglycerate**, a 3-carbon sugar with phosphate at both ends. This step generates NADH and requires inorganic phosphate. The 1,3-bisphosphoglycerate is then converted to **pyruvate** in stages which yield two molecules of ATP (four per initial glucose as each 6-carbon glucose has generated two 3-carbon sugars). The net ATP yield from glycolysis is therefore two molecules (four produced, two consumed) per molecule of glucose.

In the presence of oxygen, pyruvate is respired in the citric acid cycle (see below) in the mitochondrion. In anaerobic conditions (for instance in a root in flooded soil) it is **fermented** to yield carbon dioxide and ethanol. NADH is oxidized to NAD⁺, recycling the NADH produced in glycolysis. The total yield of anaerobic respiration is therefore limited to two ATP molecules per molecule of glucose (*Table 1*).

The citric acid cycle

The next stage of respiration, the **citric acid cycle** (sometimes known as the **Krebs cycle**) requires oxygen and occurs in the mitochondrion (*Fig. 2*). Its substrate is the pyruvate generated by glycolysis. **Pyruvate** is combined with **coenzyme A (CoA)** to generate **acetyl CoA** and liberate CO_2. The 2C compound is combined with **4C oxaloacetate** to generate **6C citrate**. CoA is released for re-use. Citrate is converted back to oxaloacetate in seven stages, with the production of one ATP, three NADH and one flavin adenine dinucleotide (reduced form; $FADH_2$). The cycle also liberates two CO_2; the pyruvate is therefore completely oxidized to yield three molecules of CO_2 and all the energy of the C-C bonds liberated.

Table 1.   The maximum overall energy yield from the oxidation of one molecule of glucose

	Cytosol	Matrix of mitochondrion	Electron transport and ATP synthesis	Total
Glycolysis	2ATP			2ATP
	2NADH		4ATP	4ATP
Pyruvate to acetyl CoA		2NADH	6ATP	6ATP
Citric acid cycle		2ATP		2ATP
		6NADH	18ATP	18ATP
		2FADH₂	4ATP	4ATP
Total				36ATP

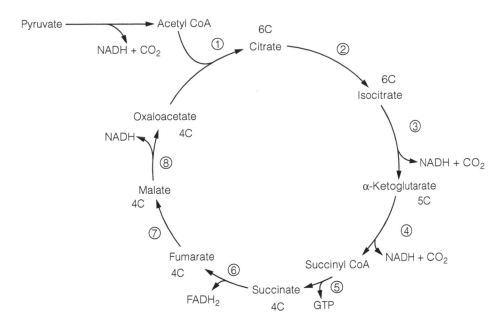

Fig. 2. The citric acid cycle. Two carbons enter the cycle as the acetyl group of acetyl CoA and two carbons are released (as CO_2). NADH and $FADH_2$ are generated. The steps involve the following enzymes: (1) citrate synthase; (2) aconitase; (3) isocitrate dehydrogenase; (4) α-ketoglutarate dehydrogenase; (5) succinyl CoA synthase; (6) succinate dehydrogenase; (7) fumarase; (8) malate dehydrogenase.

The electron transport chain

The mitochondrial **electron transport chain** generates **ATP** from the products of the citric acid cycle. In it, electrons removed from glucose are transported through a series of electron carriers located in the mitochondrial inner membrane until they react with protons and oxygen to give water. As the electrons pass through the electron transport chain, protons are pumped into the intermembrane space, generating a proton-motive force that then drives the synthesis of ATP. The key components of the electron transport chain are illustrated in *Fig. 3a*. Finally, the proton motive force is used to synthesize ATP from ADP by **ATP-synthase** (*Fig. 3b*). This process is known as **oxidative phosphorylation**.

Sources and sinks

Carbohydrate is exported from the photosynthetic tissue of the plant via the phloem. Net exporters are **source tissues**. Tissues which are net consumers or accumulating stores are **sink tissues**. Plants use a variety of storage products, including starch, protein and oils (lipids). All these storage products are produced in pathways originating from intermediates in glycolysis and the citric acid cycle. Major sink tissues include storage organs such as tubers, seeds and fruits. Newly growing tissues, including young leaves, are also sink tissues; a leaf will develop from being a sink to a source as the balance between energy requirements for growth and export from photosynthesis changes.

Phloem transport

Sugars move rapidly ($0.05–0.25 \text{ m h}^{-1}$) in the phloem (Topic D1 details the structure of phloem). Movement may be **bi-directional** in the same group of tubes. The best model to describe phloem transport is the **pressure-flow model**. It is proposed that the driving force for transport in the phloem results from an

(a)

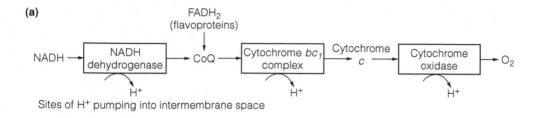

Sites of H$^+$ pumping into intermembrane space

(b)

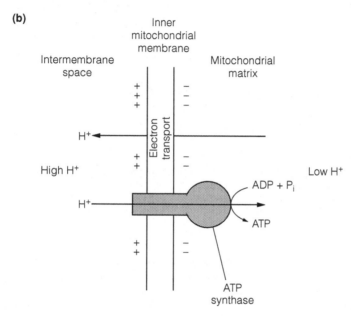

Fig. 3. ATP generation in the mitochondrion. (a) As electrons pass through the electron transport chain, protons (H$^+$) are transported out of the mitochondrial matrix and into the intermembrane space. (b) The energy stored in the resulting proton gradient across the membrane is used to generate ATP as protons flow through the ATP synthase complex.

osmotic gradient between the source and the sink ends of the tube. *Figure 4* shows the predicted development of the turgor gradient that drives transport. The presence of sieve plates and the associated **P-protein** that seals the sieve plate if the tube is damaged helps to maintain the pressure gradient.

The key processes in phloem transport are **loading** and **unloading,** which create the pressure gradient. Sugars are concentrated in the phloem by active transport; a proton-pumping ATPase establishes a proton gradient and sucrose is carried into the companion cells/phloem by a sucrose/proton cotransporter (Topic E3). Phloem unloading also requires metabolic energy; sucrose may leave the phloem passively and be converted to glucose and fructose by the enzyme **acid invertase**, in which case glucose and fructose will be transported into the sink; alternatively, sucrose may leave the phloem, either via **plasmodesmata** or via a **sucrose transporter**. If the sugars are rapidly metabolized within the sink (e.g. to form starch), a concentration gradient favoring sink loading will be maintained.

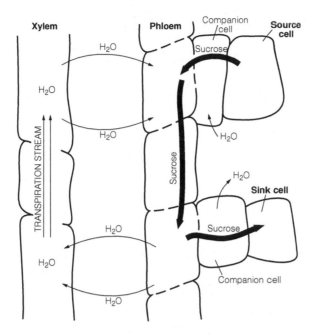

Fig. 4. Transport in the phloem. Sucrose is actively loaded into the companion cells and phloem tubules by sucrose/proton co-transporters. This increases the solute concentration resulting in an influx of water, generating a high turgor pressure at the site of loading. Sucrose is removed at the sink tissue giving a low turgor pressure.

F5 AMINO ACID, LIPID, POLYSACCHARIDE AND SECONDARY PRODUCT METABOLISM

Key Notes

Role of the citric acid cycle

Acetyl CoA, pyruvate and citric acid cycle intermediates are the starting point for the production of amino acids, lipids, polysaccharides and secondary products in plant metabolism.

Amino acid biosynthesis

Production of amino acids is linked to the assimilation of nitrogen by the plant. Nitrate is converted to ammonium by nitrate reductase and ammonium is then incorporated into glutamine and glutamate, either by the glutamine synthase-glutamate synthase (GS-GOGAT) pathway or by glutamate dehydrogenase (GDH). Ammonium is toxic and is converted to organic nitrogen compounds in the root. Nitrate may be converted to ammonium in the roots, or is carried to shoots and leaves and either stored in the vacuole or converted to ammonium. Other amino compounds are formed by transamination reactions.

Lipid biosynthesis

Plants synthesize a wide variety of lipids, including membrane lipids, cuticular waxes and seed storage lipids (mostly triacylglycerols). Synthesis of glycerolipids occurs in two stages: addition of the fatty acid chains to glycerol-3-phosphate and addition of a head group. Triacylglycerols consist of a glycerol to which three fatty acid chains have been added and are synthesized in the endoplasmic reticulum. They are stored in oil bodies, small lipid droplets with a surrounding coating of lipid and protein.

Sucrose, polysaccharides and starch

Plant cells produce monosaccharides, from 3-carbon trioses to 6-carbon hexoses. Sucrose is a disaccharide, made of glucose and fructose. Starch is a polysaccharide made up of α-(1–4) and α-(1–6) branched D-glucose residues. Cellulose is made up of β-D-glucose. Sucrose is synthesized in the cytoplasm either by sucrose phosphate synthase and sucrose phosphate phosphatase or by sucrose synthase.

Plant secondary products

A secondary product is one that is not involved in primary metabolism. They are generally produced in specialized tissues, with highly developed multi-enzyme pathways for their production. Plant secondary products include alkaloids, terpenoids and phenolic compounds. Many are involved in plant defenses against herbivory or fungal pathogens; others are of economic importance as medicinal and industrial compounds.

Related topics

Respiration and carbohydrate metabolism (F4)

Plant cell and tissue culture (K2)
Functions of mineral nutrients (E5)

Role of the citric acid cycle

Figure 1 illustrates that **pyruvate, acetyl CoA** and the **citric acid cycle** (Topic F4) are central in the production of many compounds. The activity of each of the pathways depends on substrate availability (usually monosaccharide) and on the tissue. As the pathways are often complex and involve many enzymes, initiating secondary product metabolism involves the developmental activation of many genes. Such gene expression and the production of secondary metabolites may be enhanced by stress, wounding, pathogen attack and herbivory.

Amino acid biosynthesis

Production of **amino acids** and other nitrogen-containing compounds is directly linked to the assimilation of nitrogen by the plant. **Nitrogen** is taken up as either nitrate or ammonium, the nitrate being converted by nitrate reductase into ammonium. **Ammonium** is incorporated into **glutamine** and **glutamate**, either by the **glutamine synthase-glutamate synthase (GS-GOGAT)** pathway or by **glutamate dehydrogenase (GDH)** (Topic E5). Ammonium is toxic and is converted to organic nitrogen compounds in the root. Nitrate is converted to ammonium in the roots, or carried to shoots and leaves and either stored in the vacuole or converted to ammonium there.

Roots have **glutamine synthase (GS)** in cytosol and plastids. Root plastids also have glutamate synthase that uses NADH as electron donor (**NADH-GOGAT**). If a root is supplied with nitrate, expression of **ferredoxin-dependent (Fd)-GOGAT** in plastids is induced.

Shoots have GS in the cytosol and chloroplasts. The chloroplast form takes ammonium generated by photorespiration, preventing it becoming toxic. Shoots and leaves express Fd-GOGAT in chloroplasts. Chloroplasts of shoots and leaves also contain GDH which synthesizes glutamate from ammonium and 2-oxoglutarate.

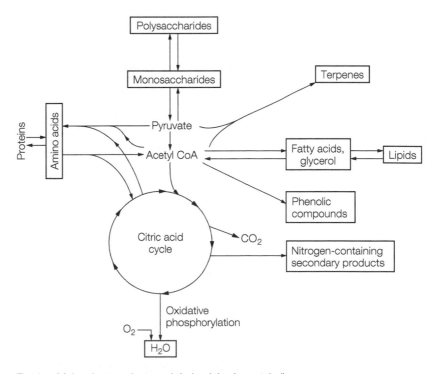

Fig. 1. *Major plant products and their origins in metabolism.*

Transamination reactions involve the transfer of an amino group from one compound to another. For instance, **asparagine synthase** (**AS**; *Fig. 2*) converts aspartate and glutamine to asparagine and glutamate. Asparagine contains a high quantity of nitrogen (2N) for each carbon (4C) present, compared with glutamate (1N:5C) or glutamine (2N:5C). Expression of the AS gene is repressed (inhibited) by high light and high carbohydrate production, so the plant makes nitrogen compounds with low quantities of nitrogen. When light is low and carbohydrate is scarce, carbon is conserved by the synthesis of compounds with a much higher nitrogen content.

Lipid biosynthesis

Plants synthesize a wide variety of lipids, including: membrane lipids, in which long-chain fatty acids, commonly 16–18 carbons in length, are esterified to glycerol; the waxes of the cuticle and suberin of the endodermis; and the storage lipids of seeds (most commonly triacylglycerols).

Plants use acetyl CoA as the basic building block for assembling **long-chain fatty acids**. Acetyl CoA is produced by **pyruvate dehydrogenase** present both in plastids and mitochondria. In the chloroplast, **pyruvate** is produced from 3-phosphoglycerate from the Calvin cycle; in nonphotosynthetic tissue, it originates from glycolysis. Synthesis of glycerolipids occurs in two stages: addition of the fatty acid chains to glycerol-3-phosphate and the addition of a head group. Glycerol-3-phosphate is derived in stages originating from glyceraldehyde-3-phosphate. In photosynthetic tissue, the chloroplast is the chief source of fatty acids, which may be used directly in the chloroplast or exported to the cytoplasm. Plant cells can synthesize **glycerolipids** at the endoplasmic reticulum (ER) and in mitochondria as well as in plastids; all three of these organelles are believed to be interconnected and lipids traffic between them and to the other cellular membranes.

Seeds make **triacylglycerols**, which are **storage oils**, and frequently very important human foods. Up to 60% of the dry weight of a seed may be in the form of storage oils. Triacylglycerols consist of a glycerol to which three fatty acid chains have been added (*Fig. 3*). They are stored in oil bodies, small lipid droplets with a surrounding coating of lipid and protein. The oil is first synthesized in the ER.

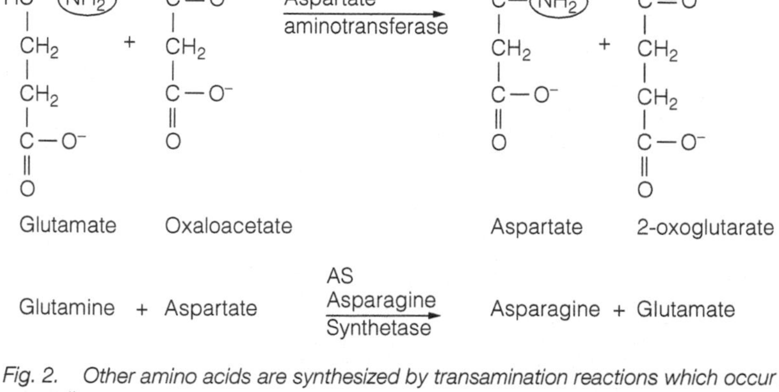

Fig. 2. Other amino acids are synthesized by transamination reactions which occur in various organelles.

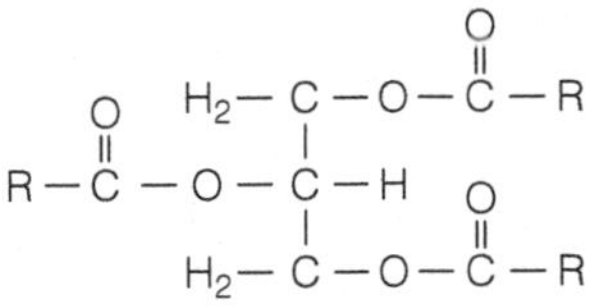

Fig. 3.   The structure of a triacylglycerol.

Sucrose, polysaccharides and starch

Plant cells produce a range of **monosaccharides**, from 3-carbon triose sugars such as dihydroxyacetone and glyceraldehyde to 6-carbon hexoses such as D-glucose, D-fructose, D-mannose and D-galactose (*Fig. 4*). **Sucrose** is a disaccharide, made of glucose and fructose. Polysaccharides are large polymers of these monosaccharides. By far the most important of these as a storage product is starch, made up of α-(1–4) and α-(1–6) branched D-glucose residues (*Fig. 4*). Cellulose, a major structural component of cell walls, is another polysaccharide, made up of β-D glucose (*Fig. 4*).

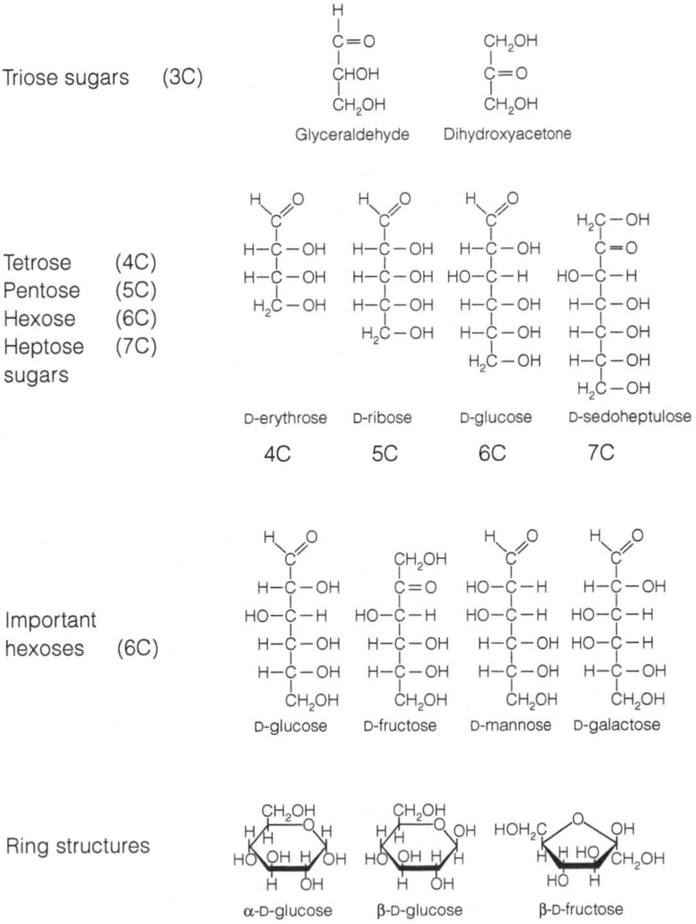

Fig. 4.   Structures of major sugars. Note that for ring structures, α or β depends on the position of the hydroxyl (OH) groups adjacent to the oxygen which closes the ring.

Sucrose synthesis occurs in the cytoplasm by one of two routes as a result of the activity of two enzymes, **sucrose phosphate synthase** and **sucrose phosphate phosphatase**:

$$\text{UDP-glucose + fructose-6-phosphate} \xrightarrow[\text{synthase}]{\text{sucrose phosphate}} \text{sucrose-6-phosphate + UDP}$$

$$\text{sucrose -6-phosphate + H}_2\text{O} \xrightarrow[\text{phosphatase}]{\text{sucrose phosphate}} \text{sucrose + P}_i$$

or by a single enzyme, **sucrose synthase**:

$$\text{UDP-glucose + fructose} \xrightarrow{\text{sucrose synthase}} \text{UDP + sucrose}$$

In leaves, **starch synthesis** occurs primarily in chloroplasts, commencing with **fructose-6-P**.

$$\text{Fructose-6-P} \xrightarrow{\text{hexose phosphate isomerase}} \text{glucose-6-P}$$

$$\text{Glucose-6-P} \xrightarrow{\text{phosphoglucomutase}} \text{glucose-1-P}$$

$$\text{Glucose-1-P + ATP} \xrightarrow{\text{ADP glucose phosphorylase}} \text{ADP-glucose + PP}_i$$

$$\text{ADP-glucose} + \alpha\ (1\text{-}4)\ \text{glucan} \xrightarrow{\text{starch synthase}} \alpha\text{-}(1\text{-}4)\text{glucosyl-glucan + ADP}$$

Cellulose is synthesized by **cellulose synthase**, an enzyme located in rosettes in the plasma membrane (Topic C2). Other polysaccharide complexes inserted in the cell wall are synthesized in the Golgi apparatus and secreted in vesicles, which fuse with the plasma membrane.

Regulating both the rates of sucrose and starch synthesis is important. If too much sucrose is produced, the chloroplast becomes depleted in the intermediates of the citric acid cycle, and photosynthesis will be inhibited. If too little starch is produced, the cell will not have sufficient reserves of carbohydrate for respiration during the night. Photosynthetic tissues also export assimilated carbon to other tissues and so regulation of the whole process must allow for this.

Plant secondary products

Plant **secondary products** are compounds generated by **secondary pathways** and not from primary metabolism. Many are toxic or give the plant an unpleasant taste and it is likely they give a selective advantage as anti-herbivory agents. Numerous plant secondary products have been used over many hundreds of years for a wide array of purposes (Topics N2 to N4). They are generally produced in specialized tissues, with highly developed multi-enzyme pathways for their production. *Table 1* summarizes major secondary products, with their origins and uses.

Table 1.   Plant secondary products

Family of compounds	Biosynthesis	Examples	Examples of key producers
Terpenoids, isoprenoids; based on isoprene: CH_3 | $CH_2=C-CH=CH_2$	Mevalonic acid pathway (originating from acetyl CoA)	Carotenoid pigments; Sterols and steroid derivatives (e.g. cardiac glycosides like digitoxin); Hormones (abscisic acid, gibberellins); Latex Flavors (e.g. menthol); Odors	*Digitalis purpurea* (foxglove) *Hevea brasiliensis* (rubber tree) *Mentha palustris* (mint)
Phenolics based on phenol: 〈 〉–OH	Shikimic acid pathway (originating from aromatic amino acids tryptophan, tyrosine, phenylalanine)	Many defence and anti-herbivory compounds including: phenolic monomers; tannins (also flavor wine and tea); lignins (also wood); flavonoids (pigments like anthocyanins)	

Alkaloids (diverse group, containing a N-group and water soluble)	Diverse origins. Name derived from slight alkalinity in solution. May be aromatic (heterocyclic) or aliphatic	Stimulants
		  Caffeine
		  Nicotine
		Sedatives
		  Morphine
		  Codeine
		Poisons
		  Atropine
		
		Chemotherapy agents
		  Vinblastine
		  Vincristine
		Narcotics
		  Cocaine
			Coffea arabica (coffee)
			Nicotiana tabacum (tobacco)
			
			Papaver somniferum (opium poppy)
			
			Atropa belladonna (deadly nightshade)
			
			Catharanthus roseus (rosy periwinkle)
			
			Erythroxylum coca (coca)

G1 THE FLOWER

Key Notes

General structure of the flower

In a typical flowering shoot there is a receptacle at the top of a stem with four whorls of floral organs: sepals, petals, stamens and carpels. In some plants one or more whorls may be absent, e.g. in unisexual flowers and reduced flowers of wind-pollinated plants.

Sepal and petal structure

Sepals are usually green and protective, resembling simple leaves. Petals are often brightly colored and attractive to animals with a specialized epidermis, often pigmented vacuoles and air-filled reflective mesophyll. The pigments are mainly flavonoids with some betalains and carotenoids. Insects have different color sensitivity from us and the colors of petals are different and more varied for insects.

Stamens

The stamens have a filament with four or two anthers. The anther wall has three layers and splits (dehisces) to release pollen. Some plants have specialized dehiscence such as only releasing pollen when insects vibrate their wings and orchids have sticky pollen masses known as pollinia. The filaments may be long and thin or flat and some are sensitive to insect contact.

Carpels

The carpel consists of an ovary enclosing the ovules, usually a stalk, the style, and a stigma with a receptive surface for pollen grains. The ovaries may be separate or fused, and on the surface, i.e. superior, or buried in the receptacle, inferior. In animal-pollinated plants the stigma is club-shaped or lobed and may be dry or sticky; in wind-pollinated plants it usually feathery with a large surface area. The ovaries contain one to many ovules. The ovules usually have two integuments and a nucellus surrounding the embryo sac, with a tiny gap, the micropyle, for pollen tube entry.

Nectar

Nectar is mainly a sugar solution of varying concentration with small quantities of other substances. It is secreted at the base of the petals or elsewhere in the flower. Many flowers have no nectar and use pollen as a food reward, more rarely oil or resin.

Related topics

Pollen and ovules (G2)
Breeding systems (G3)
Self incompatibility (G4)

The seed (H1)
Physiology of floral initiation and development (J4)

General structure of the flower

Flowers are unique to flowering plants, normally developing on vegetative shoots (Topic J4). A typical flower is **hermaphrodite**, i.e. bearing fertile organs of both sexes, and pollinated by insects (*Fig. 1*). The stalk that bears the flower has a more or less swollen top just underneath the flower called the **receptacle** and on this there are four different organs (from outside in): sepals, petals, stamens and carpels. There is usually one **whorl** each of sepals and petals and many plants have

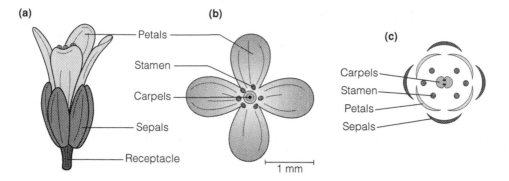

Fig. 1.   A flower of Arabidopsis thaliana: (a) side view, (b) top view, (c) diagrammatic overview of floral parts.

a fixed number, often five or three of each, though some have an indefinite larger number. Some plants have an indefinite number of more than 10 stamens, whereas in others the number corresponds to the number of petals or twice the number if there are two whorls. The carpels are usually fewer in number.

One or more of the whorls may not be present. Either sepals or petals or both may be absent, such as some wind-pollinated flowers, or the two whorls may not be fully differentiated, especially in the primitive dicotyledons. Some flowers have **unisexual** flowers without one of the whorls of fertile organs, though many retain small infertile remnants of the other parts. If male flowers and female flowers occur on the same plant it is known as **monoecious** and many of these flowers are very small without sepals or petals, such as those of many catkin-bearing trees. If unisexual flowers are on separate plants they are known as **dioecious**.

Sepal and petal structure

Sepals normally surround the flower in bud, providing a protective function. They are the most leaf-like floral organ and are often green and photosynthetic and, in structure, resemble a simple leaf. In some plants, especially among monocots, they resemble the petals. The **petals** are usually the main organs attracting insects to flowers and many are brightly colored, though they may be photosynthetic as well. They resemble leaves in their overall organization but have a specialized epidermis and sub-epidermal layer. The epidermal cells frequently have small projections that absorb the incident light and scatter the light reflected from the mesophyll layer underneath. This layer has many air spaces between its cells and reflects all the light. The **pigment** is usually in the **vacuoles** of the epidermal cells but sometimes there is some in the mesophyll too.

There are three main types of pigment in petals. The most widespread are **flavonoids**, which include at least three types of **anthocyanins** (*Fig.* 2) responsible for the pink, purple and blue colors and a number of similar **anthoxanthins** responsible for ivory-white and yellow colors which may be present in addition to the anthocyanins. **Betalains**, similar in structure to **alkaloids** (Topic F5), form intense purples and yellows in a few plant families, such as the cacti, and **carotenoids** occur in some yellow flowers such as daffodils and marigolds.

Most flower-visiting insects have good color vision but are sensitive to a **different spectrum** from us and other vertebrates, being insensitive to red but sensitive to the **near ultraviolet**. Floral pigments of insect-pollinated flowers are more varied within this color spectrum than they are for vertebrates, many flowers that are white to us absorbing in part of the ultraviolet, providing a

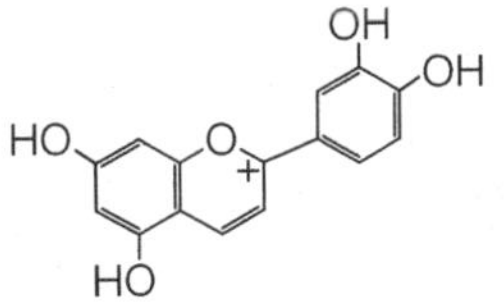

Fig. 2. The molecular structure of an anthocyanin, cyanidin.

range of shades to an insect. The lines towards the base of some petals, known as **nectar guide** marks, are exaggerated for insects since most absorb in the ultraviolet giving a darker center to the flower for an insect.

Stamens

The **stamens** consist of a stalk, the **filament**, and an **anther** enclosing the **pollen grains**. Most plants have four **anthers** on each **filament**; some have two (*Fig. 3*). The anthers have an outer epidermis, and inside that a **fibrous layer** forming the bulk of the wall, and an inner **tapetum** providing the nutrients for the developing pollen grains inside. When the anther is mature it **dehisces**, i.e. the wall splits to release the pollen. In most plants it splits all the way down along a line of weakness, but in some only the tip of the anther splits, and certain families have specialized dehiscence. In some plants, such as nightshades, the pollen is only released when a visiting bee vibrates its wings at a certain pitch (known as **buzz-pollination**). In the orchid family there is only one anther and this divides into two **pollinia**, sticky masses of thousands of pollen grains which are dispersed as a unit. The filaments vary, wind-pollinated flowers often having long flexible ones, and a few plants bearing wide filaments acting as the main attractive organs for insects. Some are sensitive to touch, contracting on contact with insects.

Carpels

Carpels consist of a basal **ovary** which contains the **ovules**, usually a stalk, the **style**, and a **stigma**, the surface onto which pollen grains will land for fertilization. The **ovaries** are usually oblong-shaped but some are spherical or disc-like and they may be separate or a few, often three or five, fused together. The ovaries may be situated on the surface of the flower, known as **superior** ovaries, or may be buried in the receptacle with only the style and stigma projecting, in

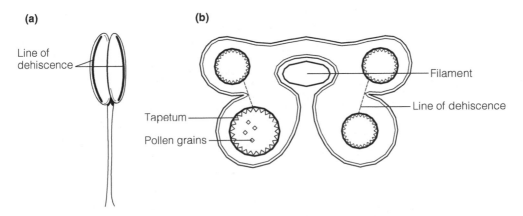

Fig. 3. A typical stamen with four anthers: (a) surface view, (b) transverse section.

which case they are known as **inferior**. There are intermediates. In most plants the stamens and carpels are of similar length and position in the open flower, but in many one sex matures before the other (Topic G3).

A **style** may or may not be present, and, where the ovaries are fused, there is only one; it is up to several centimeters long in some bulbous plants. The style is usually solid with some central **transmitting tissue** through which the **pollen tubes** will grow after the pollen grains have germinated on the stigma, but in some it is hollow and pollen tubes grow over the inner surface. The **stigma** in animal-pollinated plants is usually club-shaped or with two to five lobes, sometimes one lobe per ovary. The surface may be dry or sticky (Topic G4). In wind-pollinated flowers the stigma is often lobed and feathery with a large surface area, frequently with no style.

Each ovary may have one **ovule** or several, up to a few thousand in the orchids. The ovules are usually attached to the side of the ovary nearest the center of the flower. They have a central **embryo sac** which includes the **egg**, surrounded by a **nucellus** of parenchyma cells and with usually two protective **integuments** around that. A few plant families have lost one or both integuments and others have four. There is a small opening in the integuments, the **micropyle**, through which the pollen tubes penetrate. The ovule is usually turned through 180° with the micropyle facing the stalk, known as **anatropous**, but a few species have upright ovules, ones turned to the side or even turned through 360° (*Fig. 4*).

Nectar

Nectar is secreted by many flowers and is collected as a food by animal visitors. It is mainly a solution of **glucose, fructose** and **sucrose** in varying proportions, depending in part on which are the main pollinators – the more specialist bee-, butterfly- or bird-pollinated flowers having high sucrose content. The concentration ranges from **15%** sugar by weight to **80%** or even crystalline sugar depending on flower form and weather conditions. It may also contain small quantities of **amino acids** and other substances that may be important nutrients for some pollinators such as flies and antioxidants. It is secreted in various different parts of a flower, most commonly at the base of the petals, or in a specialized **spur** at the base of a petal or petal tube in some flowers. In other flowers it is on the receptacle and some plants have special nectar-secreting organs, either modified petals or parts of petals. **Honey** is made by bees mainly from nectar, with pollen and other impurities. It is concentrated to about 80% sugar, almost entirely glucose and fructose as the sucrose is digested.

Flowers do not all secrete nectar; some use pollen as a food reward and others, mainly tropical plants, secrete **oils** or **resins** or even **scents** (in some orchids) that are collected by bees for various uses. Wind-pollinated flowers do not normally secrete nectar and some animal-pollinated flowers produce no reward but rely on deception to attract pollinators.

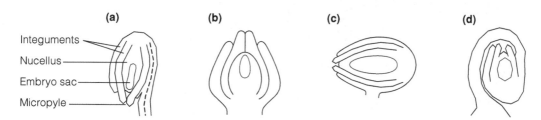

Fig. 4.   Ovules: (a) anatropous orientation (the most widespread type); (b–d) other types.

G2 POLLEN AND OVULES

Key Notes

The structure of pollen	Pollen grains consist of one haploid cell surrounded by an inner wall, intine, of cellulose and an outer wall, exine, of resistant sporopollenin. The nucleus divides to produce the pollen tube and two sperms. The tapetum of the anther provides nutrients for the developing pollen, and proteins from it impregnate the exine.
Variation in pollen grains	There is great variation in size, shape, type and arrangement of apertures and exine sculpturing in pollen. This is diagnostic to plant family, genus and sometimes species. Wind-pollinated plants vary less than animal-pollinated plants. Pollen may be dispersed singly or in groups. Little is known about the functional significance of pollen variation.
Embryo sac	The commonest type of embryo sac consists of eight cells: an egg with two synergids, three antipodal cells and two that fuse to form a diploid central cell. Other types are known with four or 16 cells in the embryo sac.
Pollination and fertilization	Pollen grains germinate and the pollen tube grows through stigma and style to reach the ovule. Two sperm cells are discharged – one fertilizing the egg, the other the diploid central nucleus. The embryo is then diploid but the central nucleus that grows to form the endosperm is usually triploid.
Related topics	The flower (G1) The seed (H1) Self incompatibility (G4)

The structure of pollen

A pollen grain is a single cell deriving by meiosis (Topic C6) from pollen mother cells in the anther. It lays down two coats, an outer **exine** and an inner **intine**. The intine is laid down first and is made of cellulose and pectins. The exine matures soon afterwards and is made of extremely hard polymers of carotenoids known as **sporopollenin**. It has numerous tiny cavities in its surface and is often beautifully sculptured. Pollen exine is one of the most resistant of plant structures and can survive long after the rest of a plant (including the inner parts of the pollen grain) decays. Inside the pollen grain the haploid cell divides to produce two nuclei, a large **vegetative nucleus**, which will form the **pollen tube**, and a **generative nucleus**. The generative nucleus divides again either before dispersal or after it has germinated on a stigma to produce two sperms that travel down the pollen tube.

Before the first division of meiosis the **pollen mother cells** of the anther are interconnected by cytoplasmic strands and this may enable the pollen grains all to develop synchronously. Meiosis in these gives a group of four haploid cells and these four pollen grains may mature and be dispersed together or separately. As the pollen grains mature, proteins from the tapetum of the anther impregnate the cavities of the wall and remain there when the pollen disperses.

It is this protein that is involved in the recognition interaction of sporophytic self incompatibility (Topic G4), and responsible for the allergic reaction of hay fever.

Variation in pollen grains

The shape and size of pollen grains varies enormously. The shape varies from spherical to narrowly oblong, tetrahedral or even dumb-bell shaped, and water-pollinated grains of the marine grass, *Zostera*, resemble spaghetti and can reach over 1 mm in length. In animal- or wind-dispersed grains the size varies from 5 μm across in, e.g. forget-me-nots, *Myosotis*, to over 200 μm in some members of the cucumber family, but is usually around 40 μm; wind-dispersed grains are all around 40 μm.

The pollen grain usually has one or more **apertures** in the exine through which the pollen tube will emerge. This aperture may be elongated forming a **furrow**, or more or less rounded forming a **pore**. A single furrow was the first form to evolve, with pores and multiple apertures, which germinate more efficiently on the stigma, appearing later in the fossil record. A combination of the number and orientation of pores or furrows and the sculpturing of the exine is diagnostic of plant families, genera or sometimes species (*Fig. 1*). The exine of pollen that is dispersed by the wind is fairly smooth and in some families, such as the grasses, it varies little. By contrast, in animal-dispersed pollen the sculpturing varies hugely from circular holes to projections of many sizes and shapes and the beautiful window-like form of some dandelion-like composites (*Fig. 1*).

Pollen grains may be dispersed singly, as four together in a **tetrad**, or as many together in a **polyad**. In orchids and asclepiads (milkweeds) they are dispersed as entire pollinia with thousands of grains (Topic G1), although most of these fragment when they reach the stigmas and can fertilize more than one flower.

To some extent pollen grain size is correlated with the length of the style down which the tube must grow, but other factors must be involved and little is known about the significance of most variation in pollen.

Embryo sac

After meiosis, three of the four daughter cells abort, and one divides to form the embryo sac. This is contained within the ovule (Topic G1) and the most common form consists of eight haploid cells (*Fig. 2*). Three of these settle at the micropyle end to form the **egg** and its two associated **synergid** cells; three others remain at the other end to form **antipodal** cells. The two remaining cells,

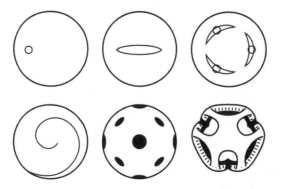

Fig. 1. Pollen grains showing the variety of apertures and sculpturing.

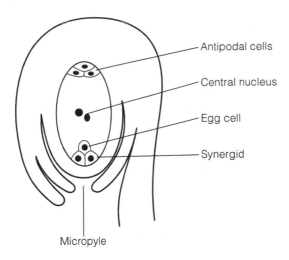

Fig. 2.   An eight-celled embryo sac.

one from each end, fuse to form a diploid central nucleus that will become the **endosperm** after fertilization.

The embryo sac described above occurs in about 70% of flowering plants but there are at least 10 other types of embryo sac known, some characteristic of certain families, others within one family or even on the same plant. The total number of cells in an embryo sac is four, eight or 16 in various orientations, but in all types there is a single haploid embryo nucleus that will form the next generation after fertilization and a second nucleus that will form the endosperm. The others play no further part after fertilization.

Pollination and fertilization

Pollination occurs when pollen grains land on the stigma. If compatible, a **pollen tube** emerges through one of the apertures of the grain and grows through the stigma and into the style. It follows the transmitting tissue in the center of the style or the inner surface of a hollow style and its progress is marked by deposits of the polysaccharide **callose** (Topics G3, M4) behind the growing tip. It can then grow through the ovary and into the micropyle of the ovule (more rarely into the other end). The rate of pollen tube growth is variable, usually around 3–4 mm h^{-1} but may be up to 35 mm h^{-1}.

Two **sperm cells** are discharged, either into one of the synergid cells or direct into the embryo sac, and they lose their cytoplasm before one fertilizes the egg cell, the other the central diploid cell that will form the endosperm. The embryo is then diploid, with its cytoplasm deriving entirely from the egg, and will grow into the new plant. The endosperm in the commonest form of embryo sac (see above) is **triploid**, since the haploid male nucleus fuses with a diploid cell. Other types of embryo sac have diploid or pentaploid (5n) endosperm. In most plants the endosperm grows after fertilization to provide the food store for the developing seed.

G3 BREEDING SYSTEMS

Key Notes

Range of breeding systems	Most plants are hermaphrodite and they may cross- or self-fertilize. Others have unisexual flowers either on the same plant, monoecious, or separate plants, dioecious, and others are intermediate or variable. A few are asexual.
Cross- and self-fertility	Many plants have a self-incompatibility system stopping self-fertilization, but others are self-fertile. Early maturation of flowers can lead to self-fertility. Some species are partially self incompatible.
Separation of floral organs	Cross-pollination is favored by a separation of fertile organs in the flower. Stamens may mature before carpels, known as protandry, or after, protogyny. Protandry is common in specialized insect-pollinated flowers in inflorescences and protogyny is associated with wind pollination and unspecialized flowers. Stamens and carpels may be separated spatially.
Asexual reproduction	A few plants have bulbils in place of some or all flowers and others produce seeds without any fertilization. These are often polyploids deriving from hybridization between sexual species and many of these clones have spread.
Control of sex expression	Floral development genes and hormones affect sex expression in all plants, but in dioecious species there is a chromosomal XX/XY system too, sometimes over-ridden in polyploids. There is an interaction with environmental conditions and sex expression can vary in plants with unisexual flowers.
Related topics	The flower (G1) General features of plant evolution Self incompatibility (G4) (Q5) Ecology of flowering and pollination (G5)

Range of breeding systems

Most flowering plant species bear **hermaphrodite** flowers with functional stamens and carpels (Topic G1). Some of these are almost entirely cross-fertilized, some mainly self-fertilized (see below). Other plants bear **unisexual** flowers, with only stamens or carpels functional, and others reproduce **asexually**. These have different implications as breeding systems (*Table 1*; Topic L5). Breeding systems are flexible in plants, e.g. in populations of many hermaphrodite species occasional individuals produce no viable pollen; individuals of dioecious species sometimes produce flowers of the opposite sex (*Fig. 1*). A few species have a mixture of hermaphrodite, male and female plants.

Cross- and self-fertility

In hermaphrodite and monoecious species there is the potential for **self-pollination** leading to **self-fertilization** (**selfing**) unless there is a mechanism to avoid it. The commonest form of inhibition of selfing is some form of physiological **self-incompatibility** (**SI**) system (Topic G4). In most plants with an SI

Table 1.   Types of flowers and breeding systems in flowering plants

Type	Description	Comments
Hermaphrodite	Both sexes in each flower	Widespread: 80% of all flowering plants
Monoecious	Flowers unisexual; male and female on same plant	Particularly wind-pollinated trees and certain families, e.g. arum lilies; 5% of plant species
Dioecious	Flowers unisexual; male and female on separate plants (*Fig. 1*)	Widespread in many families; common in tropics and on islands; 10% of plant species
Other mixed types	Some flowers hermaphrodite, others unisexual on same or different plants	Some members of certain families, e.g. male and hermaphrodite flowers on each plant in carrot family; female and hermaphrodite plants in thyme family
Asexual	Clonally produced bulbils in place of flowers, or seeds without fertilization	A few genera; mainly polyploids with odd numbers of chromosomes
Sterile	Flowers without fertile parts	Flowers for attraction only; always associated with fertile flowers in inflorescences

system that have been extensively studied a few individual plants have been shown to be **self compatible**, showing that there is always the potential for self-compatible individuals to spread if cross-pollination fails in the long term. SI systems do not work if the stigma is pollinated experimentally when the flower is still in bud (allowing easy study of the system). Situations favoring a quick life cycle, such as agricultural land, have probably led to the evolution of selfing many times, since rapid development and early maturation of the stamens or early opening of the flower (*Fig. 2*) may allow self-fertilization before the SI system becomes operational. These plants will have smaller flowers than self-incompatible relatives.

Some plants are partially self incompatible, with a plant's own pollen growing more slowly than that from another individual, allowing for self-fertilization if there is no cross-pollination. In general it seems that many hermaphrodite species are flexible in their degree of outbreeding and, among flowering plants, there is a whole range from nearly 100% outbreeding to nearly

(a)   **(b)**

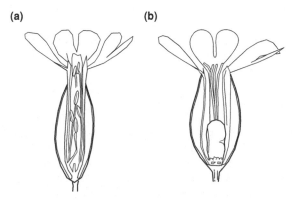

Fig. 1.   Dioecious flowers of the campion, Silene dioica *(a) male, (b) female, each showing rudimentary organs of the opposite sex.*

Fig. 2.   Self-fertilizing flower of the mouse-ear, Cerastium glomeratum. (Redrawn from M. Proctor and P. Yeo (1973). The Pollination of Flowers, Harper Collins Publishers.)

100% inbreeding. In a few plants self-fertilized seeds have slightly different properties from cross-fertilized seeds (e.g. larger and with a softer testa).

Separation of floral organs

Many species with a SI system, and some without one, separate stamens from stigmas either temporally, one maturing before the other, or spatially. This will favor the dispersal of pollen to other flowers, which is important in self-incompatible species because the stigma can otherwise become clogged with a plant's own incompatible pollen. In self-compatible species cross-fertilization will be more likely. If the stamens mature before the carpels the plant is known as **protandrous**; if the carpels mature first it is **protogynous**. Protogyny is common among some unspecialized insect-pollinated flowers and in wind-pollinated plants, whereas protandry is a common feature of the more specialized insect-pollinated flowers. Protandry is a feature of plants that bear their flowers in inflorescences in which the insects visit the older flowers first, e.g. spikes of flowers that mature from the base upwards, or composite inflorescences maturing from the outside inwards. In protogyny and protandry there may be overlap in function between the sexes, allowing for self-pollination if cross-pollination fails.

The floral organs may be separated spatially within the flower. Some flowers have their parts separated, often by a small distance, and rely on visitors crawling around the flower for pollen to reach the stigmas. Some others have parts that move as they mature or in response to insect visits; in these a spatial separation can be combined with a temporal separation. After a flower has been open for a time the anthers and stigmas may come together, allowing selfing if there has been no cross-pollination.

Asexual reproduction

Many plants can spread by vegetative means, with rhizomes, root or stem fragments, etc., not involving floral structures. These reproduce sexually with flowers as well. A few plants have **bulbils** in place of some or all of their flowers. These resemble tiny bulbs (Fig. 3) or plants and are clones of the parent plant. A few plant groups produce seeds that are formed without any fertilization of the embryo, so again they are clones of the parent plant. This is particularly associated with members of the rose family (Rosaceae) such as the brambles (Rubus), and composites (Asteraceae) such as the dandelions (Taraxacum). These **agamospermous** plants are nearly all high **polyploids**, often with odd numbers of

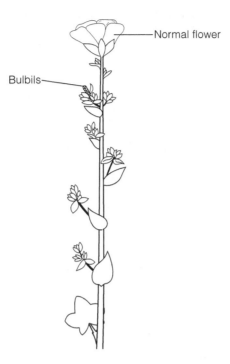

Fig. 3.   One normal flower and bulbils in place of others in the saxifrage, Saxifraga cernua.

chromosomes, derived by hybridization from sexual diploids in the center of the range of the genus. Numerous different agamospermous clones, differing in minor ways, have appeared and these present a problem of classification, each clone sometimes being described as a separate species.

Control of sex expression

In hermaphrodite flowers sex expression is controlled by floral development genes (Topic J4). In unisexual flowers some of the same genes are likely to be present and their expression is subject to hormonal and genetic control. Dioecious plants normally have identifiable sex chromosomes with an XX/XY system, XX normally, but not always, being the female. In some dioecious plants the Y chromosome is visually distinct from the X. This works for diploid plants, but numerous plant species are polyploid and must therefore have multiple sex chromosomes. Many of these polyploids are hermaphrodite even if the diploid is dioecious, although this is not always so. In some, the Y genes seem to over-ride expression of X genes even if there is only one Y chromosome and several Xs, and the polyploids remain dioecious.

In some dioecious species, such as willows, sex expression never varies, but in others production of male and female flowers can be influenced by environmental conditions, and hormonal expression within the plant interacts with the genetic makeup. This can make genetically male plants produce female flowers and vice versa. Auxin levels (Topic J3) differ along a shoot and affect which flowers grow, the auxin levels themselves being affected by day length, temperature and quantity and quality of light. In monoecious species male flowers are usually produced nearer the tip of a shoot than female flowers, this being related to the levels of auxins and other hormones.

G4 SELF INCOMPATIBILITY

Key Notes

Types of self incompatibility

In many hermaphrodite and monoecious plants there is a physiological self incompatibility (SI) in which a plant's own pollen is rejected. There are several known types of SI: gametophytic systems involving recognition of determinants of the pollen tube by the style; sporophytic systems in which proteins of the pollen surface are recognized; and ovarian systems in which the tubes reach the ovule but the ovule is aborted.

Gametophytic systems

In the commonest system incompatible pollen tubes are arrested or burst in the style. The recognition is controlled by one locus with many 'S' alleles. In the grasses the pollen tube is blocked at the stigma surface with pectins and callose. Alleles at both loci need to be the same for an incompatible reaction.

Sporophytic systems

One sporophytic system has two linked loci and the genetics may be complex involving dominance hierarchies. The pollen is inhibited from germinating or the tube is blocked before penetration. Heteromorphic systems involve two or three different forms of flower with differences in pollen and stigma structure and usually differences in style and stamen length. Pollen-stigma interaction may differ but each form can cross only with other forms. The system involves one or two loci each with two alleles.

Ovarian or late-acting systems

This is a little known, highly varied group of SI systems, in some of which there may be slow self pollen tube growth, others a signaling system. It may involve sporophytic or gametophytic recognition or a combination of both.

Molecular basis of self incompatibility

In the commonest form of gametophytic SI glycoproteins with ribonuclease activity are produced by the S gene, perhaps disabling RNA at the pollen tube tip. In the grass system wall formation is inhibited by glycoproteins and, in the poppies, small stigmatic proteins act as signalers. In sporophytic SI two linked loci produce glycoprotein and a kinase that disable each other.

Evolution of self incompatibility

A system with gametophytic recognition and RNase activity is likely to be primitive and others, including all forms of sporophytic recognition, evolved from it, some several times. It is unknown where ovarian systems fit in.

Related topics   Pollen and ovules (G2)   Breeding systems (G3)

Types of self incompatibility

Many hermaphrodite and monoecious plants have a physiological self-incompatibility (SI) system involving an interaction between the pollen and the carpel which prevents seeds forming or maturing as a result of self-fertilization (Topic G3). These interactions all involve the ability of the carpel to recognize and reject its own pollen and accept pollen from a different plant to fertilize the ovules. The site of recognition and the mechanisms appear to vary widely within plants (*Table 1*) with the

Table 1.   Types of self incompatibility (SI) in flowering plants

Name	Site of recognition	Genetic basis	Features	Distribution
Gametophytic (normal type)	Pollen tube in the style	One locus, many alleles	Stigma with wet surface and cuticle with gaps	Widespread; more than one diverse type
Gametophytic (grass type)	Pollen tube on stigma surface	Two loci, several alleles	Stigma dry with continuous cuticle	Grasses (others?)
Sporophytic	Surface of pollen grain on stigma	Two linked loci, many alleles	Stigma dry with continuous cuticle	Several unrelated families
Heteromorphic (two forms)	Surface of pollen grain on stigma or style	One (or two linked) locus, two alleles	Stigma normally dry with continuous cuticle	25 mainly unrelated families
Heteromorphic (three forms) Ovarian, or late-acting	Surface of pollen grain on stigma or style Pollen tube at ovule or signal from stigma	Two loci each with two alleles Unknown	Stigma normally dry with continuous cuticle Variable	4 unrelated families Widespread? – especially tropical woody plants

main interaction known as gametophytic, sporophytic or ovarian (late-acting) (*Fig. 1*). In a **gametophytic** interaction the pollen germinates and the growing pollen tube is recognized by the female plant, normally in the style. Since a pollen grain is the result of meiosis, it has a single set of chromosomes and it is this that is recognized. In a **sporophytic** interaction the surface of the pollen grain is recognized, normally by the stigma. The pollen surface is impregnated with the anther tapetum before being shed (Topic G2) and it is this coating, derived from the parent plant and with both sets of parental chromosomes, that is recognized. The terms 'gametophytic' and 'sporophytic' are used to indicate the origins of the structures (Topic Q4). In the ovarian interaction the pollen tube reaches the ovary but the ovules are aborted. In all SI systems the controlling genes are known as S genes.

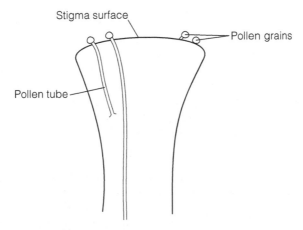

Fig. 1.   Stigma showing incompatible and compatible pollen from a typical gametophytic interaction (left) and incompatible pollen from a sporophytic interaction (right).

Gametophytic systems

These are widespread in flowering plants, occurring in more than 60 families. In most types, both compatible and incompatible pollen tubes grow down the style but only compatible pollen tubes reach the ovule where they burst and release the sperm cells. Incompatible pollen tubes grow more slowly and do not reach the ovule, either bursting in the style or being arrested and blocked. The polysaccharide callose is deposited around the tube and may block off an incompatible pollen tube. This system involves one gene locus with several alleles (i.e. different forms of the one gene), the number varying between about 12 and more than 40. If the S allele of the pollen tube is the same as either one of those in the style the tube will be stopped, thereby stopping all a plant's own pollen and other pollen with one of the same S alleles.

In the grasses the germinating pollen tube is blocked before it penetrates the stigma or just as it starts. Incompatible pollen is blocked, initially, by pectins, subsequently by callose deposition. Two gene loci are involved (S and Z loci) and only if the alleles at both loci in the pollen are the same as two of the alleles in the female parent is the incompatibility reaction triggered. The number of alleles at each locus is probably 6–20, smaller than most single-locus systems, but the two independent loci allow for more compatible pollinations.

Sporophytic systems

A sporophytic interaction with multiple S alleles similar to those in gametophytic systems has been best studied in the cabbage family, Brassicaceae, but is also known from the unrelated daisies, Asteraceae, bindweed family, Convolvulaceae, the birch family, Betulaceae, and a few others. The mode of action is normally similar to that in the grasses, with incompatible pollen blocked at the stigma surface. Two tightly linked gene loci are involved and genetic control is more complex than in gametophytic systems. There is some variation in the response, although a plant's own pollen is always stopped. The number of S alleles (the two loci considered together) is variable, as in gametophytic systems, but some have a dominance hierarchy.

Heteromorphic systems (*Table 1*) mainly have sporophytic recognition and the system, best known in the primroses, Primulaceae, occurs in several widely scattered plant families. In this SI system the flowers of different plants usually take one of two or three different forms and the site of recognition can differ between the forms of one species. The dimorphic type has one form with a short style and long stamens and the other with a long style and short stamens (*Fig. 2*). In trimorphic plants there are three different style lengths with stamens occupying the other two sizes in each plant. In most heteromorphic plants the stigmas of the different morphs have different surface textures and the pollen may be of different size or sculpturing or both. Normally in dimorphic systems each plant can cross only with a plant with the other type of flower, or either of the other types in trimorphic systems, but there are exceptions and one species, the daffodil, *Narcissus triandrus*, has a trimorphic morphology but a completely independent ovarian incompatibility.

Ovarian or late-acting systems

In some self-incompatible plants the pollen tube reaches the ovule but the ovule aborts, either because fertilization fails or because the ovule aborts at an early stage after fertilization. How this works appears to differ markedly between the few plants so far studied. In some the system appears similar to a gametophytic system with incompatible pollen tubes growing more slowly than compatible ones and only reaching the ovules in time for flower abscission. In others it appears that the embryo sacs remain partially developed until the stigma or

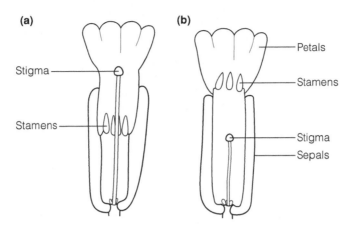

Fig. 2.   Heteromorphic flowers of the cowslip, Primula veris: *(a) long-styled, (b) short-styled.*

style signals to the ovary, perhaps with a hormone, if a compatible pollen grain has germinated. No signal comes from an incompatible grain. Other mechanisms are likely to be discovered and the recognition could be gametophytic or sporophytic in different groups and there may even be both types within one species.

In some plants it may even be simply the production of weak offspring as a result of selfing, **inbreeding depression**, that is giving rise to a loss of selfed seeds and this may be hard to distinguish from ovarian SI systems, although in some plants it is clear that there is an incompatibility system. Some form of this system is known from several families, particularly tropical woody species. It involves abortion of an ovule or the whole ovary each time, so could be wasteful of resources.

There are some other ill-defined SI systems, some with two or more loci, often giving partial SI of uncertain origin or relationship.

Molecular basis of self incompatibility

S genes of several species have been cloned. In the most widespread type, the gene gives a glycoprotein with ribonuclease (RNase) activity. In gametophytic recognition ribosomal RNA at the tip of the pollen tube is probably recognized by this protein in the style and disabled. As the pollen tube grows, enzymes from the tube digest the stylar tissue providing a path and this may activate the RNase. The mechanism is still not clear but some form of nucleic acid-enzyme recognition seems likely. In the grass system it is likely that the S allele glycoprotein interacts with carbohydrates at the pollen tube tip. This interferes with the extension of microfibrils of carbohydrate that would normally occur to make the pollen tube wall. The precursors of these wall molecules, microfibrillar pectins, then accumulate and block the tube. In the sporophytic SI system of the cabbage family, one locus produces a glycoprotein and the other a receptor kinase. These are structurally similar and are expressed on both the stigma surface and the surface of the pollen grain. If the kinase is disabled there is no incompatibility reaction so it must involve some interaction between the two proteins. Pollen grains in sporophytic SI hydrate by enzymic breakdown of cells on the stigma surface and this stage may be inhibited, so that pollen germination is stopped through insufficient hydration. In wet or very humid conditions more incompatible pollen germinates, but this is stopped from penetrating, so the interaction is clearly not confined to germination.

A completely different set of protein interactions control the gametophytic recognition in the poppy family, Papaveraceae, involving small stigmatic proteins acting as signaling ligands. These interact with a receptor on the plasma membrane of the pollen. This incompatibility reaction appears to be quicker than the RNase system.

Evolution of self incompatibility

The possibility of SI through a pollen:stigma interaction has been frequently cited as a major factor in the success of the angiosperms. Full or partial SI is widespread and occurs among the primitive dicotyledons. It is likely that a system with gametophytic recognition and involving RNase is primitive as this has been found in some form in a large group of plant families. The different gametophytic systems in poppies, grasses and perhaps other families are thought to be derived. Sporophytic systems appear to have evolved many times, perhaps from the primitive gametophytic system as similar proteins are often involved, although the recognition is different. Some ovarian systems may prove to be similar too, but it seems that the ovarian systems are highly variable and they are still too little known to make conclusions about their relationships with other systems. Much research is going on in this area of plant science.

G5 ECOLOGY OF FLOWERING AND POLLINATION

Key Notes

The dual function of flowers	Cross-pollinating hermaphrodite flowers must disseminate pollen and receive pollen from other flowers, these two functions involving different adaptations. More insect visits may increase pollen dissemination but not fruit set.
Seed and fruit set	There is great variation in size and resource input into fruits and seeds by different plants. In some the resources of the parent plant limit seed or fruit set whereas in others the amount of pollination is limited. Many plants are limited by a combination of factors.
The dissemination of pollen	Wind-pollinated plants produce large quantities of pollen, but pollinating animals vary in their effectiveness to carry pollen both through their structure and behavior. Some plants have precise adaptations for particular pollinators. Pollen is deposited on the stigmas of up to the next 30 flowers visited in decreasing quantity.
Effective fertilization	Once pollen has reached a stigma there may be competition between pollen grains, with some pollen tubes growing faster than others and potentially complex interaction with the styles. There may be selective abortion, particularly of selfed seeds.
Self-fertilization	Hermaphrodite flowers always have the potential for self-pollination and some plants are habitually self-fertilized. Advantages of self-fertilization are assured seed set, less resources needed for flowers and, after some generations with high mortality, identical seedlings with no new unfavorable combinations. Disadvantages are lack of variation allowing no adaptation to changing environments or pathogen attack and inbreeding depression. Inbreeding depression is manifest in nearly all plants studied at some stage in the life cycle.
Related topics	The flower (G1) Polymorphisms and population Breeding systems (G3) genetics (L5)

The dual function of flowers

Hermaphrodite flowers perform two functions: disseminating pollen to other flowers, and receiving pollen to fertilize their own ovules. These functions involve slightly different adaptations. There is a selective advantage for a flower to disperse its pollen as widely as possible to give it the maximum chance of siring seeds. In its capacity as a female parent the flower must receive enough pollen for maximal fertilization, but the plant must also provide enough energy and nutrients for the formation of seeds and fruits.

If nectar is the food reward for insect visitors this must be produced in suffi-cient quantity and of suitable concentration to attract but not satiate insect polli-nators, and one plant species may have to compete with coexisting species for the attentions of visiting insects. More insect visits will generally lead to greater pollen dispersal but not necessarily greater seed set (see below).

Seed and fruit set

The success of a plant as a female parent is easier to measure than its success as a pollen donor, since the plant forms its seeds on the plant. Plants differ greatly in the size and number of seeds and fruits produced and how many pollina-tions they require for maximal seed formation. Plants with many seeds per fruit may require few insect visits for full seed set if they are effective pollinators. The most extreme example is the orchids, which produce pollen in sticky pollinia (Topic G1) and have thousands of microscopic seeds per fruit. They require just one successful insect visit per fruit so need to attract much smaller numbers of pollinators than other plants. In contrast members of several fami-lies, such as the daisy family and grasses, have one-seeded fruits, so each flower or floret must be pollinated separately. The amount of resources required from the parent plant is highly variable. Small-fruited species such as grasses and orchids require few resources compared with those with large fruits such as coconuts (here mainly the seed) or water melons that set only a few fruits, each with a large amount of resources. On some plants with large fruits, such as many fleshy-fruited trees, only a small proportion of flowers can mature into fruits because of shortage of space.

In most flowers seed set is limited by a combination of the amount of pollina-tion and resources available for the developing seeds. In large-fruited plants the resources from the plant limit fruit set and it is well known that domestic fruit crops often have a poor fruiting year after a good one, suggesting that there are insufficient resources for two consecutive years of high yield. Pollination may also limit fruit set, and in many studies in which the amount of pollination has been increased experimentally, more fruit has been set. This has often led to poorer growth or less fruit set the following year even in quite small-fruited plants, so clearly these two factors interact.

In dioecious plants (Topic G3) female plants frequently produce fewer flow-ers than male plants, giving a smaller floral display, suggesting that limitation of resources is important.

The dissemination of pollen

For pollen to spread to its maximum many pollinator visits or effective spread of the pollen by wind are required. Few resources are needed for the pollen grains compared with fruit and seed set, and attraction of numerous insects to a flower will mainly enhance pollen dissemination. This suggests that, in many animal-pollinated species, much of the floral display is in relation to a flower's male function. In wind-pollinated flowers, wide dissemination of pollen is assured by the plants producing an enormous quantity on fine days, as hay fever sufferers know only too well.

In animal-pollinated flowers the quantity and nature of the pollen depends on the animal. The furry bodies of many bees, moths and bats and the feathers of birds can carry much more pollen than the smooth bodies of flies, some bees or the thin proboscis of a butterfly. They are likely to be more efficient as polli-nators as a result, but their efficiency will also depend on their behavior. For instance, bees forage systematically, visiting adjacent flowers, often on the same plant and sometimes crawling, with occasional long flights. They frequently groom themselves to remove the pollen on their bodies. Butterflies take mainly longer flights between flower visits so potentially take pollen further, and vertebrates are active fliers but require more nectar resources. Different species within each pollinator group behave differently with, e.g. some birds and some bees being territorial and some pollinators making a hole in the flower by the nectar, potentially avoiding the anthers. Different flowers are adapted to particular pollinators in the way their flowers are presented and the quantity and position of the pollen and nectar.

Once pollen has been picked up from a flower it may be deposited on many subsequent flowers in diminishing quantity, depending on the pollinator, with pollen being recorded traveling to up to 30 or more flowers visited (*Fig. 1*). In a self-incompatible plant the first flowers visited may be on the same plant or an adjacent sibling or offspring which will be incompatible with it, so pollen flow further than that may be essential for effective fertilization.

Effective fertilization

Once pollen has reached a stigma it may be in competition with other pollen grains for fertilization. Many plants have a self-incompatibility (SI) system (Topic G4) but, even within apparently compatible pollen grains, there is variation in their effectiveness in growing pollen tubes. This may be through genetic differences and connected with the SI system or because of where they land on the stigma and through interaction with other pollen tubes growing down the style. There may be different interaction between certain pollen grains and the stylar tissue affecting which pollen fertilizes the ovules in particular plants. Vigorous pollen tube growth may lead to vigorous seedlings. Once the ovule is fertilized, if resources are limiting seed set, there may be selective abortion of the ovules. Some plants abort self-fertilized seeds selectively, perhaps similar to ovarian SI, and may abort weaker cross-fertilized seeds, though evidence of this is limited.

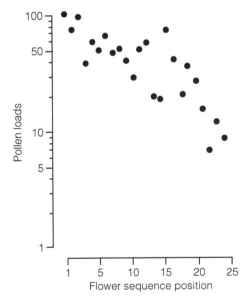

Fig. 1.   The number of pollen grains deposited by bee pollinators on stigmas of successive flowers visited in the larkspur Delphinium nelsonii *(from Waser NM, Functional Ecology 1988; 2, 41–48).*

Self-fertilization Flowers originally evolved for cross-pollination, but there is always the potential for self-pollination in a hermaphrodite flower and some species can produce full seed set from self-fertilization (Topic G3). A much larger number of plants are self-compatible and can self-fertilize but normally cross-fertilize. In these, self-fertilization frequently occurs only if cross-fertilization has failed, often with the pollen tubes of a plant's own grains growing more slowly than those from another plant, or the pollen only reaching receptive stigmas as the flower ages.

The advantages of self-fertilization are that no external pollinating agent is needed so flowers can be small, providing no attractive parts and producing only enough pollen to fertilize their own ovules. Seed set is assured, and with fewer resources being used in flowers, it may be able to flower at an earlier developmental stage (Topic G3). This has particular advantages in colonizing situations and in ephemeral habitats.

Constant selfing will mean that the plant loses much of its genetic variation and, after a few generations, will be **homozygous** at almost all its genes, i.e. each gene copy will be identical. This can lead to deleterious genes being manifest and much mortality in the early generations, but once any deleterious genes have disappeared the plants will be all identical. This means that only proven successful plants will be produced and no seeds are 'wasted' on poor gene combinations. Many habitual selfers are polyploid so have several copies of each gene anyway, avoiding some of the problems of homozygosity.

The disadvantages of self-fertilization are the lack of new gene combinations leading to much reduced variation, dependent only on new mutations. This means that they may not be able to adapt to a changing environment, particularly important in developing resistance to herbivores or pathogens or in colonizing new sites. It also means that they never have the advantages of hybrid vigor, which probably arises from a plant having two different forms of certain genes in each cell (Topic L5).

In plants that may cross- or self-fertilize there is nearly always a disadvantage to selfing. This may be manifest in lower seed set, smaller or less resistant seeds, difficulties of germination, ability of the seedling to establish, vigor and survival of the offspring or their ability to reproduce or any combination of these features. This is known as **inbreeding depression**. Even in most habitual self-fertilizing plants studied, experimental cross-pollination has led to more vigorous plants than selfing and almost all plants do occasionally cross-fertilize.

H1 THE SEED

Key Notes

Definition	The seed is the mature ovule and consists of a seed coat, endosperm and embryo. In angiosperms it is enclosed within a fruit.
The seed coat	Known as the testa, this is normally formed from one or both integuments. Frequently it is tough and survives unfavorable conditions; a few species have a fleshy outer integument.
The endosperm	This is normally triploid and the cells start dividing immediately after fertilization, sometimes forming an acellular mass first. It is the food store of the seed and may be rich in carbohydrate, protein or lipid. It remains as the main part of the seed at maturity in some plants.
The embryo	After the first transverse cell division the upper cell forms the shoot, cotyledons and root, the lower cell a suspensor of no clear function. Most flowering plants have two cotyledons but some have one and grasses have an extra sheathing coleoptile. Some mature seeds, such as beans, are almost entirely filled with the embryo. On germination the cotyledons may remain in the seed and wither with shoot growth, or become the first leaves.
Seed development	Seeds undergo three phases: maturation, post abscission and desiccation. As they mature there are high levels of RNA activity, storage products are produced and ABA levels are high; post abscission ABA and RNA decline until metabolism slows and the seed becomes desiccated.
Related topics	Pollen and ovules (G2) Seed and fruit dispersal (H3) Fruits (H2) Seed dormancy (H4)

Definition

The ovule matures into a seed before it is dispersed from the parent plant. Seeds consist of three morphologically and genetically distinct parts: a seed coat, derived from maternal tissue, the new embryo and, in flowering plants, endosperm derived from the second fertilization and usually triploid. In other seed plants the remains of the haploid female gametophyte are present and there is no endosperm. Seeds are enormously variable in size, shape and structure, from coconuts weighing several kilograms to the microscopic seeds of orchids. In flowering plants the seeds are contained within **fruits** at maturity. In other seed plants they are exposed.

The seed coat

As the seed matures, the outer layers of the ovule develop into the seed coat or **testa**. The structure of the testa is highly variable and it may be formed from one or both integuments and sometimes the nucellus as well (Topic G1). The testa hardens with the deposition of cutin or lignin and phenols and, in many plants, it is extremely tough (Topic H4). In some seeds, such as those of yew trees, *Taxus*, the outer integument or part of it develops into a fleshy coat, or **aril**, attractive to animals, with a hard testa inside. Plants with dry indehiscent fruits may have almost no seed coat at all, its function being replaced by the fruit, as in grasses. The seed coat hardens at different stages of development in different plants.

The endosperm

After the double fertilization of the ovule the first part to grow is the endosperm, normally consisting of triploid cells. Initially in many plants the endosperm nucleus divides freely giving an **acellular** tissue only later developing cell membranes and cell walls, although others are cellular from the start. Some remain part cellular and part acellular, such as the coconut, in which the milk is acellular. The endosperm grows rapidly and may absorb part of the nucellus or inner integument, stimulating their cells to break down. It incorporates within it a large supply of nutrients, a mixture of carbohydrate, lipid and protein and remains in the ripe seed of some plants, which may be mainly starchy as in cereals such as wheat or oil-rich as in castor oil, linseed or oilseed rape (canola). It also contains plant hormones which may stimulate the growth of the embryo (Topic H3). In some acellular endosperms nuclei with **high ploidy** levels have been found where, it seems, the chromosomes have divided but never separated. In some plants the endosperm has been all or mostly absorbed by the growing embryo by the time the seed is mature, such as in beans and Brazil nuts, both of which have protein-rich seeds.

The embryo

The embryo starts to grow either shortly after fertilization or after some delay, up to a few months (*Fig. 1*). The first cell division is transverse, with just a few exceptions, giving a terminal and a basal cell. After this details vary somewhat but the terminal cell always gives rise to the **stem apex** and the **cotyledons**, which are the embryo's first leaves and sometimes food store and, in some, to the **radicle**, consisting of **root apex** and root cap, as well. The basal cell provides the **suspensor**, along with the root tip and cap in some. Topic J1 illustrates this for arabidopsis. The suspensor becomes a short string of cells and the lowest cell swells into a vesicle, but the function of the suspensor is not clear and it disintegrates with the enlargement of the embryo. At this point cell division stops and the second stage of seed development begins, resulting in the formation of a seed capable of protecting the embryo.

The earliest flowering plants almost certainly had two cotyledons and in most living plants the shoot tip develops two lobes that will become the two cotyledons

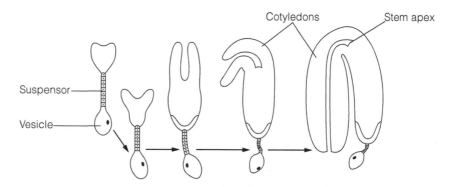

Fig. 1. Early growth of the embryo in a dicotyledon, Capsella bursa-pastoris.

with the shoot apex between them. Beneath this, in the **hypocotyl**, vascular tissue develops. The developing root acquires a distinct cap. In some plants one of the cotyledons appears to be suppressed. This is a feature of one of the major divisions of the flowering plants, called the **monocotyledons** because of this (Topic Q4), although some **eudicotyledons** suppress one cotyledon in a similar way. Grasses, which are monocotyledons, have their own unique embryo formation with a swelling at the tip of the embryo forming two different outgrowths, the **scutellum** which may represent the cotyledon, and the **coleoptile** which grows beside it and forms a kind of sheath enclosing the shoot apex (*Fig. 2*). A similar sheath appears over the root apex.

On germination the cotyledons may remain in the seed and wither away as the nutrients pass to the developing shoot, known as **hypogeal** germination, or emerge as the first photosynthetic organs, known as **epigeal** (*Fig. 3*). Small seeds are mostly epigeal, since, with low nutrient content, immediate photosynthesis may be a priority; large seeds are mainly hypogeal. They normally have a simple form, quite different from the typical foliage leaf and wither soon after the first foliage leaves have formed. In a few plants they persist.

Seed development

The final stage of seed development involves three phases: **maturation, post abscission** and **desiccation** (*Table 1*). Genetic analysis of mutants of seed development in

Table 1. Stages of seed development

Stage	Events
Maturation	Embryo completes growth; dehydration of embryo begins; ABA levels reach maximum in dicots. High levels of storage reserve synthesis (protein, lipid, carbohydrate, depending on species)
Post abscission	Seed separated from connection with mother plant; ABA levels peak in monocots. Browning of seed coat (testa). mRNAs for storage proteins decline and disappear. Late embryogenesis abundant (LEA) mRNAs for highly hydrophylic proteins increase which protect during dehydration
Desiccation	Seed metabolically inactive. Very little DNA turnover or mRNA synthesis; embryo inactive until germination initiated

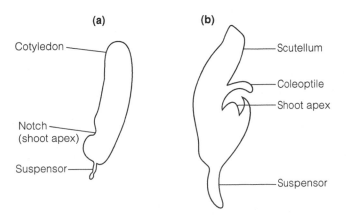

Fig. 2.   Embryos of monocotyledons: (a) onion, Allium, *(b) a grass.*

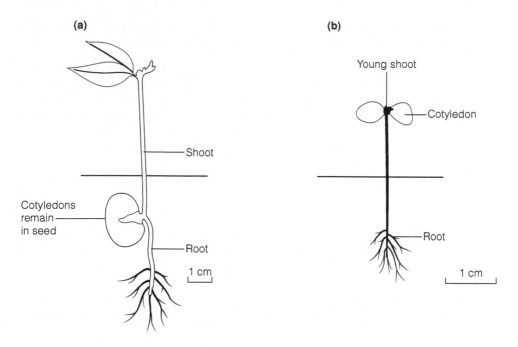

Fig. 3. Seed germination: (a) hypogeal in the bean, Faba sp., (b) epigeal in Arabidopsis.

arabidopsis indicates that regulation of the events in the different phases is complex, with a number of genes being involved. One, ***abi3***, is involved in sensitivity to ABA as, when mutated, the plant is ABA-insensitive. It is only expressed in the seed and it controls expression of some of the genes of the maturation and post abscission stages. *abi3* mutants germinate on the mother plant and do not undergo desiccation. Other mutants are also known that result in this type of early (precocious) germination. These include *fus3* mutants of arabidopsis which have cotyledons which develop to be like the primordia of true leaves and which do not synthesize seed storage proteins.

H2 FRUITS

Key Notes

Fruit structure
Fruits are formed from the ovary and sometimes other parts of the carpel or other parts of the flower. Dehiscent fruits usually have a hard wall made partly of sclerenchyma and break along a line of weakness where fusion has happened.

Indehiscent fruits
These may be dry and resemble the seed coat in the one-seeded fruits of grasses and some other plants and dispersed with the seed. Fleshy fruits are variable, some with a complex structure with a separate rind, layers of different cells and juice-filled parenchyma. When immature the fruits are often photosynthetic, becoming colored and often with an increase in sugar as they mature.

Related topics
The flower (G1)
The seed (H1)

Plants as food (N1)

Fruit structure

The true fruit is a feature only found in the flowering plants since it is formed, at least in part, from the **carpel** (the so-called 'fruits' of conifers such as yew or juniper are fleshy outgrowths from the seeds or cone scales). There are numerous different shapes, sizes and types of fruits – from water melons and pineapples to dehiscent pods and the dry chaff of a grass. They have been classified in several different ways according to the structures that give rise to them. These always involve the **ovaries**, which may be free or fused, superior or inferior. Sometimes the rest of the carpel develops into part of the fruit as well and other floral parts such as the **bases** of **sepals** or **petals**, or the **receptacle** may be involved. A few fruits such as the fig and pineapple are derived from a whole **inflorescence**.

Dehiscent fruits, i.e. those which open to disperse their seeds (*Fig. 1*), usually have several seeds and the fruit itself often has a protective function as the seeds mature. It may derive from a single carpel as in the pods of peas and other legumes or from two or more fused together, as in *Arabidopsis* and other members of the cabbage family. The outer fruit wall is normally hard and contains one or more layers

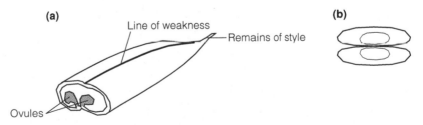

(a)
Line of weakness
Remains of style
Ovules

(b)

Fig. 1. Dehiscent fruits: (a) bean derived from one carpel, (b) transverse section of fruit of Arabidopsis derived from two fused carpels.

of sclerenchyma cells mixed with parenchyma, sometimes with the cell layers at different angles. As they dry out the cell layers dry differentially. Lines of weakness occur along the carpel margin or where several carpels fused and these break to disperse the seeds. In many legumes differential drying in the fruit wall leads to an explosive break and twisting of the two halves of the fruit, dispersing the seeds.

Indehiscent fruits   Many fruits do not dehisce and these may be **dry**, as in grasses, or **fleshy**. In dry indehiscent fruits the fruit normally resembles the seed testa and the fruit is dispersed with the seed. The fruit consists solely of one or more layers of sclerenchyma and parenchyma cells, this type being characteristic of grasses, composites (daisy family) and some other families. It may include the ovary wall and part of the receptacle or base of the flower that has become indistinguishable from it. In wheat this wall, along with what remains of the seed integuments, is the bran.

Fleshy fruits have a more complex structure. Some have a **rind** quite separate from the flesh (e.g. citrus fruits, banana) and most change color, texture and chemical structure as they mature. As one example, citrus fruits (*Fig. 2*) have an outer cuticle and epidermis over compact parenchyma with oil glands. Inside the epidermis, the white tissue consists of parenchyma with large air spaces and a vascular network and another epidermal layer lies inside that. The edible segments consist mainly of juice-filled sacs derived from the inner epidermal cells. Other fleshy fruits have no separate rind, such as cherries and tomatoes, and here there is an epidermis and elongate fleshy cells underneath and, in some, an inner lignified layer around the seed. In apples and pears the flesh is derived from the receptacle and other fruits are derived from the base of the sepals and petals. These grow around the ovary, the ovary itself becoming the **core**. In the fruits deriving from inflorescences, petals, sepals and flower stalks all swell to aggregate together as the dispersal unit with fleshy cells mixed in with harder epidermal cells and other types.

Many fleshy fruits are green and photosynthetic either when immature or at maturity (e.g. cucumbers) from chloroplasts in the outer parenchyma or epidermis. These may change to become colored **chromoplasts** (Topic C3) and produce anthocyanins or carotenoids (Topic G1) as the fruit matures. Other chemical changes involve the production of sugars from more complex polysaccharides. In many plants, the maturation of the fruit involves a fusion of the floral parts that have given rise to them.

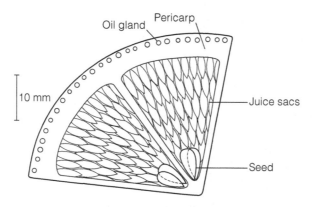

Fig. 2.   Cross-section of a citrus fruit.

H3 FRUIT AND SEED DISPERSAL

Key Notes

Size and ecology

Seed size is a compromise between the germination benefits of large nutrient- or carbon-rich seeds and the number and greater dispersal of small seeds. Large seeds are associated with plants of low light conditions in all habitats and size increases towards the equator. Some species produce seeds of constant weight but in others it varies considerably. Some plants produce more than one type of seed which differ in dispersal and dormancy characteristics.

Dispersal by the elements

Many dehiscent fruits simply drop their seeds but chance dispersal through wind, flood or mud on animals' feet may be effective with long-lived seeds. Wind-dispersed seeds may be tiny, sometimes with specialized germination requirements, or have an attached sail. Some freshwater and coastal plants have seeds dispersed by water.

Fruit and seed eating

Vertebrates disperse the seeds of any fleshy-fruited plants either internally or by discarding the seeds. Fruits relying on specialist frugivores often have nutrient-rich hard fruits, whereas smaller sugar-rich fruits are typical for opportunist frugivores. Some nutrient-rich seeds, particularly of trees, are hoarded or buried by rodents and birds. Ants disperse some seeds a short distance, mainly into good germination sites.

Other modes of dispersal

Seeds with burs or hooks are dispersed externally on mammals or birds. Some plants have explosive fruits that scatter the seeds, though there may be other adaptations too. Some rainforest trees have large short-lived seeds with almost no dispersal powers.

The measurement of dispersal

Effective dispersal is hard to measure. Many seeds do not travel far from the parent plant but some travel long distances. Effective dispersal requires the seed to land in a safe germination site. The floras of oceanic islands have many bird-dispersed seeds, with fleshy fruits on wet islands, mainly burs on dry islands, and few wind-dispersed, suggesting that bird dispersal is the most effective for long distances.

Related topics

The seed (H1)
Fruits (H2)

Seed dormancy (H4)

Size and ecology

Large seeds contain more nutrient or carbon stores for germination and rapid establishment than small seeds and they can continue to depend on stored nutrients for a considerable period after germination. Frequently they are more flexible in the timing of leaf production by the seedlings and this means that they can take advantage of early light periods before canopy closure. Small seeds or fruits are likely to disperse further than large ones and can be produced in larger numbers at similar cost to the plant, but they have little stored nutrient and seedlings from small seeds generally take longer before the first leaves appear and then produce fewer. Without light or, for some, mycorrhizal infection, they will die. Seeds are a compromise between these opposing selection pressures.

Seed size and weight differ enormously between species and, within any one habitat, seed weight can vary by several orders of magnitude. In general, mean seed size increases with decreasing light intensity and those plants that survive in dense shade have a few large seeds. In addition, in spite of all other factors, mean seed weight increases towards the equator by a remarkably constant factor of 10 for every 23° of latitude. There is no obvious reason for such a constant increase, though large nutrient-rich seeds are likely to be advantageous in tropical plants in stable environments.

Within some species seed size remains fairly constant despite different sizes of adults growing in different conditions, but in others it varies with conditions experienced by the adult, with the weight frequently declining through a flowering season or dependent on day length or growth conditions. In any one species, smaller seeds frequently remain viable for longer than larger ones and plants growing in dry places often have larger seeds than those in wet places, perhaps because rapid growth is important to avoid drought. In some species there may be up to a 100-fold difference in weight between seeds on one plant. In grasses and composites, and perhaps others, the position of the seed within the inflorescence determines its size, shape and longevity, and some species have two totally different types of seed (*Fig. 1*). In these the seeds with the greater dispersal powers are shorter lived and it seems that these species can disperse in time and space using the different types of seeds.

Dispersal by the elements

Most dehiscent fruits open with a **slit** or **pores** and the seeds simply drop out, although they may be transported at least a short way by the **wind**, particularly if the fruiting stalk is long and flexible or the seeds are shed at the tip of dry fruit. Such fruits and seeds, without any specialized structures for dispersal, are the commonest type. Dispersal may happen through chance events, such as high winds, occasional water currents or in mud on the feet of passing animals, particularly if the seed is long-lived.

Some seeds have clear adaptations for dispersal by the wind, e.g. the tiny seeds of orchids, heathers and some parasitic plants that rely solely on their size for effective dispersal. These plants often produce enormous numbers of seeds but many have specialized **germination requirements** involving mycorrhizal fungi (Topic M1) or, for the parasites, a host plant, and the seed has few stored

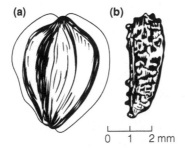

(a)   **(b)**

0   1   2 mm

Fig. 1.   Two different types of seeds of the composite Dimorphotheca pluvialis, *(a) from ray florets, (b) from disc florets. (From Fenner M.* Seed Ecology. *1985 Kluwer Academic Publishers, redrawn with kind permission.)*

nutrients. Other wind-dispersed seeds or small fruits (Topic H2) have a cottony **sail** on the seed, e.g. willows and members of the Malvaceae such as cotton, or a sail formed by a ring of persistent modified sepals in many members of the daisy family. Some larger seeds are attached to a sail-like fruit, or **samara**, as in maples, *Acer*, and birches, *Betula*, that may be blown for some distance, the fruit spinning as it descends.

Many freshwater plants have seeds dispersed by **water**, in contrast to pollen which is rarely water-dispersed, and many coastal plants have seeds that are partially resistant to salt water so they can drift with **ocean currents**. Many species that occur by coasts are widespread in distribution and colonize islands easily.

Fruit and seed eating

Many indehiscent fruits are dispersed by animals. Mammals and birds can ingest the smaller **fleshy fruits** or arillate seeds (Topic H1) whole and the seeds will be defecated or regurgitated elsewhere; with larger fruits or those with larger seeds vertebrates will often discard the seeds. Many of these seeds have a thick resistant seed coat and may germinate more freely after passage through a vertebrate gut (Topic H4). Passage through a gut can take from a few hours to days or even weeks and dispersal may be long distance. Fruits dispersed by specialist **frugivorous birds** frequently have a different chemical composition from those attractive to opportunists, with much more protein and fat and often a larger seed. Opportunist frugivores take softer sugar-rich fruit and some damage the seeds. This is compensated for by large numbers of fruit produced with much lower nutrient investment in each.

Some **nutritious seeds**, particularly those of trees, are themselves a food source for vertebrates, particularly rodents and some birds such as jays and other members of the crow family. Some of these animals store the seeds, often burying them as a long-term food supply, after which they may not need them or they may forget where they have put them or be killed before they return.

Ants can be important seed dispersers and seeds dispersed by ants often have a nutrient-rich **elaiosome** attached to them which the ants remove once the seed is in the nest, before discarding the seed to a refuse pile or outside. This is common in temperate deciduous woodland and dry shrub communities in Australia and elsewhere but occurs everywhere. The distances of dispersal are short, normally less than 1 m, though can be up to 80 m, and the main advantage is thought to be that the seeds cease to be so vulnerable to predation and/or are taken to a good germination site.

Other modes of dispersal

Dry indehiscent fruits may disperse by becoming attached to mammals or birds. Some plants have seeds with spikes or hooks, formed by the fruit wall or a persistent style, which attach themselves to fur (or clothing), such as those forming **burs**. The dispersal unit of some grasses is the fruit with surrounding persistent bracts, which enlarge and attach themselves with sharp points or teeth. Some of these may be dispersed mainly by one particular species of mammal, and may be adapted in some way to it. This can be most effective as a dispersal mechanism, with long distances being covered and some colonizing oceanic islands.

Some plants have an **explosive** fruit which throws the seeds out, such as some legumes and balsams, although the distances covered are likely to be short and the seeds may have other mechanisms of dispersal as well, e.g. by ants.

Some plants, particularly rainforest trees, have short-lived large seeds with no clear dispersal mechanism. In these, dispersal is limited and they often have a restricted natural distribution and do not colonize islands.

The measurement of dispersal

Most studies have concentrated on initial dispersal, since this is easy to measure, but may mask the true picture. True effective seed dispersal is notoriously hard to measure since the area to be studied rapidly expands with distance and, if a seed is potentially long-lived or dormant, chance dispersal may happen at any time over a long period. Studies on plants that are not dispersed by animals suggest that a majority of seeds land close to the parent but there may often be a long 'tail' of longer distances (*Fig. 2*). If a plant is in a new area, short dispersal distances and dense colonization may be possible, but for others the **effective dispersal** distance may be great. For light wind-dispersed seeds, an initial updraft, such as a thermal or eddy, can disperse seeds long distances. A seed must be dispersed to a **safe site** for germination to be effective and, for many, this will not be immediately beside the adult. Safe sites may be common or infrequent and, if infrequent, the production of many small seeds will be an advantage. Colonizing species in particular must have effective long-distance dispersal.

We can infer the effectiveness of different dispersal types by examining the floras of remote islands. One remarkable fact is that, although some island plants are wind-dispersed and probably arrived by wind, the number is quite small and wind dispersal is rarer than on continents. This is underlined by the fact that the most remote islands have the fewest wind-dispersed species. Much the commonest means of seed dispersal on oceanic islands is by birds and they appear to be the most effective long-distance seed vectors. Many plants on wetter islands have small fleshy fruits. On dry islands there are many burred plants, perhaps through shortage of water for fleshy fruits, and colonization appears to have come through external transport by birds. Sea drift seems to be important for coastal plants. Chance events such as storms probably play an important part in driving the birds there, since successful colonization needs to have happened only once in every few thousand years on average to account for the present floras, though much of it is likely to have happened in the early years.

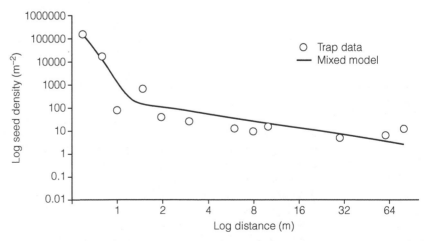

Fig. 2. Dispersal distances of seeds of heather, Calluna vulgaris, *with a line fitted from a model of the system. (Redrawn from J.M. Bullock and R.T. Clarke,* Long distance seed

H4 SEED DORMANCY

Key Notes

Dormancy and longevity

Seeds differ in how long they will persist until they reach a suitable site for germination. True dormancy is a property of the seed and can be morphological, where the embryo is not fully formed; physiological where a chemical change in the seed is needed; or physical where the seed coat is impermeable.

Physical dormancy

This is characteristic of seeds of shrubs in fire-prone habitats and some vertebrate-dispersed fruits. The seed is dry and will germinate only after a fire or abrasion has broken through the testa.

Morphological and physiological dormancy

Both these types require water to be imbibed and many plants have both types. Physiological dormancy is reversible. Over time conditions to break the dormancy may change. There may be annual cycles in dormancy characteristics and microbes may interact with seeds to affect it.

Seed banks

These vary enormously from small seed banks in woodlands to large in disturbed land. The species composition may differ from that of the adult plants, especially in mature stable communities. They also differ seasonally and may be patchy. They allow dispersal in time and mixing of generations. A few plants retain a seed bank on the plant.

Related topics   The seed (H1)   Fruit and seed dispersal (H3)

Dormancy and longevity

Once it is shed from the parent plant a seed must reach a suitable germination site and suitable conditions for germination. Seeds of different species vary enormously in how long they can remain alive until they find a suitable germination site, some lasting only a few weeks or even days whereas others potentially last for centuries. Trees of the tropical rainforest characteristically have short-lived seeds that will die after a few weeks, though there are exceptions. The seeds of some mangrove trees germinate before they disperse from the trees; these possess a radicle that plants itself in the mud on release from the parent. By contrast, pioneer plants have seeds that remain viable for years if no suitable site is reached. The longest-lived seed reliably recorded (by carbon dating), a sacred lotus, *Nelumbo*, from China, germinated after about 1200 years. The description here has been known as **enforced dormancy** but if the plants germinate as soon as conditions become suitable it is best regarded as seed longevity. Many seeds can be kept in a cold dry atmosphere for much longer than in other conditions, a feature used by plant nurseries and seed banks.

Most recent definitions of dormancy regard it as an inherent property of the seed, preventing it from germinating in suitable conditions until a set of prior conditions has been met. There are three different types of dormancy: **morphological** dormancy in which the seed continues to mature after being shed from the parent; **physiological** in which the chemical structure of the seed needs to be modified; and **physical** in which a physical factor such as high temperature, a fire or serious abrasion is needed on the seed coat before it can germinate. The commonest type is physiological dormancy and this may be reversible. Many plants have a combination of physiological and morphological dormancy. In general fewer plants from a predictable climate and environment show any dormancy than those in unpredictable climates, and the dominant plants in a mature community have less dormancy than colonizers.

Physical dormancy

This is characteristic of seeds in areas with a marked dry season or those prone to fires and extremes of temperature. It is particularly associated with a number of Australian woody plants such as some mallee trees, *Eucalyptus* species, and *Banksia*s, and with some conifers, all plants prone to periodic bush fires. With physical dormancy the seeds remain dry until the germination barrier, usually a tough seed coat, is removed. Removal of the seed coat permits germination, and dormancy is due to lack of oxygen and water for the embryo.

Seeds of some fleshy fruits require physical abrasion by passage through a vertebrate gut, though this has probably been overestimated in the past. The famous case of the tambalacoque, *Sideroxylon*, from Mauritius needing to pass through a dodo's gut for germination is discredited; the seeds will germinate if freed from fruit pulp and not infected with fungi. They do not have physical dormancy. In contrast the guanacaste tree, *Enterolobium*, does have physical dormancy, needing abrasion by the gut of a mammal or by soil before it germinates.

Morphological and physiological dormancy

Morphological dormancy, perhaps the earliest type, requires the embryo to grow and it can only do this when the seed can imbibe water. Physiological dormancy, frequently coupled with it, also requires some change, with temperature changes usually the dominating influence. The best known example is of seeds of a temperate plant requiring a chilling treatment, sometimes known as **vernalization**, so that they do not germinate in autumn but remain as seeds until similar conditions in spring and frosts are avoided as the seedlings grow.

Sometimes several low temperature treatments are needed. Seeds that have experienced conditions of release from physiological dormancy can become dormant again if unfavorable conditions return. This has been known as **induced** dormancy, although it is only known from those seeds that have some physiological dormancy anyway. Conditions for a second release from dormancy may be different from the first and the conditions required to break dormancy can be demonstrated to change with the age of the seed. Temperature in particular can have different effects at different times and there may be **annual cycles** in the degree of dormancy of certain species, often giving maximal germination in particular seasons.

The physiological dormancy regulator, at least in some plants, has been shown to be **abscisic acid** (**ABA**), produced by the embryo during the late stages of seed maturation (Topic J2).

The importance of microbes in seed dormancy and germination is little known. Some seed coats have antifungal or antibacterial properties, though this may reduce with age. Microbial attack may also break dormancy, either through physical decay of the seed coat or through chemical changes breaking a physiological dormancy. There is much still to be learned in this area of plant–microbial interaction (see Topics M1, M2).

Seed banks

Seeds that remain in a soil build up a seed bank. Habitats vary enormously in their seed banks from fewer than 100 to 100 000 seeds per square meter of surface. The habitats with the greatest numbers of seeds in the seed bank are those prone to periodic major disturbance, such as agricultural land, wetlands or fire-prone shrub habitats. Those with smaller seed banks include many woodlands and arctic environments, both places with long-lived perennials and little disturbance. Usually a seed bank consists largely of **pioneer** species and in disturbed conditions most species will be the same as in the standing vegetation. In mature communities the seed bank is quite different in species composition from the existing vegetation and there may be almost no overlap in the species present since many perennial species typical of mature plant communities have short-lived seeds.

A seed bank will be constantly shifting and be different in constitution at different times of year, depending on seed longevity, differences in dormancy and timing of seed production. In temperate environments it will normally be much lower in early to mid summer than in winter. It will also be patchy, with seeds congregating in certain places. In most soils the viable seeds are nearly all near the surface although ploughing will bury some deeper. Most studies on seed banks have involved collecting soil and recording what germinates, but this can miss some seeds that may have different germination requirements. The few studies that have also investigated the seeds themselves in the soil have recorded different results from germination tests, but these also can miss seeds, especially the very small ones, and identification can be a serious problem.

Seed longevity and dormancy allow **dispersal of seeds in time** as well as in space. The seed bank of a species with potentially long-lived seeds can consist of the seeds from many different generations of the plant and, when favorable germination conditions arise, these generations will mix. This mixing will prevent any short-term genetic change from occurring.

A seed bank can be retained on the plant, especially those with physical dormancy, in a few places where fire is a regular feature of the community. Some conifers and plants of Australian savannahs, such as some *Eucalyptus* trees and other genera, retain dormant seeds in dry cones or fruits on the plant. They can remain there for decades until a fire breaks the dormancy and they germinate.

H5 REGENERATION AND ESTABLISHMENT

Key Notes

Seed germination	Seeds must reach a suitable germination site, often a vegetation gap. They may be sensitive to quality of light, temperature fluctuations, water regime in the soil, soil chemistry and other factors. Different species germinate under different sets of conditions.
Seedling growth	Initial growth is dependent on seed size with larger seeds growing more quickly than small. Seedlings from species with small seeds are often light demanding and epigeal, those from large seeds shade-tolerant and hypogeal. Epigeal seedlings often grow faster than hypogeal in light conditions.
Gene expression in germination	Germination begins with water absorption (imbibition). Next the seed food reserves are mobilized. In grasses, the starch in the endosperm is broken down by enzymes synthesized in the aleurone layer in response to gibberellic acid from the embryo. Production of a photosynthetic apparatus involves: Phase I, cell division and division of etioplasts; Phase II, plastids go on dividing; plastid and nuclear genes for the photosynthetic apparatus and Calvin cycle are active; Phase III, maintenance of the photosynthetic apparatus.
Vegetation gaps	Gaps vary enormously in size, shape and how they are formed, and different plant species are favored in particular conditions. Conditions in small gaps are buffered and many seedlings are likely to appear, resulting in intense competition. Some seedlings are facilitated directly by other plants. In large gaps conditions will be harsher and there will be fewer seedlings. The timing of gap formation can influence which species establish. Early germination and seedling growth have great advantages and this stage is probably the most critical in a plant's life cycle.
Related topics	Fruit and seed dispersal (H3) Plant communities (L3) Seed dormancy (H4)

Seed germination
Seeds will only germinate if they reach a suitable place. Germination conditions differ greatly for different species, many requiring a gap in the vegetation, though others are facilitated by the presence of other plants.

Light is one of the main stimulants for germination. If a seed is under a vegetation canopy the light is reduced and its quality changed. Leaves filter out red light and many seeds respond to the ratio between red and 'far-red' light, which is filtered much less, using phytochrome (Topic I1). Light-demanding seeds germinate when the proportion of red light increases, i.e. when not under a leaf canopy, with different species responding to different levels. Many agricultural weeds germinate following light stimulation. Sometimes only a few seconds are needed as the trigger; ploughing at night can reduce the numbers of weeds germinating by up to four times. Shade-tolerant species do not respond like this.

Some seeds respond to daily fluctuations in temperature, which will be greater by 5°C or more at the soil surface within a gap, although certain species will only respond to temperature under particular light regimes. This will also be affected by whether the seed is buried, since soil will buffer fluctuations; there will be more germination near the surface.

Conditions in the soil vary greatly. The water regime and humidity and nutrient conditions must be favorable for each species, many seeds being particularly sensitive to nitrate concentration, only germinating where it is high enough. The type of soil and nature of the disturbance will affect how a seed lies on the soil, e.g. to allow it to imbibe water for germination. If a seed is buried there is likely to be a higher CO_2 concentration than in the atmosphere and this can stimulate germination if levels are slightly raised but inhibit germination at high levels. All the factors mentioned vary continuously and plant species differ enormously in their sensitivity to the different factors and the interaction between them, leading to great differences in germination conditions for different species.

Seedling growth

The speed of initial growth of a seedling will depend on seed size, with larger seeds having several advantages over small (Topic H3). Quicker growth at this early stage can be critical for the seedling to get established ahead of others. Most small seeds are epigeal (Topic H1) and the cotyledons are needed as the plant's first leaves. Many of these species are light-demanding pioneers. Many large seeds are hypogeal, more reliant on the seed's food stores. In light conditions, seedlings from epigeal small seeds may have a higher **relative growth rate** than those from large seeds so can catch up.

The stage between initial seedling growth and establishment as an adult is the most critical stage in a plant's life cycle and is key to understanding diversity in plant communities. It is in this stage that it will face greatest competition from other plants and be most vulnerable to attack by herbivores and pathogens of all kinds, and the role of mycorrhiza and microbial conditions in the soil is only beginning to be explored. Many of the dominant plants in a community such as the trees or the grasses that form the sward in a grassland are long-lived, some living for centuries once established, so study of these plants in their critical stage is often difficult. In at least some **mast-fruiting** trees, i.e. those that produce large quantities of fruits in certain years with lean years between, there is only successful recruitment from mast years; in other years all seeds are lost to seed predators, herbivores or pathogens.

Gene expression in germination

The first stage of germination is **imbibition**, in which the seed takes up water. Next **seed storage reserves** are mobilized and the embryo grows. Mobilization of food reserves varies depending on the nature of the major reserves of the seed. The process has been studied in detail in barley.

Cereal seeds (**grains**) contain a large reserve of **starch** that is broken down to sugars that supply the developing embryo. A barley seed is made up of three components: an **embryo**, an **endosperm** containing storage reserves and a layer surrounding the endosperm called the **aleurone** (*Fig. 1*). When a dormant seed imbibes water, **gibberellic acid** (**GA**; Topic J2) is produced by the embryo and diffuses to the aleurone layer, resulting in the synthesis of new proteins, including hydrolytic

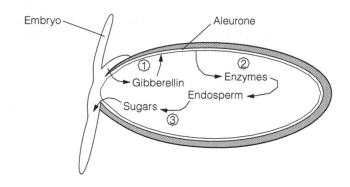

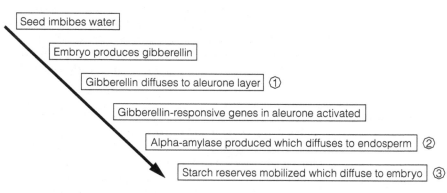

Fig. 1. *The mobilization of storage reserves in barley seeds. After imbibition, the embryo produces gibberellic acid (GA; 1) which initiates gene expression in the aleurone (2) including the starch degrading enzyme α-amylase. This breaks starch down to yield sugars that are transported to feed the growing embryo (3).*

enzymes like **α-amylase** that break down the starch of the endosperm. GA greatly enhances the rate of transcription of the α-amylase gene. A small number of **GA response elements** (**GAREs**) have been identified, 200–300 bp from the transcription start site of the gene, which form a **GA response complex** (**GARC**). The GARC is activated by the binding of **transcription factors** – proteins produced in response to GA that bind to DNA. One such factor, **GAMYB**, is one of a family of transcription factors (**MYBs**) known to be involved in development and is up-regulated by GA well before α-amylase gene expression begins.

After emergence of the seedling, the shoot can photosynthesize and supply the nutrients required for growth. Production of a fully functional photosynthetic apparatus (described for a grass) occurs in three phases:

- The coleoptile (Topic H1) and primary leaf (plumule) contain etioplasts (precursors of chloroplasts which cannot photosynthesize; Topic C3).
- Phase I. The cells of the leaf divide as the leaf grows, together with the etioplasts within them.
- Phase II. Cell division stops, but plastids go on dividing and increase in abundance. Genes are activated which generate the photosynthetic apparatus (Topic F1) and enzymes of the Calvin cycle (Topic F2). This involves both plastid and nuclear genes. The photosynthetic system becomes more efficient as more components are added.

● Phase III, the activity of the photosynthetic apparatus is maintained, and damage repaired. Finally, the apparatus begins to senesce and efficiency is lost as the leaf withers and activity increases further up the plant.

Vegetation gaps

Gaps are constantly present in plant communities, caused by the death of established plants, by climatic factors like frost, wind or flood and by animals through herbivory, trampling or burrowing, defecating, etc. Even in a tropical rain forest, gaps are frequent and diverse (*Fig. 2*), one published figure being approximately one per hectare with a mean size of 89 m². Environmental conditions within the gap will vary and which species will establish in a gap will depend on numerous properties of the gap.

Small gaps, perhaps created by the death of a single large plant, will remain mainly surrounded by vegetation and heavily influenced by shade and by root growth of the surrounding plants. They are usually colonized by many individuals and species as the conditions for growth are often favorable, without extremes of temperature or other weather conditions. There will be a period of intense competition. Many plants benefit from less extreme conditions provided by nearby plants and some are more directly facilitated in their germination or establishment by the presence of other plants. Fallen logs, the roots of some plants and bryophytes supply a potential medium for catching seeds and can provide a germination site for certain species. Mycorrhiza and other microbial interactions in the soil may have a profound effect too. Large gaps, normally created by human clearance or a natural disaster, may have areas in their centers in which there is no influence of any surrounding vegetation, grading to their edges in partial shade and with roots from neighboring plants. In a large gap, depending on its nature and formation, fewer seeds may be present and conditions more extreme, particularly drought, so if the seedling can survive its initial growth, competition and other interactions will be less intense. As a gap ages, conditions will change and there will be succession (Topic L3) with the new conditions favoring different species.

The timing of the creation of the gap will be critical. In nearly all communities the plants flower and set seed at particular seasons and there will be a different species composition of seeds at different seasons. In addition many of the long-lived plants of a community do not set fruit every year or fruits vary greatly in their abundance so years will differ.

For successful establishment it is vital for a plant to have seeds in place as soon as a site becomes suitable for germination, and for these to grow quickly. Early growth may be critical for survival since the first to grow will overtop later seedlings and thereby have a crucial competitive advantage. As stronger seedlings they will also be less susceptible to herbivore attack, e.g. from molluscs, and more resistant to pathogens. Differences of one or a few days in germination time can be critical for survival over the next several years.

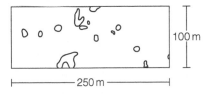

Fig. 2. Gaps in a primary lowland tropical rain forest in Sumatra (redrawn from Richards PW, The Tropical Rain Forest, 2nd Edn, 1996. Cambridge).

I1 PHOTOPERIODISM, PHOTOMORPHOGENESIS AND CIRCADIAN RHYTHMS

Key Notes

Photoperiodism

Photoperiodism is the response of a plant to length of day. It governs processes such as dormancy and flowering. Species may be either long-day, short-day or day-neutral in their response to day length.

Photomorphogenesis

Photomorphogenesis is the direct influence of light on growth and development. It involves responses to certain wavelengths of light perceived by photoreceptor pigments.

Phytochrome

Phytochrome is a photoreceptor protein. It is synthesized as Pr that absorbs red light (666 nm) and is converted to Pfr which absorbs far-red light (720 nm), and initiates cell signaling events. Many phytochrome responses initiated by red light are reversed by far-red light. Five phytochrome genes *PHYA–PHYE* have been identified. *PHYA* is expressed at high levels in etiolated tissue. In red light its expression is switched off and the protein is rapidly degraded. *PHYB–E* are constantly expressed at low levels in tissue and are involved in other phytochrome responses. Phytochrome responses include: etiolation/de-etiolation, circadian rhythms like leaf and petal movement and seed germination. Phytochrome regulates processes which involve changes in cell turgor (like leaf movements) by altering proton and potassium transport at the cell membrane. Longer-term phytochrome responses involve phytochrome-activated genes.

Blue light receptors

A variety of processes including stomatal aperture, chloroplast movement and phototropism are regulated by blue light. Cryptochromes and phototropins are proteins associated with flavins and respond to blue light.

Circadian rhythms

Processes with a 24-h cycle, including gene expression, stomatal aperture and growth and leaf and petal movements. An internal oscillator (clock) is generated by the expression of a number of genes, including *CCA1*, *LHY* and *TOC1*.

Related topics

Features of growth and development (J1)

Tropisms (I2)

Photoperiodism

Photoperiodism permits seasonal responses in temperate species that are independent of temperature. Plants may be classified according to whether flowering is induced by short days (**short-day plants – SDPs**); long days (**long-day plants –LDPs**) or is day-length independent (**day-neutral species**). Flowering in

LDPs occurs only when day length is greater than a given value (depending on species); that in SDPs when it is less than a given value. LDPs and SDPs in fact respond to the length of the night, as introducing a brief period of illumination (a **night-break**) overcomes the effect of long nights. Photoperiodism results from two processes – **perception of light** by the photopigment phytochrome in leaves, and an endogenous circadian rhythm (see below). Genetic control of flowering has been indicated by the description of mutant LDPs which are day-neutral and mutant day-neutral plants which are SDPs.

Photomorpho-genesis

Plants use light for photosynthesis. They also respond to light in other ways. For instance, a germinated seedling grows rapidly to reach light suitable for photosynthesis before its food reserves are exhausted. During this rapid elonga-tion growth, chlorophyll and some chloroplast proteins are not synthesized and the seedling remains **etiolated** – pale and lacking developed chloroplasts – until light is reached. Then immediately, patterns of gene expression are initiated and the seedling begins to form mature, photosynthetically active chloroplasts and its growth form alters from rapid elongation to the production of leaves and a stem capable of supporting them. This is **photomorphogenesis** – change of form in response to light. There are many other examples, including seed germination (some seeds will only germinate after being exposed to red light), leaf morphology and control of flowering. Photomorphogenesis involves specific responses to certain wavelengths of light, different to those needed for photosynthesis. These are usually either responses to blue or red light; responses to blue light include growth to unilateral light (**phototropism**), the mechanism for which is discussed in Topic I2. Growth in response to red light involves the pigment **phytochrome**.

Phytochrome

Phytochrome is synthesized as a protein, **Pr**, able to absorb red (666 nm) light. When it absorbs red light, it converts to a **Pfr**, able to absorb far-red light at **730 nm** (that converts it back to Pr; *Fig. 1*). Many phytochrome responses show 'red–far red reversibility' – when a process has been activated by a short period of red light, it will be stopped or reversed by a subsequent pulse of far-red light.

Phytochrome is a soluble protein made up of two identical subunits, in total sized 250 kDa. Each monomer (subunit) has a pigment (**chromophore**) molecule attached to it through a -S- (thioether) bond to the amino acid cysteine. When the chromophore absorbs red light, its structure alters slightly (see *Fig. 2*) and this alters the conformation of the protein, initiating events which ultimately result in altered gene expression.

A multi-gene family of phytochromes has been identified in arabidopsis (Topic B1), with five members, *PHYA*, *PHYB*, *PHYC*, *PHYD* and *PHYE*. These can be subdivided into two types of phytochrome; **PHYA** encodes **type 1 phytochrome**, which is the most abundant form in etiolated seedlings; **PHYB–E** encode **type II phytochrome** which is synthesized at much lower rates.

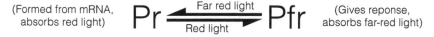

(Formed from mRNA,   Pr ⇌ Far red light / Red light ⟶ Pfr   (Gives reponse,
absorbs red light)   absorbs far-red light)

Fig. 1.   Formation of phytochrome and its interconversion between Pr and Pfr by red and dark red light.

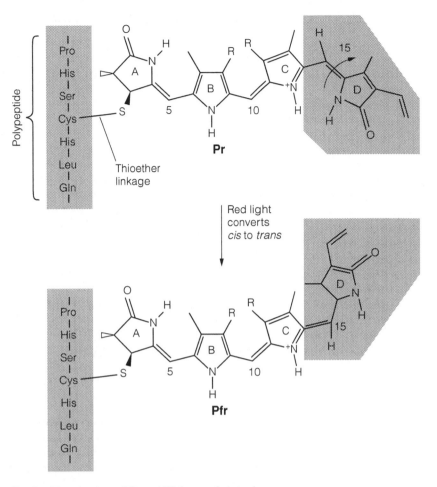

Fig. 2. The structure of Pr and Pfr forms of phytochrome.

Transcription of the *PHYA* gene is regulated by **negative feedback** in red light (which causes the formation of Pfr); so when an etiolated seedling (with high levels of type 1 phytochrome) is exposed to light, production of type 1 is greatly reduced as one part of photomorphogenesis (*Fig. 3*). In addition, type 1 Pfr phytochrome is very sensitive to proteolysis, so the level of the protein quickly reduces when it is not being newly synthesized. Transcription of the *PHYB–E* genes is not sensitive to light, and type II phytochrome is much less sensitive to proteolysis, so it remains more or less constant in the plant.

The **red/far red ratio** of light changes in different environments and through the day. Daylight, for instance, has an R/FR ratio of 1.19, while at sunset it is 0.96 and under a leaf canopy it can be 0.1. Light intensity also varies throughout the day. Phytochrome is involved in a wide range of plant responses to light including:

- **Etiolation**, in which a seedling or organ rapidly elongates without the production of chloroplasts until it receives red illumination, whereupon **de-etiolation** occurs and functional chloroplasts are produced.
- **Circadian rhythms**. A number of plant processes, including metabolism and

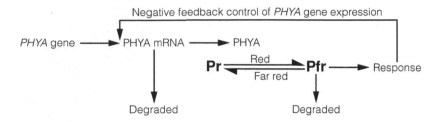

PHYB-E: low level of production, not sensitive to degradation, no negative feedback from response

Fig. 3.   PHYA is the form of phytochrome active in the etiolation response of dark-grown seedlings. The PHYA gene is expressed at high levels, producing Pr; when the plant is exposed to red light and Pfr is produced, the plant responds by producing functional chloroplasts and altered growth. PHYA production is stopped by a negative feedback control of the expression of the gene and any Pfr already formed is degraded. PHYB–E are involved in other phytochrome responses and produced at much lower levels. They are not sensitive to degradation and do not show decreased gene expression after exposure to red light.

leaf positioning follow a periodic cycle of 24 h. The phytochrome response ensures synchrony of the rhythm with day length.

- **Seed germination.** Many seeds are stimulated to germinate by light in a phytochrome-mediated response. This may require only brief irradiation or prolonged illumination, depending on species. Other seeds (like wild oat) show germination inhibited by light, though this requires intense irradiation over long periods, and is unlikely to involve phytochrome.

In some processes, changes induced by phytochrome involve changes in **cell turgor**. The circadian (day/night) movement of leaves and petals in some plants are examples (see below and Topic I3). Such **turgor** movements are rapid and involve altered proton pumping and K^+ movement through the plasma membrane (Topic E3).

Phytochrome also regulates changes in **gene expression**. The gene for the small subunit of ribulose bisphosphate oxygenase (Rubisco) is a phytochrome-regulated gene involved in the de-etiolation of dark-grown seedlings. Red light perception by Pr activates a protein activator of the Rubisco small subunit gene. This activator (called a *trans* **acting factor**) binds to the promoter region of the gene (the *cis* **regulator region**) and activates it. Phytochrome regulates a number of genes in de-etiolation and this results in the change from nonphotosynthetic plastids (etioplasts) to fully functional, photosynthetic chloroplasts.

Blue light receptors

Plants also show responses to blue light detected by pigments. Blue light responses include stomatal aperture, sun tracking by leaves, chloroplast movement, inhibition of stem elongation growth, phototropism (Topic I2) and gene expression. Cryptochrome receptors are sensitive to blue, green and UVA light. Cryptochrome is a protein with two chromophores attached, one sensitive to green and the other to blue light. There are two cryptochrome genes in arabidopsis. Phytochrome and cryptochrome interact. Other blue light photoreceptors include phototropin, involved in phototropism and chloroplast movement, which has two flavin-binding domains and is a self-phosphorylating protein kinase. In blue light, it binds flavins and self-phosphorylates. Zeaxanthin, a carotenoid pigment, is the photoreceptor for stomatal opening.

Circadian rhythms

Many plant processes, from gene expression to the movement of leaves or petals, occur in a 24-h cycle (for examples, see *Table 1*). When the regular succession of darkness and light are replaced by constant darkness or constant illumination, these rhythmic processes continue for some days, although frequently they will occur with a slightly altered length. Circadian (Latin *'circa diem'* or almost a day) rhythms are therefore an internal property of the plant, maintained at 24 h by the influence of day and night. There is therefore an internal oscillator (a clock) that is set (entrained) by light perceived by light receptors including phytochromes and cryptochromes.

Control of the circadian clock in *Arabidopsis* involves the action of clock genes that produce regulatory proteins. The interactions of the genes are shown in *Fig. 4*; light increases the expression of two, CCA1 and LHY, which reaches high

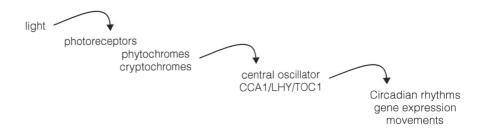

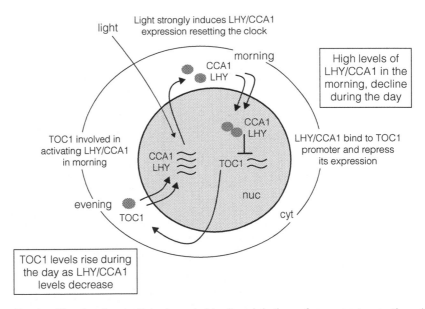

Fig. 4. The circadian oscillator is central to diurnal rhythms of movement, growth and gene expression. It is entrained by red and blue light and involves the action of key genes which act in a feedback system. Expression of two genes, LHY and CCA1 is highest around dawn, when light greatly stimulates their expression. During the day, levels of the two proteins decline, allowing a third protein, TOC1, to be synthesized. LHY/CCA1 inhibit expression of TOC1. In the evening, levels of TOC1 decline allowing new CCA1 and LHY to be produced until the cycle resumes with a burst of LHY and CCA1 expression at dawn. The oscillator ('clock') controls the expression of a variety of proteins and of processes.

Table 1.   Genes and processes showing circadian rhythms

Process	Examples
Growth and movement	'Sleep' movements of leaves and petals due to turgor changes in specialized pulvini Hypocotyl elongation
Stomatal aperture	Regulated on a diurnal rhythm by guard cell turgor. Topic E2

Gene expression	Numerous, including:
	Photosynthesis
	  Chlorophyll *a/b* binding proteins
	  Rubsico small subunit
	Photorespiration
	  Catalase
	  Serine hydroxymethyl transferase

Hormone production	Ethylene levels show a circadian rhythm in many species
Flowering time	Mutations that affect circadian rhythm also affect flowering time response, indicating that the systems are linked

levels after dawn. These two proteins inhibit the expression of another protein, TOC1. As CCA1 and LHY decrease during the day, they no longer inhibit TOC1. These genes then act to regulate a wide variety of other genes, some of which are most highly expressed in the evening and others in the morning.

12 TROPISMS

Key Notes

What are tropisms?	Tropisms are plant responses to environmental stimuli including unilateral light and gravity which involve altered growth. They may be positive involving growth towards the stimulus, or negative, away from the stimulus.
Phototropism	Phototropism is the response of the plant to unilateral light. The photoreceptor is likely to be a plasma membrane protein (NPH-1 in arabidopsis) which is phosphorylated in the presence of blue light. The signal is transmitted to the growing region by auxin redistribution between the light and dark sides of the tissue. Auxin stimulates elongation growth resulting in curvature.
Gravitropism	Gravitropism is the response of the plant to gravity. Roots are positively gravitropic, while shoots are negatively gravitropic. The stimulus is perceived either by amyloplasts in specialized statocytes in root caps and around vascular tissue or by stretch-activated ion channels and transmitted by auxin redistribution to the growth zone.
Related topics	Biochemistry of growth regulation (J2) Molecular action of plant hormones and intracellular messengers (J3)
	Photoperiodism, photomorphogenesis and circadian rhythms (I1)

What are tropisms?

Tropisms are responses to environmental stimuli which involve altered growth. Plants are able to respond to a range of environmental stimuli in order to make optimal use of their environment. *Table 1* describes a range of tropisms and their functions. Tropisms may be **positive** (towards the stimulus) or **negative** (away from the stimulus). All tropisms involve a **receptor** to sense the stimulus involved (a **gravireceptor** or **photoreceptor**, for instance) and a mechanism to transmit that stimulus to the region of the plant in which the altered growth will take place.

Roots are generally positively **gravitropic** (i.e. grow downwards with gravity), but nonphototropic. Shoots are generally positively **phototropic**, but either non- or

Table 1. Tropisms

Tropism	Function
Phototropism	Orientation towards/away from a unilateral illumination
Gravitropism	Growth towards/away from gravity
Hydrotropism	Growth towards water
Thigmotropism	Growth towards touch (e.g. in a tendril curling around a support)
Chemotropism	Growth towards a chemical stimulus

negatively gravitropic. Leaves, orientated at an angle to catch sunlight are **pla-giotropic**.

Tropisms have attracted a lot of interest since they were first studied by Charles Darwin in the 1870s. Much of the work has concentrated on the responses of the coleoptile of grasses. The **coleoptile** is a sheath of cells which covers and protects the primary leaf, and shows rapid extension growth (Topic H1).

Phototropism

The photoreceptor

When a **coleoptile** is exposed to unilateral light for a short period (1–2 h) it bends towards the light. The action spectrum of phototropism shows sensitivity to blue light. Early experiments showed that it was located at the tip of the coleoptile, as removal or covering the tip prevented the response. A protein photoreceptor has been identified by studying arabidopsis mutants that do not show phototropisms. These mutants lack **NPH-1**, a plasma membrane receptor protein phosphorylated in response to blue light. The action spectrum of NPH-1 and phototropism correspond almost exactly (Topic I1).

Transmission of the signal

Auxin (Topic J2) has been suggested as the transmitted signal in phototropism. The **Cholodny-Went hypothesis** states that unilateral light causes **auxin redistribution** near the apex, with more auxin on the shaded side. The Cholodny-Went hypothesis was criticized for many years, as it was hard to prove auxin redistribution using the techniques that were available. However, transgenic plants expressing **promoter-reporter gene constructs** (Topic B2) have been used to show that measurable changes in auxin concentrations do occur (*Fig. 1*). It is therefore likely that auxin is one of the regulators that transmits the signal from perception to the growing tissue.

The growth response

Auxin causes increased growth on the dark side of the shoot or coleoptile stimulating acidification of the cell wall, resulting in loosening of the **wall matrix**. Turgor pressure against the loosened wall results in elongation (Topic J2).

Gravitropism

The gravireceptor

The perception of gravity requires the presence of an object that responds (e.g. by movement) to the gravitational field. Such objects are named **statoliths**. Starch-filled plastids (**amyloplasts**) are good candidates, as their density means that they move readily within the cytoplasm in a gravitational field. Amyloplasts move against endoplasmic reticulum (ER) at the cortex (outer edge) of specialized cells termed **statocytes**. *Figure 2* illustrates how statocytes function. The statocytes are in the root cap (in primary roots) or in a layer (the **starch sheath**) adjacent to the vascular tissue. Removal of the root cap results in a loss of sensitivity to gravity in primary roots. The *scr* (*scarecrow*) mutant of arabidopsis, which has no layer of starch-containing cells around its vascular tissue, does not respond to gravity.

There are some plants (the alga, *Chara*, for instance) which appear to have no starch grains and still show responses to gravity. An alternative hypothesis is that plant cells have specialized ion channels that are activated by stretching. Such **stretch-activated channels** would sense movement of the protoplast under gravity and respond by allowing a calcium gradient to be set up. It may also be that stretch-activated channels in the ER respond to moving amyloplasts.

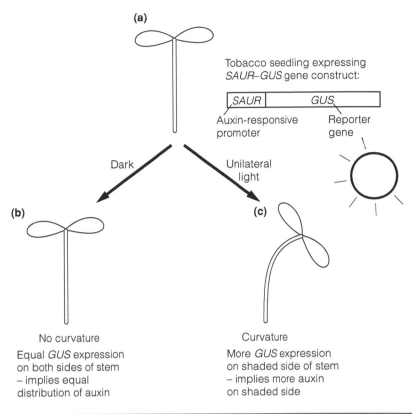

(a)

Tobacco seedling expressing
SAUR–GUS gene construct:

| *SAUR* | *GUS* |

Auxin-responsive   Reporter
promoter   gene

Dark   Unilateral
                     light

(b)   **(c)**

No curvature   Curvature

Equal *GUS* expression   More *GUS* expression
on both sides of stem   on shaded side of stem
– implies equal   – implies more auxin
distribution of auxin   on shaded side

Time (h)	Curvature (°)	% *GUS* expression on the dark side
0	0	50
12	30	65
8 + auxin transport inhibitor	0	50

Fig. 1. Use of tobacco seedlings transformed with a construct of SAUR (small auxin up-regulated) promoter (which responds to auxin) and a β-glucuronidase (GUS) gene to detect altered auxin concentrations in phototropism (a). The transgenic plants created were either kept in darkness (b) or exposed to unilateral light (c) and the concentration of auxin measured by estimating the amount of purple-colored product formed when the seedlings were soaked in the substrate for GUS. As the GUS gene is regulated by the SAUR promoter, its expression is proportional to the concentration of auxin; so high auxin ultimately results in a more intense purple coloration.

Transmission of the signal and growth response

Auxin is suggested to transmit the signal from the statocytes to the growth zone. In roots, high levels of auxin on the lower side inhibit growth, while lower levels on the upper side permit it; in shoots, higher levels on the lower side stimulate growth while lower levels on the upper side inhibit it. Experiments using plants with promoter-reporter gene constructs (*Fig. 1*; see also Topic B2) have confirmed a redistribution of auxin in seedlings experiencing altering gravitational forces. The mechanism of action of auxin in stimulating growth is described in Topic J2. Calcium has also been suggested to be involved in the graviresponse, as transport of calcium from the upper side to the lower side of roots has been shown, and roots will bend towards an agar block containing calcium.

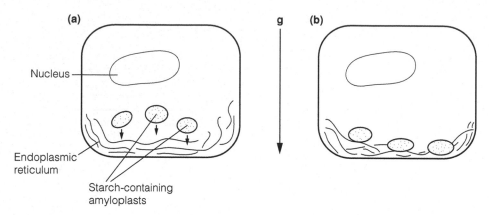

(a)

Nucleus

Endoplasmic
reticulum

Starch-containing
amyloplasts

g

(b)

Fig. 2.   *The starch sedimentation theory of graviresponse. Statocyte cells within the root cap or in a starch sheath around the vascular tissue contain amyloplasts (starch filled plastids) which sediment in the cytoplasm under a gravitational field (g) (a). When the amyloplasts interact with ER in the cell cortex, they initiate a signaling chain resulting in altered growth (b).*

I3 NASTIC RESPONSES

Key Notes

Nastic responses	Nastic responses are movements in which the direction of the stimulus is unimportant. Most nastic movements involve turgor changes in specialized cells, such as thermonastic movements in response to temperature or thigmonastic curling in response to contact.
Nyctinasty	Nyctinasty is movement of leaves or petals in a diurnal rhythm resulting in opening during daytime and closure at night. Movement results from turgor changes in the motor cells of specialized pulvini, where the flexor cells are turgid while the extensors are not and vice versa. The movements frequently occur in a circadian rhythm.
Seismonasty	Seismonastic movements are rapid movements of leaves and petioles in response to touch. They involve turgor changes in specialized pulvini triggered by a depolarization of the cell membrane propagated through the phloem tissue.

Related topics	Molecular action of plant hormones Tropisms (I2)
	    and intracellular messengers (J3)
	Photoperiodism,
	    photomorphogenesis and
	    circadian rhythms (I1)

Nastic responses Nastic responses are plant movements in response to stimuli in which the direction of movement is not related to the direction of the stimulus; this differentiates them from the tropisms (Topic I2). Most nastic movements do not involve growth (i.e. they are not permanent), though the terms **epinasty** and **hyponasty** are used to describe bending of an organ which does involve growth. The others are commonly also called **turgor movements**, as the mechanism of movement is usually a change in turgor of the tissue involved.

A wide range of nastic movements have been described which include:

- **epinasty**, downward curvature of an organ;
- **hyponasty**, upward bending of an organ;
- **thermonasty**, plant movements in response to temperature changes. The petals of some flowers show thermonasty, opening or closing the flower, in response to temperature;
- **thigmonasty**, curling in response to contact with a support. Tendrils of some plants curl around supports with which they contact;
- **seismonasty**, rapid movements in response to touch. The leaves of the sensitive plant, *Mimosa pudica*, fold rapidly when touched;
- **nyctinasty**, 'sleep' movements of petals and leaves in which leaves fold or close at night and open again next morning.

Nyctinasty

Flowers of many species close at night and leaves of some, including the prayer plant (*Maranta*), *Coleus* and French bean (*Vicia faba*) fold together at night and open out in daylight. This rhythm is maintained by an endogenous clock, as it persists even when the plant is placed in continuous light or continuous darkness for several days. The rhythm is therefore said to be **circadian** – an endogenous rhythm of about 24 h reinforced by regular exposure to light and darkness.

Leaf movement occurs at a **hinge region** in the petiole, the **pulvinus**. Pulvini appear as swellings in the petiole that contain **motor cells** surrounding central vascular tissue which drive the movement. Motor cells may be divided into two groups: **extensors** and **flexors** (*Fig. 1*) which lie opposite to one another. Turgor driven swelling and shrinkage result in movement; thus when the extensors are fully turgid, the flexors are flaccid and *vice versa*. These turgor changes are driven by movement of K$^+$ and associated cations (Cl$^-$, organic acids). The ion movements are driven by the *trans*-membrane electrical gradient set up by the plasma membrane proton pump and occur through ion channels in the plasma membrane (Topic E3).

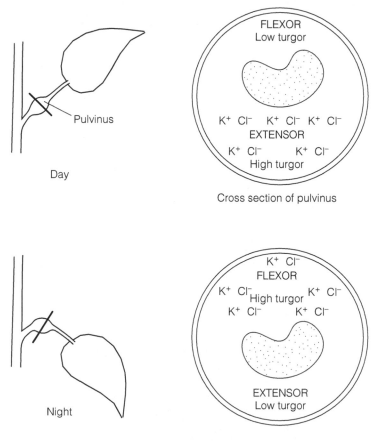

Fig. 1.   The role of extensor and flexor motor cells in the pulvinus of a plant (e.g. a bean) undergoing nyctinasty. Note the movement of K$^+$ and anions into the highly turgid extensor cells in daytime and from these cells into the flexor cells at night, resulting in leaf folding.

While the mechanism of movement is well understood, the mechanism governing the circadian rhythm is less so, although it is known that phytochrome is involved in entraining ('setting') the endogenous clock to coincide with light and dark in the 24 h period (Topic I1). It is likely that intracellular messengers are also involved. Understanding the clock is likely to be achieved through analysis of mutants of arabidopsis which show altered circadian rhythms.

Seismonasty

Seismonasty is most dramatically seen in sensitive plants, such as *Mimosa pudica*, in which touch will cause leaflets and then leaves to progressively collapse and fold (*Fig. 2*). It is believed that this deters herbivory, as insects either fall off, or are presented with a less palatable meal. The mechanism of folding is identical to that of nyctinasty – turgor changes in motor cells in pulvini in the petioles of the leaves. However, the trigger is different and a circadian rhythm is not involved. As the leaf collapse is progressive, i.e. it occurs near the point touched initially, then spreads to other pulvini as the leaf is touched more vigorously, there must be a mechanism for the signal to be transmitted across the leaf. An electrical signal has been measured flowing from the point of contact to the petiole. This resembles the action potential of a nerve and may involve the opening and closing of potassium channels that depolarize the plasma membrane of the phloem.

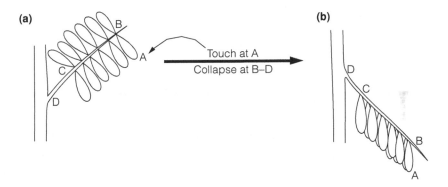

Fig. 2.   The leaf structure of the sensitive plant, Mimosa pudica, *includes many pulvini. (a) Touch at point A results in progressive collapse due to turgor changes in these pulvini (b). A measurable electrical potential flows along the phloem between each pulvinus propagated by potassium channels.*

14 ABSCISSION

Key Notes

Biological importance	Abscission is a controlled process resulting in the removal of plant parts either when they have completed their development or function (e.g. flowers) or in response to adverse conditions (e.g. leaves). The process permits the plant to achieve efficient fruit dispersal and to survive an unfavorable period, particularly drought.
Mechanism	An abscission zone forms at the point of attachment of the organ to the plant. Initially, high auxin levels from young tissue keep the zone inactive; later, the abscission zone develops and becomes sensitive to ethylene as auxin levels decline. Finally, ethylene triggers the release of cell wall degrading enzymes into the wall and the tissues separate at separation layers within the abscission zone.
Related topics	Features of growth and development (J1) Molecular action of hormones and intracellular messengers (J3)
	Biochemistry of growth regulation (J2)

Biological importance

Abscission describes the removal of plant parts (leaves, flowers, fruit) either in response to environmental stimuli or at specified points in the life cycle in a controlled and ordered manner. In different species, abscission permits survival in temporary adverse conditions, e.g. drought, cold, survival during regular or adverse conditions, fruit and seed dispersal; shedding of damaged organs or those which have completed development such as flowers and fruit.

Leaf abscission occurs in a number of circumstances. Most familiar in temperate climates is the abscission of leaves before winter in perennial deciduous species. Many of the nutrients within the leaf are reabsorbed by the plant and removal of the leaf reduces water loss during the winter when photosynthetic gain would be low and leaf damage (due to frost or pathogens) great. Controlled loss of the leaf and the sealing of the point of separation prevents pathogens from penetrating. Leaf abscission also occurs in many species in drought, thus reducing water loss and enhancing survival. It occurs in older leaves in plants in temperate zones throughout the growing season, and throughout the year in the tropics.

Fruit and flower abscission occur at the end of their development. Loss of the remains of the flower prevents necrosis and pathogen accumulation at the plant surface and removes organs no longer needed after pollen shedding or when no possibility of pollination remains. Fruit abscission results in the dispersal of seeds.

Mechanism

The abscission process ensures two things: the separation and sealing of the point of separation of the abscized organ and the appropriate timing of the abscission event.

The **abscission zone** is recognizable as a slight swelling in the petiole or stalk attaching the organ to the plant (*Fig. 1*). Within it, one or two layers of cells, the **abscission layer**, forms as the zone develops. These layers become sensitive to ethylene and develop the secretory machinery necessary to deposit cell-wall-digesting enzymes into the wall space between them. It is the separation of these layers that results in abscission at a well-defined location resulting in a 'clean break' of minimal surface area.

Young leaves produce large amounts of auxin (Topic J2); this diminishes as the leaf matures, or if it is damaged (*Fig. 1*). Drought (wilting), damage, short days or declining temperatures stimulate ethylene production and diminish auxin production in the leaf. With high auxin and low ethylene, the cells of the abscission zone remain inactive. As the leaf production of auxin diminishes, the abscission zone becomes sensitive to ethylene and begins to develop. Finally, when ethylene levels are sufficient, the layer is activated and enzymes (e.g. polygalacturonases, cellulases) are secreted and digest the cell walls holding the leaf to the stem. The break is then sealed preventing water loss and pathogen penetration. While abscisic acid (ABA; Topic J2) was named because of a supposed involvement in abscission, it is now known that it acts only indirectly in this process by accelerating ethylene production.

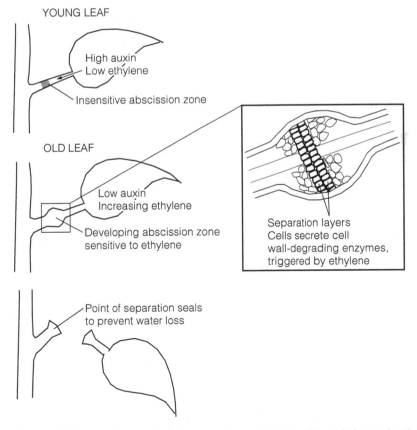

Fig. 1.   Development of an abscission layer. In young leaves, high auxin keeps the abscission layer insensitive to ethylene. Later, auxin production declines and the abscission layer develops and becomes sensitive to ethylene. Finally, rising ethylene triggers the secretion of cell wall hydrolytic enzymes into the separation layers and the two layers separate.

I5 STRESS AVOIDANCE AND ADAPTATION

Key Notes

Plants as sessile organisms	Plants are sessile – they cannot move to avoid adverse environments or predators. They show adaptations to permit survival without movement. These include plasticity of development, ability to regenerate new organs and a range of defenses to deter herbivory and pathogenesis.
Herbivory and pathogenesis	Plants deter herbivores by physical barriers such as a cellulose cell wall, waxy cuticle, lignification, hairs and stinging hairs, and also by chemical deterrents including compounds which make the plant unpalatable or poisonous or disrupt digestion.
Toxic ions	Many nutrient ions become toxic at high concentrations, others are almost universally toxic. Plants tolerate toxicity by one of four mechanisms: tolerance permits growth and metabolism in the presence of the toxin; exclusion where barriers at the root surface or root/shoot interface prevent uptake; amelioration, the dilution or chelation of the toxin; and phenological escape, where growth only occurs in favorable seasons.
Gaseous toxicity	Gaseous toxins include oxides of sulfur and nitrogen and ozone. Gaseous pollutants have the direct effects of cell and membrane damage and the indirect effects of soil acidification and inhibition of nutrient uptake.
Waterlogging	Plant roots growing in waterlogged soil become anoxic and growth is inhibited. Some species form aerenchyma, either constitutive or induced by ethylene. The root meristem develops anaerobic metabolism.
Salinity	Salinity causes both toxicity and osmotic problems. Most plants have little or no salt tolerance. Salt tolerant species may produce nonprotein amino acids as compatible solutes to overcome osmotic problems, some form salt glands that excrete the salt.

Related topics	Features of growth and development (J1)	Movement of ions across membranes (E3)
	Abscission (I4)	Uptake of mineral nutrients by plants (E4)
	Plants and water (E1)	

Plants as sessile organisms

The vegetative form of many higher plants is sessile; they cannot respond to changing environments by moving to new locations. Therefore, they must be able to tolerate adverse conditions. The range of adaptations varies between species: some characteristics are shown by almost all plants, but many others are only present in a few plants that tolerate particular environments. Plant development is plastic; in other words, it is not predetermined to a rigid path as mammalian development is. The presence of persistent embryonic tissue (meristems; Topic D1) and the fact that new meristems can be induced, means that plants can respond to a changing environment, wounding, etc., by altered growth. The form of growth may also alter, for instance to a lower, more compact form in response to high wind speeds. Plant hormones (Topic J2) are essential to plasticity as they integrate environmental stimuli with development.

Herbivory and pathogenesis

Plants are exposed to many herbivores, ranging from mammals to insects. Defenses include:

- **Physical barriers and defenses.** Cellulose cell walls; waxy cuticles; spines, hairs and stinging hairs; bark and secondary thickening; silica deposition.
- **Chemical defenses.** Formation of toxic or unpalatable secondary products (*Table 1*; see also Topic F5). Many of the compounds indicated are synthesized rapidly in response to herbivory or pathogen attack and plants may acquire resistance to further attack. **Salicylic acid** is produced by infected tissue and induces **defense genes** elsewhere in the plant. Fragments of the pathogen and the plant cell wall produced during the initial stages of infection by hydrolytic enzymes from the pathogen, called **elicitors**, also act to initiate cell signaling pathways resulting in the activation of defense genes, including those for the enzymes responsible for the production of many of the compounds included in *Table 1*.

Table 1. Compounds involved in defenses against herbivory

Type of compound	Examples	Effects
Phenolic compounds	Coumarins	Cytotoxic and irritant
	Lignin	Indigestible
Terpenes	Pyrethrin	Insecticidal
Tannins	Various	Bind proteins and inhibit digestion
Isoflavonoids	Phytoalexins	Fungicidal/bacteriocidal
Alkaloids	Cocaine, morphine, nicotine, caffeine	Toxic and/or physiologically active drugs
Non-protein amino acids	Various	Inhibit protein digestion
Proteinase inhibitors	Various	Inhibit proteinases in the herbivore, preventing digestion of the plant material

Toxic ions

Most nutrient ions, at high enough concentrations, become toxic to plant growth. Others, such as **Cd** and **Al** are almost universally toxic, even at relatively low concentrations. *Table 2* lists some major toxic ions in soil. Toxic ion concentrations may arise when the toxin is added to the soil (e.g. by atmospheric pollution or in industrial waste) or when soil conditions change (e.g. acidification will release Al from nontoxic complexes to free solution).

In general terms, toxicity occurs when growth is inhibited or a plant is prevented from completing its life cycle. Toxicity may result from (i) inhibition of **resource acquisition** (e.g. water uptake, the uptake of essential nutrients or of photosynthesis), or (ii) inhibition of the **utilization of resources** (e.g. inhibition of enzymes, damage to cell membranes, etc.).

Table 2. Some toxic ions and their effects

Toxic ion	Circumstances in which toxicity occurs	Effect
Aluminum	Acid soils below pH 4	Inhibition of root growth; binds to phosphate, DNA, RNA, disrupts membranes and ATP metabolism
Boron[a]	Soil contaminated with fuel ash	Chlorosis and necrosis of tissue
Copper[a]	Mine-spoil contaminated land	Damaged root cell membranes; inhibition of growth
Magnesium[a]	Soils with high Mg/low Ca	Causes Ca deficiency
Manganese[a]	Acid soils	Causes Ca and Mg deficiency; inhibits shoot growth
Sodium	Saline soils; irrigated soils	Competes with potassium for uptake and osmoregulation; osmotic effects; stomata remain open
Chloride	Saline soils; irrigated soils	Osmotic effects; competes with other anions for uptake giving deficiency

[a] Essential at low concentrations, toxic at high.

Tolerance to toxicity

Plants may respond to the presence of a toxic ion in a number of ways. For many plants, severely inhibited growth and death will occur, but others show adaptations that permit them to survive. Such plants may only grow slowly, but because they can occupy an environment that other plants cannot, they are able to benefit from that lack of competition. Plants may tolerate toxicity in one of four ways presented in *Table 3*. Hyper-accumulator species that are able to tolerate and accumulate large amounts of toxic ions are considered in Topic N5 (Bioremediation).

Table 3. Adaptations to toxicity

Tolerance	Adaptations which permit normal growth and metabolism in the presence of the toxin, even when in the tissues. This includes modified enzymes which are not inhibited by the toxin. Species which take up the toxin are termed 'includers'
Exclusion	The toxic ion is not taken up by the plant. This may be because of an effective barrier at the root surface (through the possession of very specific ion transport systems) or internally, particularly at the root–shoot interface. Species which take up the toxin into the root tissue and prevent it from moving to the shoot (e.g. at the root endodermis) are termed 'includer/excluders'
Amelioration	Some 'includer' plants minimize the effects of a toxic ion by modifying or storing it away from the key enzymes of growth and metabolism, usually in the vacuole. Amelioration may involve: compartmentation of the ion in the vacuole; chelation, in which the ion is complexed with an organic compound (for instance citrate or malate) and then deposited in the vacuole; and dilution in which the ion is diluted to below toxic concentrations
Phenological escape	A species may escape a seasonal stress by growing in the other seasons when the stress is minimal, for instance species growing in zones only covered by extreme tides in a salt marsh

Gaseous toxicity   Gaseous toxicity may occur when plants are exposed to toxic gases from industry or volcanoes. Pollutants include **ozone**, **sulfur dioxide**, **nitrogen oxides** and **carbon monoxide**. Their effects may be direct, e.g. inhibited stomatal action, damage to surfaces, inhibition of enzymes; or indirect, e.g. altered soil properties preventing nutrient uptake. *Table 4* gives the direct and indirect effects of some major atmospheric pollutants.

Table 4.   Effects of atmospheric pollutants

Pollutant	Direct effect	Indirect effect
Sulfur dioxide (precipitates from the atmosphere in rain as sulfates, including sulfuric acid)	Interveinal chlorosis (yellowing between leaf veins); death of leaves of forest trees	Acidification releases toxic aluminum in soils giving aluminum toxicity; Uptake of other nutrients inhibited
Oxides of nitrogen (precipitates from atmosphere in rain as nitric acid)	Black, necrotic lesions on leaves	Acidification releases toxic aluminum in soils giving aluminum toxicity; Uptake of other nutrients inhibited; Formation of NO in light from NO_2 results in release of O which reacts with O_2 to form ozone (O_3)
Ozone	Weakens palisade cell walls and results in necrosis, particularly near stomata; damages plasma membranes and thylakoids; inhibits photosynthesis	Produces reactive chemicals that cause genetic mutation

Waterlogging   **Waterlogged soils** contain low concentrations of oxygen which inhibit the growth of roots. Some species of plants contain aerenchyma, gas spaces in the root which both reduce the overall root oxygen demand and may supply oxygen to the growing tip. Aerenchyma may be formed constitutively, i.e. is always present, for instance in paddy rice. In other species, such as maize, it is induced by ethylene, produced in response to low oxygen. It is formed by the death of cells of the root cortex. Cells at the meristem do not die, but respond to lack of oxygen by the induction of anaerobic metabolism (fermentation). This involves induction of a number of enzymes, including alcohol dehydrogenase, which removes ethanol, a toxic product of fermentation, from the tissues.

Salinity   Toxicity due to NaCl has considerable impact on agriculture. Both Na^+ and Cl^- are toxic and occur in excess where land is exposed to sea water, in many arid zones and on irrigated land where evaporation exceeds precipitation or irrigation. Declining water tables due to deforestation and intensive agriculture and resulting salinization are an increasing agricultural problem.

    The chief causes of damage by salt are: osmotic stress and dehydration; inability to regulate stomata (as Na^+ replaces K^+); nutrient ion deficiency; and Cl^- toxicity. Plants show a wide range of salt tolerance, from **glycophytes**, able to tolerate small amounts of NaCl, to **halophytes** able to tolerate $\geq$ 500 mM. Plants able to grow in high salt conditions show a number of adaptations which are summarized in *Table 5*.

Table 5.   Salt tolerance in plants

Type and examples	Sodium concentration tolerated	Examples of mechanisms shown by some species
Glycophyte (beans, maize)	<50 mM	None
Salt-tolerant glycophyte (barley, wheat)	<200 mM (tolerated but growth retarded)	Limited salt accumulation to shoot; Production of solutes to provide osmotic balance with external solution of NaCl
Halophytes (sugar beet) (*Suaeda maritima*)	<500 mM (tolerate, but growth retarded) <700 mM (tolerated and growth stimulated up to 350 mM)	Make solutes like nonprotein amino acids and accumulate ions in the vacuole to balance external osmotic effects; Endodermis restricts transport from root; Absorption of ions from xylem flow; Succulence; Salt glands actively secrete salt from the leaves

J1 FEATURES OF GROWTH AND DEVELOPMENT

Key Notes

Growth and development

Growth involves cell division followed by cell enlargement. Primary meristems produce files of cells in concentric rings, which form the major tissues of the plant. Development occurs when cells and tissues change form and function to give the organs and structures required during the life cycle of a plant. Growth originates with new cells formed by meristems.

Cell growth

Cell growth occurs when the cell wall is made plastic by enzymes. The driving force for cell expansion is turgor pressure, which pushes the plasma membrane out against the cell wall. The direction of growth is governed by the orientation of cellulose fibers in the wall.

Embryogenesis

The fertilized ovule first divides to give an apical and a basal cell. The basal cell forms the suspensor and the root cap; the apical cell gives the root, shoot and cotyledons of the seedling. Cell lineages can be traced from the seedling through the various stages of cell division, the octant stage, the dermatogen stage and the heart-shaped embryo.

Development of tissues

The cells laid down in the meristem form all the tissues of the plant. The first stage of development is determination, in which the cell becomes established on a pathway of change. The cell then becomes differentiated to its new function. Determination and differentiation involve altered gene expression.

Tissue culture and totipotency

In tissue culture, tissue explants are de-differentiated to form a callus and then redifferentiated by varying hormone or other growth conditions. Single cells in culture can be shown to be totipotent as they can regenerate to form an entire plant.

Cell-to-cell communication

Cell-to-cell communication occurs through plasmodesmata connecting rows or blocks of cells symplastically.

Plant and animal development compared

Cell walls prevent cell movements that are characteristic of animal development. Plant embryonic tissue is maintained through the life of the plant, whereas animals have a distinct embryonic stage. This gives greater plasticity of plant development. Plant cells show totipotency, the ability for single cells to regenerate an entire organism. Cell-to-cell communication in plants is limited to plasmodesmata.

Related topics

Meristems and primary tissues (D1)
Biochemistry of growth regulation (J2)

Molecular action of plant hormones and intracellular messengers (J3)

Growth and development

Growth occurs when new cells and tissues are formed by **cell division** (Topic C6) followed by **cell enlargement**. **Development** is the process whereby those cells change form and function to form the specialized tissues, organs and structures required during the life cycle of a plant. It commences with the first cell division after fertilization of the ovule and continues through seed development, seed germination, the development of the seedling to the mature plant, flowering and production of the next generation of ovules. It also includes the processes of cell death and plant senescence.

Plant growth is **accretionary**, new cells being constantly added in meristems (Topic D1), regions that essentially remain **embryonic** throughout the life of the plant. Plant growth begins with **cell division**. Cell enlargement and change in form and function follows subsequently. In the meristem, dividing cells surround the **quiescent center** where no cell divisions occur. As the new cells are surrounded by a cell wall, migration of cells to new locations is impossible. Therefore, the rows (or files) of cells formed (usually in concentric rings) predict the future tissues of the root or shoot. As primary growth occurs by the formation of new tissues at growing tips, different cells and tissues in the same plant are of different ages. The growth of a herbaceous dicotyledon may be considered to occur as successive **phytomeres** consisting of stem, bud and a leaf (*Fig. 1*).

Cell growth

Cell growth in plants can only occur when the cell wall (Topic C2) is made **plastic** by the action of enzymes that break the cellulose cross-linkages. The direction of cell expansion is governed by the orientation of the major fibers

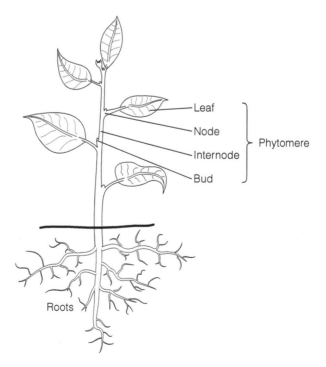

Fig. 1.   The plant may be considered as a number of repeating units (phytomeres) which in this dicotyledon comprises a leaf or leaves, a node, an internode and a bud.

(cellulose fibrils) within the wall. The driving force for cell expansion is **turgor pressure**, which pushes the plasma membrane out against the cell wall. Cell growth largely occurs separately from cell division.

Embryogenesis

The basic plan of the plant is established soon after the ovule is fertilized, in the early stages of the development of the embryo (**embryogenesis**) in the formation of the seed (Topic H1). Embryogenesis in arabidopsis (Topic B1), a dicot, is shown in *Fig. 2*. The fertilized ovule divides to give two cells: the **apical cell** and the **basal cell**. The basal cell forms the **suspensor**, connecting the embryo to maternal tissue and also the **root cap meristem**. The apical cell undergoes many cell divisions. The first stage is the **octant stage** (named from the eight cells in two tiers formed). This is followed by the **dermatogen stage** (where tangential cell divisions have occurred creating tissue layers). Finally the **heart-shaped embryo** is formed. This contains the origins of all the major structures of the seedling. The lobes of the heart shape are the cotyledons; between them lies the shoot meristem. The center of the heart forms the hypocotyl and the lower layers form the root. As plant cells cannot migrate during development, it is possible to trace **cell lineages** back to the dermatogen and octant stages. These lineages are illustrated in *Fig. 2*.

Development of tissues

The concentric rings of cells laid down in meristems are initially similar in form: nonvacuolate, isodiametric (roughly cuboid) and with thin cell walls. Subsequently they form all the tissues of the plant. This involves stages during which major changes in gene expression occur. The initial stage of this process is **determination**, in which the cell becomes established on a pathway of change but physical changes are not yet detectable. At this stage, the cell is **committed** to a pathway of development. The cell then becomes **differentiated** to its new function, losing some characteristics and gaining others. In plants, both determination and differentiation are frequently reversible given suitable treatments. A major question about development and differentiation is: what causes the altered gene expression that results in ordered patterns of differentiated tissue?

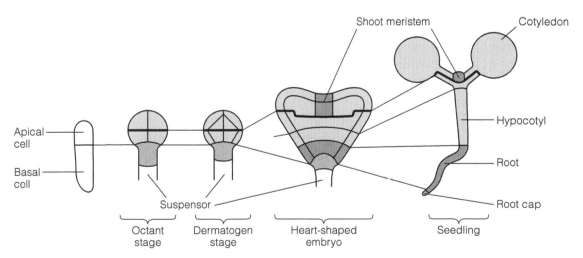

Fig. 2. Embryogenesis in a typical dicotyledon. (Redrawn from T. Laux and G. Jurgens. Embryogenesis, a new start in life. Plant Cell 1997; 9: 989–1000. American Society of Plant Physiologists.)

The position of the cell is likely to be important and endogenous chemical signals within the tissue are involved (Topics J2 and J3).

Tissue culture and totipotency

Plant tissue may be cultured in either liquid or solid media containing an energy source (sucrose), plant hormones (auxins and cytokinins; Topic J2) and a range of other minor components. In tissue culture, tissue explants (small pieces of plant) are first **de-differentiated** to form **callus**, an amorphous mass of cells, and then **redifferentiated** to form roots, shoots and other organs by varying hormone or other growth conditions. In suspension cultures, it has been shown that single cells can regenerate to form an entire plant, going through all the normal stages of embryo development. As the parent cell did not originate from reproductive cells it demonstrated that plants show **totipotency** – the ability for a differentiated cell to retain all the genetic material in a form required to form an entire organism.

Cell-to-cell communication

While plant plasma membranes are separated by the presence of the cell wall, cell-to-cell contact is made by **plasmodesmata** (Topic C2). Rows or blocks of cells are therefore connected as if in colonies. Macromolecules such as RNA and smaller signaling molecules can move between cells.

Plant and animal development compared

- Plant cells are not mobile during development due to a cell wall.
- In animals, determination and differentiation occurs in the embryo. In plants, cells in the meristems keep dividing and the newly formed cells keep differentiating throughout the life of the plant. This means that different parts of the same plant are of different ages.
- Determination and differentiation of plant cells is much more **plastic** than animals. Application of hormones, wounding or other treatments result in plant cells altering pathways of development to form different tissues and organs.
- Plant cells are **totipotent** (in other words a single, nongerm-line cell can be induced to regenerate to a whole organism), a property not generally seen in animals. This implies that the entire genome of that cell is intact and functional and that it can be brought back to an embryonic state.
- In spite of the cellulose cell wall, plant cells may communicate via plasmodesmata.

J2 BIOCHEMISTRY OF GROWTH REGULATION

Key Notes

Hormones in plants	Plant hormones or 'growth substances' are compounds that act specifically to regulate growth and development at low concentrations. Each plant hormone regulates a variety of processes; the range of concentrations over which they act is broad and there is frequently no clear separation between point of synthesis and point of action.
Auxins	Auxins have a variety of effects including elongation growth, cell division and differentiation, and apical dominance. They frequently work with other hormones, principally cytokinins. Plants show polar (directional) auxin transport. Nonpolar transport also occurs in phloem.
Ethylene	Ethylene (ethene) is a gaseous hormone first identified as a regulator of fruit ripening. It also stimulates senescence and abscission. In seedlings, it initiates the triple response: epinasty, lateral growth and inhibition of elongation. It is synthesized from S-adenosyl methionine (SAM) via 1-aminocyclopropane-1-carboxylic acid (ACC).
Gibberellins	Gibberellins are a large group of compounds formed from isoprene units. Gibberellins stimulate elongation in dwarf plants and mediate the transition from rosette form to flowering in response to temperature or daylength. They also modulate processes involved in seed and bud dormancy.
Cytokinins	Cytokinins promote cell differentiation and division, often acting in association with auxin. They delay senescence, and promote chloroplast maturation in etiolated seedlings. Their synthesis, based on isoprene units, occurs in various tissues and organs. They are transported in the xylem as cytokinin conjugates and inactivated by oxidation.
Abscisic acid	Abscisic acid is a regulator of dormancy and germination of seeds and of plant responses to stress. It is synthesized in roots and shoots, at high rates in stressed tissue. It is transported in xylem and phloem.
Polyamines	Polyamines, like putrescine and spermidine, are compounds with more than one amine group, synthesized from lysine and arginine. They have effects in cell growth and development and in stress responses.
Brassinosteroids	Brassinosteroids are derived from the sterol campesterol. They are present in plants at low levels but stimulate cell division and elongation.
Oligosaccharides	Oligosaccharides, carbohydrate fragments released from plant cell walls, have been shown to elicit plant defenses against fungal attack and may be regulators of development.

Related topics	Features of growth and development (J1) Methods in experimental plant science (B2)	Molecular action of plant hormones and intracellular messengers (J3)

Hormones in plants

A **hormone** is defined as a naturally occurring, organic substance that, at low concentration, exerts a profound influence on a physiological process and is not a part of a major metabolic pathway. The plant hormones coordinate processes as diverse as development of the embryo and response to stress. As plant development shows some major differences from animal development (Topic J1), it is not surprising that plant hormones have many differences in mode of action and nature from mammalian hormones. To avoid this confusion, other terms have been used, such as **plant growth substance** (**PGS**) and **phytohormone**. A summary of the key features of plant hormones, and how they differ from animal, is presented in *Table 1*.

Auxins

The major plant auxin is **indole-3-acetic acid** (**IAA**). A number of other compounds with auxin activity include **phenoxyacetic acid** and **indole 3-butyric acid** (*Fig. 1*).

Auxin effects
Elongation growth. The primary effect of auxin is to regulate stem growth. It does this by stimulating the growth of cells in the direction of elongation. Shoot

Table 1.   *Differences and similarities between plant and animal hormones*

	Animal	Plant
Naturally occurring organic molecule exerting profound effect on physiological process	Yes	Yes
Active at low concentrations	Yes (10× range usual between inactive and fully active)	Yes (may be a 1000× range between inactive and active)
Synthesized in a discrete organ or tissue remote from point of action	Yes	Not necessarily; synthesis may be diffuse through the plant or at, or near, point of action
Transported in a circulatory system	Yes	No circulatory system; transport in a specific direction (e.g. in xylem, or cell to cell) may occur
Has one, or a few, functions	Yes	Often multiple responses, depending on tissue, age and other factors
Require specific receptors in the cell to function	Yes	Yes. In view of the multiple effects of plant hormones, the presence of specific receptor proteins is essential in determining the final response

Naturally-occurring auxins

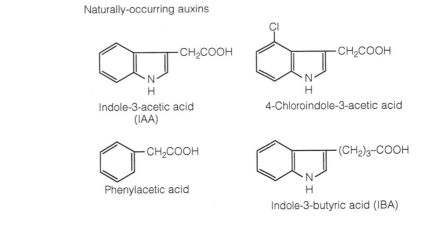

Indole-3-acetic acid
(IAA)

4-Chloroindole-3-acetic acid

Phenylacetic acid

Indole-3-butyric acid (IBA)

Synthetic auxins

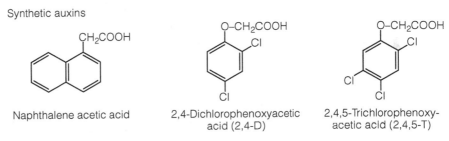

Naphthalene acetic acid

2,4-Dichlorophenoxyacetic
acid (2,4-D)

2,4,5-Trichlorophenoxy-
acetic acid (2,4,5-T)

Fig. 1. Chemical structures of auxins. Indole-3-acetic acid (IAA) is the major active auxin found in plants. Naphthalene acetic acid (NAA), 2,4-dichlorophenoxyacetic acid (2,4-D) and 2,4,5-trichlorophenoxyacetic acid (2,4,5-T) are all synthetic auxins with commercial applications.

growth is stimulated by 10^{-6}–10^{-7} M auxin. Root elongation, on the other hand, is much more sensitive, with maximum stimulation of growth at 10^{-9}–10^{-10} M auxin, and inhibition at higher concentrations.

Cell division and differentiation. When **callus**, an amorphous mass of undifferentiated cells (Topic J1) is grown on an agar plate containing nutrients, the degree of cell division and differentiation to form roots and shoots can be varied by altering the **auxin:cytokinin ratio**. Both hormones are required; *Fig. 2* summarizes the results of such an experiment. These effects are also found in plants where auxins induce lateral root formation in stem cuttings.

Apical dominance. A characteristic of the growth of many plants is the dominant growth of the apical bud. When this bud is removed, growth of axillary buds formed a little way behind the apex is stimulated, until one of them becomes dominant and the growth of the others is suppressed. Replacement of the apical bud with auxin inhibits the axillary buds, suggesting that high auxin concentrations generated at the apex inhibit axillary buds. Application of cytokinin to axilliary buds releases them from inhibition and therefore auxin-cytokinin interactions are responsible for the phenomenon.

Other auxin effects. Auxin has a range of other effects, either alone or with other hormones, including fruit development. Some plants, e.g. strawberry,

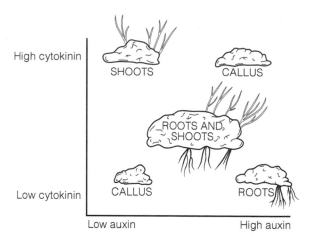

Fig. 2. Regulation of growth and development by auxin:cytokinin ratio. Explants grown on a nutrient-containing agar may be induced to form amorphous callus, or roots, shoots, leaves and buds by varying the auxin:cytokinin ratio.

tomato, cucumber, pumpkin, citrus fruits, produce **parthenocarpic** (seedless) fruits if they are treated with auxin. Senescence and abscission of mature leaves, fruits and flowers is inhibited by auxin; however, abscission of young fruits is enhanced by auxin treatment.

Commercial applications

Synthetic auxins find widespread application in agriculture and horticulture. At high concentrations, **2,4-dichlorophenoxyacetic acid** (**2,4-D**) and **2,4,5-trichlorophenoxyacetic acid** (**2,4,5-T**) (*Fig. 1*) are used as **herbicides**, particularly on broad-leaved plants, which are much more sensitive to them than monocots. Naphthalene acetic acid (NAA) is used to **stimulate rooting** of cuttings ('hormone rooting powder'), while other synthetic auxins are used to reduce fruit number early in the season in apples and to promote fruiting in tomatoes and citrus fruits.

Synthesis

Auxins are mostly synthesized from the amino acid **tryptophan**, predominantly in young leaves, shoot meristems and developing fruits, wherever cells are dividing rapidly. IAA is also made by bacteria (see *Agrobacterium tumefaciens*, Topics B2, K3) and several pathways exist, including one in which IAA is synthesized from indole or indole-3-glycerol phosphate rather than tryptophan.

Auxin transport

Auxin shows **polar transport** (unidirectional). It moves **basipetally** (from the apex to the base) in isolated coleoptiles (the sheath encasing the primary leaf in a grass) and stems. Small amounts of auxins produced at the root apex may also move basipetally (in this case from the root tip up the root) but this is limited in comparison with that from the shoot. Polar transport in stems occurs in the parenchyma surrounding the vascular tissue involving specific **auxin** transport proteins. Its transport can be inhibited by **auxin transport inhibitors** such as **1-N-naphthylphthalamic acid (NPA)**. Auxin synthesized in the leaves is also transported in a nonpolar fashion in the phloem; this process is about 10 times faster than polar transport. Studies in *Arabidopsis* have revealed a gene, *AUX1*, expressed in the root apex, which encodes an auxin transport protein which is involved in directional elongation, for instance in gravitropism (Topic I2).

Auxin conjugation and degradation
The amount of auxin available in a cell or tissue depends on three processes. The rate of auxin biosynthesis or import from other cells, the rate of auxin degradation and the amount that is conjugated (chemically bound) to other molecules. Conjugated auxin is not biologically active. Most auxin within a plant is covalently bonded to organic compounds (e.g. esters of myo-inositol and glucose, and high-molecular weight compounds such as glycoproteins) and is inactive. Transport of IAA in the phloem is predominantly in the form of these complexes, and their breakdown to release IAA supplies it to tissues like the coleoptile tip. There are several pathways for IAA breakdown, involving peroxidation of IAA to 3-methyleneoxindole and nondecarboxylation to oxindole-3-acetic acid.

Ethylene

Ethylene (ethene) was discovered in the early 1900s as a gas that regulated **fruit ripening**. It had been realized that the close proximity of ripe fruit, such as oranges or apples, speeded up the ripening of other fruits, such as tomatoes and bananas. Regulating ripening, and therefore ethylene, has become an important part of the storage, transport and marketing of fruit worldwide. Ethylene has a variety of other roles in plants, including senescence of leaves and fruit, elongation of roots, and responses to waterlogging and other stresses. Although a simple molecule, its effects are highly specific.

Ethylene effects
Fruit ripening. Many ripening fruits show a rise in ethylene production that precedes the onset of ripening. Fruits that produce and respond to ethylene in ripening are the **climacteric fruits** (apples, tomatoes and bananas); the climacteric is a characteristic burst of respiration that occurs just before the final stages of ripening take place. Ethylene production in climacteric fruit is **autocatalytic**, i.e. ethylene stimulates its own production, the rapidly rising ethylene concentration then triggering the rapid burst of respiration.
The triple response. Ethylene-treated shoots (e.g. pea seedlings) show three characteristic growth responses simultaneously: **epinasty** (downward curvature of the leaves); **decreased elongation** and **lateral cell expansion** (i.e. increase in stem width); and **loss of gravity response** to give horizontal growth. Ethylene-induced epinasty gives the apex of young dicot seedlings a 'hook'-like appearance.
Ethylene and waterlogging responses. Whereas ethylene normally inhibits elongation growth, in some wetland species, including rice, ethylene induces rapid elongation growth, allowing the plant to reach air. The formation of **aerenchyma** (air spaces in the root cortex, formed by programmed cell death; Topic D1) is also induced by ethylene, which is synthesized in response to low oxygen and accumulates in waterlogged roots.
Other roles of ethylene. High concentrations of ethylene (>10 μl l^{-1}) induce

adventitious rooting and root hair formation. In leaf abscission, ethylene accelerates the synthesis of cell-wall degrading enzymes in the abscission layer, a specialized layer of cells at the leaf pulvinus which separate from adjacent cells permitting the leaf to fall from the plant.

Synthesis and degradation

Ethylene is produced from the amino acid **methionine**, via **S-adenosyl methionine (SAM)** and **1-aminocyclopropane-1-carboxylic acid (ACC)**. The enzyme producing ACC is key in regulating ethylene production; it has a very short half-life and the expression of its gene is stimulated by factors known to induce ethylene responses. Ethylene is inactivated by oxidation (e.g. to ethylene oxide or CO_2) or it can diffuse from the plant. Rates of ethylene production rise rapidly in tissues subject to stress or wounding and subsequently decline to normal levels. Ethylene is active at very low concentrations (around 1ppm or $1\ \mu l\ l^{-1}$).

Gibberellins

Gibberellins were discovered in the 1930s by Japanese scientists investigating a disease of rice caused by the fungus *Gibberella fujikuroi*, that results in tall, seedless plants. Nearly 100 gibberellins have been identified in plants though many do not have biological activity. The most studied, and probably most significant gibberellin is **GA$_3$**. In common with the other gibberellins, it has a structure based on *ent*-gibberellane (*Fig. 3*).

Gibberellin effects

Environmental responses. Many species remain as **rosette plants** until they have been exposed to either low temperatures (**vernalization**) or a number of **long days**. Spinach, for instance, retains a short, squat form until day-length increase, when it begins to grow upwards and flower. Gibberellin levels are low in rosette plants, but increase dramatically in response to the changed environment and initiate the growth response. **GA$_1$** is most significant in elongation responses and **GA$_9$** in flowering.

Seed germination

In some seeds which show dormancy, gibberellin application will break dormancy. In other seeds, gibberellins are essential in coordinating the processes of germination, increasing in activity upon rehydration of the seed and initiating the activity of the hydrolases which mobilize seed storage reserves. The role of gibberellins in germinating barley is of economic importance in the malting process, part of beer brewing.

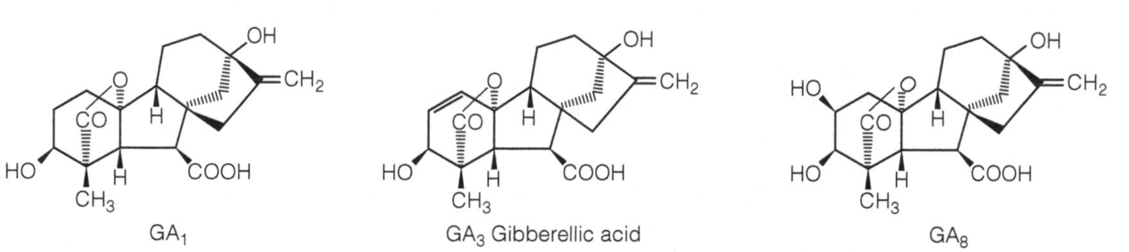

Fig. 3.   Some active and inactive gibberellins. They are all based on a structure known as ent-gibberellane.

Other effects of gibberellins

Gibberellins are involved in regulating the transition from **juvenile** to **mature** growth form in some perennial species such as ivy (*Hedera helix*); in **initiating flowering** and promoting **fruit formation**.

Synthesis

Gibberellins are **diterpene acids** synthesized by the **terpenoid pathway**. Terpenoids are compounds built of repeating **isoprene units**:

$$CH$$
$$|$$
$$—CH_2—C=CH—CH_2—$$

The location of the early stages of gibberellin synthesis is the **plastid**, where isoprene synthesis occurs from glyceraldehyde-3-phosphate and pyruvate. Later stages occur in plastids in meristems and in enzymes of the endoplasmic reticulum (ER) and cytoplasm.

Transport

Highest levels of active gibberellins in plants are found in young rapidly growing tissues like young leaves and buds, and developing seeds and fruits. Transport of gibberellins in the plant occurs predominantly in the **phloem** and is **nonpolar**.

Cytokinins

Cytokinins were described as compounds that regulate cell division in plants (Topic C6) in experiments in which compounds were screened for their effects on tissue cultures. From these experiments came two compounds: the first, **kinetin**, was isolated as the active ingredient in herring sperm DNA that caused massive cell proliferation in plant cell culture. **Zeatin** was the first natural plant cytokinin, isolated from the **liquid endosperm** of the coconut. The cytokinins constitute a small number of compounds, which are derivatives of adenine or amino purine.

Cytokinin effects

Cytokinin effects are generally associated with promotion of growth and development and delay of senescence. Cytokinins applied to leaves will delay senescence; they speed up chloroplast maturation in etiolated (dark-grown) tissue and promote cell expansion in young leaves. Cytokinins applied to lateral buds in plants showing strong apical dominance will overcome growth inhibition by auxin, causing the bud to grow out. Cytokinins, with auxins, are involved in plant tumor formation and in morphogenesis, the development of roots and shoots. *Figure 2* illustrates the effects of varying the ratio of auxin to cytokinin on the growth and morphogenesis of plant material in tissue culture.

Cytokinin synthesis

Like gibberellins, cytokinins contain isoprene subunits. The first stage involves the reaction of isopentenyl pyrophosphate with adenosine monophosphate (AMP), catalyzed by the enzyme **cytokinin synthase** to yield isopentenyl adenine ribotide. From this compound, cytokinin ribotides, ribosides and cytokinins are formed. Cytokinins are also synthesized by gene products resulting from the insertion of bacterial genes from *Agrobacterium tumefaciens* (Topics B2, K3). The **TI** (**tumor-inducing**) **plasmid** from *A. tumefaciens* introduces the gene for **isopentenyltransferase**. This enzyme generates **isopentenyl adenine**, which is converted to *trans*-**zeatin** and **dihydrozeatin** in the plant. These hormones, together with auxin produced by another TI-plasmid gene product, cause a tumor or crown gall to form at the site of infection.

Cytokinin transport

Cytokinins are synthesized in various tissues and organs, though the root apical meristem (Topic D2) is a major site of their production. They have been identified in the xylem flow from cut roots, suggesting that this may be a route for long-distance transport of cytokinins through the plant. Cytokinins in the root xylem are predominantly **zeatin ribosides**, which are rapidly converted to **free cytokinin** in leaves. Cytokinin inactivation occurs when they are oxidized to adenine by the enzyme **cytokinin oxidase**.

Abscisic acid

Abscisic acid (**ABA**) is found in all higher plants and mosses. ABA regulates **dormancy** and is central to plant responses to stress. The nature of the molecule means that it can exist in several forms. First, the carboxyl group at the end of the side chain may be *cis* or *trans* (*Fig. 4*), while the C at position one of the ring is asymmetric and gives either (+) or (–) (*S* or *R*) **enantiomers**. The **active form** of ABA is (+) *cis* **ABA**; (+)-2-*trans*-ABA also exists in plants and is active in some long-term ABA responses; it may be converted to the (+) *cis* form in tissues.

Abscisic acid effects

ABA has a variety of effects related (i) to **seed dormancy** and (ii) to **stress responses**. ABA levels rise initially during **embryo development** within the seed and then decline. ABA regulates the expression of genes for proteins in the embryo that prepare it for the final stages of seed development in which the seed desiccates and becomes dormant; it also activates genes for seed storage proteins (Topic H4). ABA also keeps some seeds dormant until the environment becomes suitable for growth. Controlling dormancy is very important in temperate climates since precocious germination may lead to the death of the seedling. ABA also accumulates in the dormant buds of woody species, although control of dormancy here is likely to be the result of the action of several hormones.

ABA also regulates several plant stress responses. Rising ABA levels in water stress initially cause stomatal closure (Topic E2) and subsequently increases the ability of root tissue to carry water; it also promotes root growth and inhibits shoot growth.

Abscisic acid synthesis

ABA biosynthesis begins in chloroplasts and amyloplasts. Synthesis, like that of gibberellins and cytokinins, involves the isoprene subunit in isopentenyl

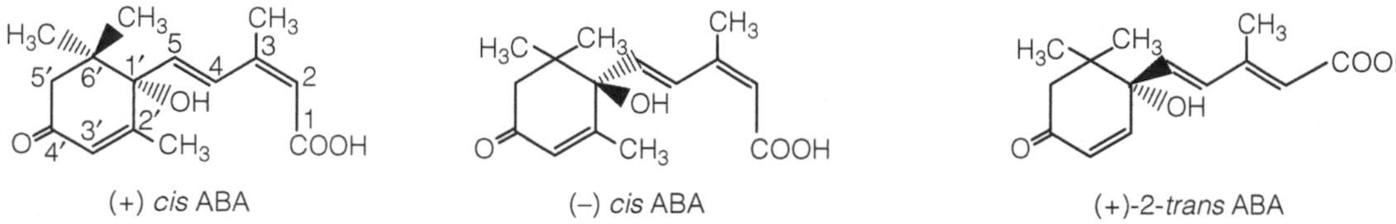

(+) *cis* ABA   (–) *cis* ABA   (+)-2-*trans* ABA

Fig. 4.   Chemical structures of (+) and (–) forms of ABA. (+) cis ABA is active; (–) cis ABA is active in slow ABA responses, but not rapid ones like stomatal closure. (+) trans ABA is inactive, but may be converted to (+) cis ABA.

pyrophosphate, which is used to produce an oxygenated carotenoid compound, **zeaxanthin**. Zeaxanthin is modified in a multi-stage process to **neoxanthin**, which is cleaved to the C15 compound **xanthoxin**; xanthoxin is then modified in two stages to produce ABA. ABA is degraded, either by oxidation or by conjugation, to form **ABA-glucosyl ester**.

Abscisic acid transport
ABA is synthesized in roots and shoots, and at much higher levels in tissues undergoing stress. Water-stressed roots, for instance, produce up to 1000 times more ABA that is transported through the xylem to the shoot. ABA is also transported from shoot to root in the phloem.

Polyamines   **Polyamines** are compounds containing two or more amine groups. Typical examples with biological activity are:

- Putrescine: $H_2N\text{-}(CH_2)_4\text{-}NH_2$
- Spermidine: $H_2N\text{-}(CH_2)_3\text{-}NH\text{-}(CH_2)_3\text{-}NH_2$
- Spermine: $H_2N\text{-}(CH_2)_3\text{-}NH\text{-}(CH_2)_3\text{-}NH\text{-}(CH_2)_3\text{-}NH_2$

Biosynthesis originates from the amino acids lysine and arginine. Levels of polyamines increase where rapid cell division is occurring; putrescine levels increase in response to some forms of stress and they may be involved in some aspects of embryo and fruit development.

Brassinosteroids   **Brassinosteroids** (or **brassins**) are a recently discovered, complex group of lipids synthesized from the sterol **campesterol**. They are present at low levels, but have strong growth promoting effects, stimulating both **cell division** and **cell elongation**. The structure of one brassinosteroid is shown in *Fig. 5*. It appears that brassinosteroids act with the other plant hormones to regulate growth and differentiation. Mutants of *Arabidopsis* and pea which are deficient in brassinosteroid biosynthesis are dwarf; application of brassinosteroid restores them to a normal phenotype, indicating that they are essential for cell elongation in normal plants.

Oligosaccharides   The complex polysaccharide cell wall of plants is a dynamic structure, which may be modified by both endogenous and exogenous enzymes. Wall changes lead to the release of fragments of long-chain polysaccharides, **oligosaccharides**, into the apoplast, some of which have been shown to have effects on development in plant tissue cultures and others of which are released during **fungal pathogen** attacks and elicit the **defense responses** of the plant (Topics I5 and M4).

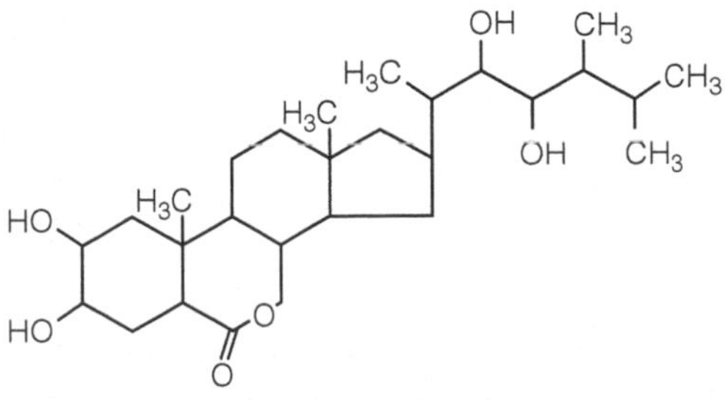

Fig. 5. Brassinolide, a brassinosteroid.

J3 MOLECULAR ACTION OF PLANT HORMONES AND INTRACELLULAR MESSENGERS

Key Notes

How do plant hormones act?	Plant hormones act in a variety of types of responses that may be long-term, like growth and development, or rapid. Hormones alter the activity of enzymes or other cytoplasmic components directly or alter gene expression and the production of new cellular components, or both.
Receptors and target tissues	In order for a hormone response to occur, the presence of the hormone must first be perceived by receptor protein. Receptors for ethylene (ETR1) and auxin (ABP1) have been identified and characterized. Target tissues respond to a given hormone because they possess the necessary receptors and pathways for response.
Hormones and the control of gene expression	Many genes are regulated by plant hormones. Some respond very rapidly, in a matter of minutes; others require hours to days. Genes that respond to hormones have a region termed a response element in the promoter region that is regulated by a protein (transcription factor), which in turn is regulated by the hormone.
Intracellular messengers	Intracellular messengers alter in response to a stimulus causing a coordinated response within the cell. Ca^{2+} and inositol trisphosphate (IP_3) are examples. IP_3 is produced when phospholipase C is activated and releases Ca^{2+} from intracellular stores. Intracellular messengers frequently activate protein kinases which phosphorylate other proteins thereby producing the cellular response.
Related topics	Features of growth and development (J1) Biochemistry of growth regulation (J2) Tropisms (I2)

How do plant hormones act?	Plant hormones influence both long-term processes like growth and development and short-term responses like the closure of stomata or curvature in unilateral light or gravity. These effects may involve altered gene expression and altered activity of cellular components like enzymes or the cytoskeleton. To do this, there must be a cellular component called a **receptor** that alters in function or properties in response to the presence of the hormone. Between the receptor and point of action there may be an inter-linked series of chemical or ionic signals, termed **intracellular messengers**, within the cell that transmits and amplifies the initial signal. Plant hormones act in a variety of different

ways, with the same hormone having several different effects using several different mechanisms in the same plant, and even in the same tissue. *Figure 1* illustrates some of the potential pathways involved, while *Table 1* details some examples of plant hormone actions and their mechanisms.

Receptors and target tissues

A **receptor** is a protein which binds the hormone. Binding is reversible and with high affinity and specificity. In other words, hormone is bound at low concentrations; similar compounds not effective as hormones are not bound and the hormone is readily released, 'switching off' the response. Binding the hormone produces a change in the receptor that results in it being able to activate other processes. Receptors may be located in membranes or may be in the cytoplasm.

Only a few plant hormone receptors have been identified with certainty. **ABP1** is a soluble **auxin-binding protein** that normally resides in the lumen of the ER. It binds auxin reversibly, with high specificity and affinity and is a strong candidate as an **auxin receptor**. Recently, a protein called 'transport inhibitor response protein 1' has been shown to bind auxin and control the degradation of a family of auxin-regulated proteins. An **ethylene receptor**, **ETR1**, has been cloned in *Arabidopsis* and subsequently in other plants, which has a high affinity and specificity for ethylene. The receptor is a **protein kinase** containing a copper atom as part of the ethylene binding site. It passes on the signal by phosphorylating other proteins in a **signal transduction chain** (see below).

It is evident that some tissues respond to a hormone while others do not. Such hormone-responsive tissues are known as target tissues because they contain the receptors and signal transduction machinery necessary to respond in a particular way.

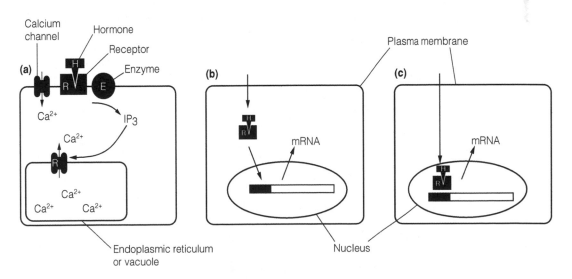

Fig. 1.   *Pathways for hormone action in plants. In pathway* (a) *the hormone (H) interacts with a receptor protein (R) in the pm. This activates an ion channel, giving an influx of Ca²⁺, or activates enzymes in the pm producing intracellular messengers like inositol trisphosphate (IP₃). This initiates subsequent events in the cell. In* (b) *and* (c) *the hormone (H) influences gene expression, either by interaction with a cytoplasmic receptor (pathway* (b) *or a nucleoplasmic receptor (pathway* (c)).

Table 1.   Examples of evidence for mechanisms of hormone action

Hormone	Effect	Mechanism
Auxin	Cell elongation	Activates the plasma membrane proton pump. Evidence for a soluble auxin receptor protein (ABP) that may result in activation of the pump and increased expression of the *ABP* gene
Auxin	Altered protein synthesis	A wide range of auxin-responsive genes have been identified in many species
Abscisic acid	Stomatal closure	PM ABA receptor; regulation of ion channels, and altered cell turgor; mechanism involves second messengers including Ca^{2+} and IP_3
Abscisic acid	Dormancy and stress	ABA-response elements which are transcription factors have been identified which regulate gene expression; a GA-activated, ABA-repressed gene has been identified in seeds which regulates one of the enzymes (α-amylase) involved in germination
Ethylene	Abscission and ripening	A number of ethylene-responsive genes have been identified and a putative ethylene receptor protein identified
Cytokinins	Delay of senescence; chloroplast maturation; development	A possible cytokinin receptor has been identified and cytokinins have been shown to alter mRNA abundance for key proteins, possibly post-transcription. Cytokinin action may also involve second messengers
Gibberellins	Seed germination; development	Strong evidence for second-messenger involvement. Gibberellin causes altered gene expression (e.g. of (α-amylase in germinating seeds). Gibberellin-specific gene promoters and a transcription factor stimulated by gibberellin have been identified

Hormones and the control of gene expression

Altered gene expression is an essential component of many hormone-regulated processes. Altered gene expression results in the production of new proteins involved in the changes in cell function that hormones regulate (*Fig. 2*). This may occur in addition to hormone action via intracellular messengers (see

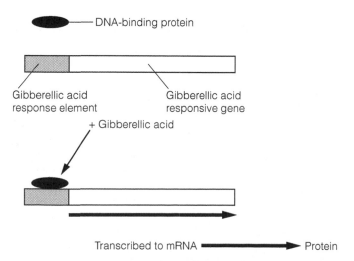

Fig. 2.   Model of regulation of a gene by a hormone (GA). The grey hatching is the gibberellic acid response element (GARE), regulated by a GA-responsive DNA-binding protein (⬬).

below) or by directly modulating cell processes by interaction with proteins already present. In this section, two examples will be given of gene regulation by plant hormones: **auxin and elongation growth** and **ethylene and fruit ripening**. Hormone responsive genes have identifiable sequences in the **promoter region** (Topic C5) termed **response elements** that interact with **protein modulators** (**transcription factors**) which in turn are regulated by the hormone.

Auxin and elongation growth

The mechanism of the stimulation of **elongation growth** by auxin has been the subject of scientific controversy for many years and is still not fully understood for all plants. When oat coleoptiles are exposed to auxin, their rate of elongation increases as a result of **cell elongation** (Topic C5). The effect of auxin can be mimicked by weak acid, and it can be shown that when auxin is added to the coleoptile, the cell wall space becomes acidified. This acidification results from increased pumping of protons (H^+) into the cell wall by the plasma membrane **proton ATPase** (**proton pump**; Topic E3). The **Acid Growth Theory** states that the following stages occur:

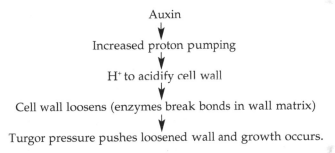

Proteins (**expansins**) present in the cell wall catalyze pH-dependent wall extension growth, by loosening bonds between components of the wall matrix. How does auxin increase the acidification of the cell wall? Two possibilities exist; the first is the activation of proton ATPase already present at the plasma membrane; the second is that auxin causes an increase in the amount of proton ATPase present by up-regulating gene expression. In addition to the effect of auxin on the proton pump, other **auxin-responsive mRNAs** have been found, some of which also increase during elongation. Auxin regulation of gene expression involves the breakdown of proteins that inhibit (repress) the transcription of auxin-responsive genes; *Table 2* summarizes some of these and their possible functions.

Many auxin-responsive genes are inactive (repressed) in the absence of auxin, due to the presence of transcriptional repressor proteins called Aux/IAA proteins. When auxin levels increase, a system for degrading proteins called the ubiquitin-proteasome pathway is activated and promotes the degradation of the Aux/IAA proteins. Expression of the auxin-regulated genes increases as their repressors are removed.

Ethylene and fruit ripening

Tomato fruit ripening follows a well defined series of events typical of climacteric fruits (fruits which show a burst of respiration during ripening; see *Fig. 3*). Ripening involves softening of the cell walls, increase in sugar content and

Table 2. Examples of auxin-responsive mRNAs

Messenger RNAs identified that are up- or down-regulated by auxin	Time take to respond	Notes
Aux/IAA family	15 min–1 h	A family of proteins, many of which are abundant in elongating tissue. Range in size from 20 to 35 kDa; present in low amounts, several targeted to the nucleus
GST family	15 min–3 h	A family of enzymes which conjugate glutathione to many substrates. Believed to be involved in detoxification of xenobiotics and cytokinins
SAUR family	2.5–5 min	Abundant in elongating coleoptiles incubated in the presence of auxin
ACS family	20 min–20 h	ACC-synthases, which are an essential component of the ethylene biosynthetic pathway
Others	30 min–48 h	A wide range of other mRNAs are either increased or decreased by auxin

color changes which ultimately result in an attractive red fruit. Studies of the proteins involved in ripening reveal that a well-regulated series of changes in gene expression occurs. Ethylene synthesis increases rapidly before the climacteric, as genes for the enzyme ACC oxidase (Topic J2) are expressed, and declines after it. The ethylene signal is then sensed by the ethylene receptor (ETR1) and transcription of a range of mRNAs for proteins involved in ripening is enhanced. The level of transcription of the gene for ETR1 also increases greatly during ripening. These ethylene-regulated genes encode mRNAs for cell wall softening and pigment synthesis. One well studied gene, which is transcribed in response to ethylene, encodes the wall-loosening enzyme polygalacturonase (PG). Tomatoes in which expression of PG has been reduced by antisense technology show enhanced storage as they ripen more slowly than conventional tomatoes. In this technique a single strand of DNA is present that binds specifically with the PG mRNA strand, thereby inactivating it. Such fruit also have better qualities for processing and suffer fewer losses in harvest and transport to the consumer or processing plant.

Intracellular messengers

Hormones generally convey signals from cell to cell and tissue to tissue. **Intracellular messengers** carry out signaling within cells. They may respond to a signal from a hormone first perceived at the plasma membrane, or they may respond to a stimulus (like light or temperature) first perceived inside the cell. A number of intracellular signaling molecules have been found in plants; two, Ca^{2+} and the **inositol trisphosphate (IP$_3$)** pathway, will be described here.

Calcium as an intracellular messenger
Cells maintain a very low concentration of Ca^{2+} in the cytoplasm (< 1 µM), by pumping it out across the plasma membrane to the apoplast using Ca^{2+} ATPases. They also pump Ca^{2+} into intracellular stores like the vacuole and

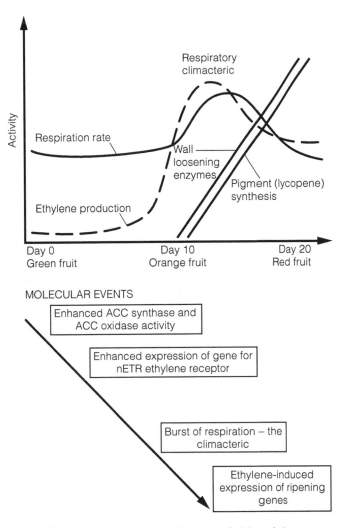

Fig. 3. Regulation of the ripening of a tomato fruit by ethylene.

endoplasmic reticulum. This results in a very steep Ca^{2+} gradient across several cell membranes. When Ca channels (Topic E3) in these membranes open, Ca^{2+} floods into the cytoplasm, giving a Ca^{2+} 'wave'. Channel opening may be regulated by a hormone receptor or by some other stimulus. This amplifies the signal (one molecule of hormone bound can keep a channel open long enough to permit tens of Ca^{2+} ions to enter) and can integrate (coordinate) several signals. The Ca^{2+} wave is then perceived by receptor proteins. The best described of these is calmodulin – a protein with four Ca^{2+}-binding sites. When calmodulin binds Ca^{2+}, it changes conformation and activates a range of proteins, including Ca^{2+}-calmodulin-dependent protein kinases (CaM kinases). Plants also have a range of Ca^{2+}-dependent (but calmodulin-independent) protein kinases (CDPKs). Protein kinases are important in signal transduction pathways as they phosphorylate other proteins at specific sites, altering their activity.

Inositol trisphosphate (IP$_3$)

The **IP$_3$** pathway begins with the conversion of a plasma membrane lipid, **phosphatidyl inositol**, into **phosphatidyl inositol bisphosphate (PIP$_2$)** by **kinase** enzymes in the plasma membrane. PIP$_2$ is then hydrolyzed by **phospholipase C (PLC)** to give **IP$_3$** and **diacylglycerol (DAG)**. PLC activity is regulated by a signal transduction pathway initiated by the binding of a hormone to a receptor, so IP$_3$ and DAG levels respond to stimuli. IP$_3$ causes Ca^{2+} channels to open at the tonoplast (vacuole) and endoplasmic reticulum – resulting in a Ca^{2+} signal being initiated (see above). DAG activates **protein kinase C** that regulates other processes. *Figure 4* summarizes the Ca^{2+} and inositol trisphosphate signaling pathways.

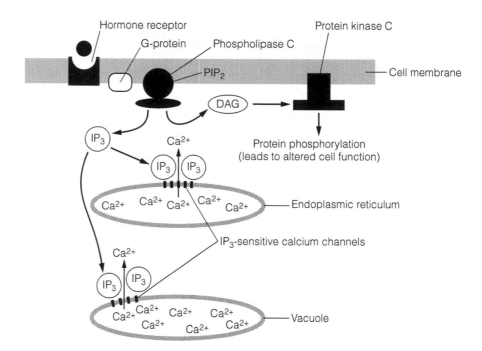

Fig. 4. *Intracellular messengers. Release of Ca2+ from the ER and vacuole results in a transient rise in cytoplasmic Ca^{2+} concentration. The Ca^{2+} binds with proteins like calmodulin and Ca^{2+}-dependent protein kinases (CDPKs) to alter cell function.*

J4 PHYSIOLOGY OF FLORAL INITIATION AND DEVELOPMENT

Key Notes

Floral meristems

Flowers originate from the shoot meristem. The change in the vegetative meristem to develop a flower is termed induction. This is followed by evocation, the development of the floral meristem. The flower then forms and becomes functional when the reproductive structures are mature. Flowers are formed in concentric whorls of sepals, petals, stamens and carpels.

Floral evocation

Floral evocation usually requires an external stimulus such as cold (vernalization) sensed by the meristem, or appropriate day length sensed by phytochrome, species being long-day, short-day or day-neutral. Hormones are important in the control of flowering.

Floral development genes

Heterochrony (flowering-time) genes regulate the conversion of the vegetative meristem to a floral meristem. Floral meristem identity genes then regulate the formation of a flower. When flowering has been initiated, cadastral genes govern the formation of the whorls. Finally, homeotic genes control the structure of the flower, influenced by the cadastral genes. The ABC model of flower development predicts that the four whorls of the flower are controlled by three homeotic genes A, B and C. Mutations of these genes form flowers in which the organs are misplaced.

Related topics

Methods in experimental
  plant science (B2)
Features of growth
  and development (J1)

Ecology of flowering and
  pollination (G5)
Self incompatibility (G4)

Floral meristems

Flowers originate from the shoot meristem (Topic D1) which normally generates leaves and shoots. The meristem stops vegetative growth when flowering commences and either produces a single flower (**determinate** or **closed inflorescence**) or a succession of floral meristems, each of which will become a flower (**indeterminate** or **open inflorescence**). Like a shoot meristem, a floral meristem is divided into layers: the **tunica** producing the outer cell layers, and the **corpus** the inner cell layers (Topic D3). The first stage of the flowering process is termed **induction**, the change in form and function of the vegetative meristem to develop a flower. This is followed by **evocation**, the development of the floral meristem. The flower itself is then formed and becomes functional when the reproductive structures are mature. The basic structure of an arabidopsis flower is shown in Topic G1.

Floral evocation

Forming a flower requires the differentiation (Topic J1) of the vegetative shoot meristem into a floral meristem when it has reached an appropriate stage of development. Flowering usually requires one of a range of external stimuli that bring about floral evocation. In temperate plants, this may be **vernalization**, a chilling period preceding flowering, possibly by weeks or months, or day length, species being classified as long-day, short-day or day-neutral species (Topic I1). Vernalization appears to be sensed by the meristem itself, as chilling the plant while the meristem is warmed does not induce flowering. Day length is sensed by **phytochrome** (Topic I1) in the young leaves, suggesting that a **hormone** is involved in transmitting the signal to the meristem. This flowering hormone was originally named florigen, but its existence has never been proven.

Floral development genes

Genes for many aspects of the process of floral development have been described in two species, **arabidopsis** and *Antirrhinum majus*. It will be helpful to have read Topic B2 before studying this section. The arabidopsis flower is actinomorphic (Topic Q5) and mainly self-pollinated, while the *Antirrhinum* flower is zygomorphic (Topic Q5) and insect-pollinated.

The earliest stages of flower formation involve the activity of genes known as **heterochrony** or **flowering-time** genes which regulate the conversion of the vegetative meristem to a floral meristem. Once this has happened, **flower meristem identity genes** regulate the formation of the flower. The arabidopsis mutant known as *leafy* (*lfy*), for instance, that has a mutation in a flower meristem identity gene, forms shoots where flowers should be. Once flowering has been initiated, a third group of genes known as **cadastral genes** are initiated which govern the formation of the whorls of the flower. Finally, the structure of the flower is governed by **homeotic genes**, which cause the right structure to appear in the right place. The function of homeotic genes is influenced by the cadastral genes expressed before them. The **ABC model** of flower development predicts that the four whorls of the flower are controlled by the action of three genes A, B and C. By studying floral mutants affecting each of these genes (*Table 1*), the way in which they control development has been established.

Each whorl is specified by the activity of one or two of the three homeotic genes, A, B and C, where: A alone → sepals; A and B → petals; B and C → stamens; C alone → carpels. This will occur regardless of where A, B and C are active in the flower; so a mutant, where B is inactive, makes two whorls of sepals and no petals (*Table 1*). A and C inhibit each other: if A is inactive, C becomes more active and *vice versa*, so a plant without C will form petals in whorls 2 and 3 and sepals in whorls 1 and 4. Sepals form in whorl 4, because the action of the A gene in this whorl was being inhibited by the activity of the C gene.

Table 1. Mutants of arabidopsis and Antirrhinum *and the ABC model for formation of the structures of the flower*

Genotype	Gene function	Whorl 1	Whorl 2	Whorl 3	Whorl 4
Wildtype		Sepals	Petals	Stamens	Carpels
apetala 2 squamosa	A	Carpels	Stamens	Stamens	Carpels
apetala 3, pistillata deficiens	B	Sepals	Sepals	Carpels	Carpels
agamous plena	C	Sepals	Petals	Petals	Sepals

The arabidopsis mutants described are: *ap2* (*apetala 2*), *ap3* (*apetala 3*); *pi* (*pistillata*) and *ag* (*agamous*); the *Antirrhinum* mutants are *squa* (*squamosa*), *def* (*deficiens*) and *ple* (*plena*).

K1 PLANT BREEDING

Key Notes

Historical perspective	Modern crops are the result of intensive plant breeding and have many characteristics different from their wild ancestors. The process of plant breeding has resulted in increased yield and removal of undesirable characteristics, but with this is the risk of increased disease susceptibility of clonal populations and loss of biodiversity.
Plant breeding methods	Selection has resulted in recognizable varieties or landraces of crops. These varieties are heterozygous for crops that cross-pollinate, but homozygous for those (like wheat) which do not. Cross hybridization gives hybrid vigor. Cross-pollination between lines can be achieved by removing the anthers from one line and pollinating with a second line. Recently male sterile lines that do not make pollen have been produced. Desirable characteristics are introduced into plants by cross-fertilizing lines to produce hybrids. Back crossing allows a desirable trait to be introduced into an existing useful line.
Limits to conventional plant breeding	Plant breeding has been very successful in generating high-yielding crops. However, as well as being time consuming and labor intensive, it is also limited by natural barriers of pollination between species.
Related topics	Breeding systems (G3) Self incompatibility (G4) Plant genetic engineering (K3)

Historical perspective

All the major crops world-wide are the result of a repetitive process of breeding, selection and further breeding to alter characteristics and improve yield. Alongside this process have come great increases in human population as agricultural productivity has attempted to meet the needs of that population. For some crops, the process has been carried out only recently (e.g. for oil-seed rape, Canola) while for others it has been going on for many thousands of years and in many civilizations. Wheat was domesticated first in the near-East. Bread wheat appears to have resulted from crossing primitive einkorn wheat with goatgrass to generate emmer wheat, which was crossed again with goatgrass to yield bread wheat. Corn (maize) was domesticated in Mesoamerica. One of its closest relatives is teosinte, which produces small, corn-like seed ears with hard outer husks. Rice was domesticated in Indo-China from a wild rice, *Oryza rufipogon*. Apart from yield, domestication has altered many characteristics of crop plants, including loss of dormancy from the seed and loss of dispersal mechanisms. There are also some negative effects of domestication; for instance, domesticated varieties are frequently demanding of nutrients and soil conditions. Intensive breeding programs resulting in near clonal crops may result in loss of biodiversity and susceptibility to disease.

Plant breeding methods

Selecting plants for desirable traits resulted in the development of identifiable **varieties** or landraces of crops, each with slightly different characteristics. As some species can cross-fertilize, their landraces tend to be heterozygous (genetically mixed) while those that do not normally cross-fertilize (like wheat) will form genetically pure (homozygous) lines.

Selection breeding is the process of choosing plants showing desirable characteristics and generating seed from them. It is straightforward if the crop is self-pollinated but produces increasingly inbred lines and may result in loss of yield. Cross-pollinators can also be bred in this way, though creation of a homozygous line will not be possible. Loss of yield in inbred lines can be overcome by deliberate cross hybridization to achieve what is known as hybrid vigor. This is achieved by removing the anthers from one line and planting it adjacent to a second line with anthers. All the pollination of the first line will then be by the second line, and the seeds produced by that line will be hybrids of the two lines. Some lines are naturally male sterile (i.e. do not produce viable pollen). These plants are very valuable in plant breeding, as the anthers do not have to be removed by hand. More recently, genetic engineering (Topic K3) has been used to create male sterile lines, by linking ribonuclease gene expression with a promoter sequence (Topic B2) controlling an anther-specific gene. Ribonuclease produced in the anthers degrades the messenger RNA (mRNA) they produce. This results in plants that do not make pollen.

Desirable characteristics are introduced into plants by **cross breeding**. Lines are cross-fertilized with others (or with wild ancestors or related species) to produce hybrids containing a mixture of characteristics from both parents, some useful and some not. To eliminate the undesirable characteristics and develop useful ones, **back crossing** is carried out. In this process, the progeny of a cross (for instance between a high-yielding strain and a disease-resistant but low yielding strain; *Table 1*) are repeatedly hybridized with the high-yielding line.

Limits to conventional plant breeding

Conventional plant breeding has been very successful in generating the varieties of high-yield crops we use today. It is time consuming and labor intensive. It is also limited by the natural **pollination barriers** between species that mean desirable traits cannot be easily introduced from one species to another. **Artificial mutagenesis** (by X-rays or chemicals) has been used to generate new characteristics, together with **tissue culture** techniques (Topic K2). However, these have largely been replaced by **genetic manipulation** (Topic K3). This technique allows single characteristics to be transferred into a crop in a far more controlled and specific manner than was previously possible.

Table 1. Back crossing

Season	Action
1	Select two lines (one high-yielding, one disease-resistant); cross; collect seed
2	Grow seed from (1) and select plants that have good disease resistance and the best yield; cross with high-yielding line and keep seed
3	Grow seed from (2) and select plants that have good disease resistance and the best yield; cross with high-yielding line and keep seed
4–	Keep repeating stages 2–3 until a stable new line with acceptable yield and good disease resistance is produced (this may take eight or more repeats)
	Produce sufficient seed for large-scale trials and agricultural production

K2 PLANT CELL AND TISSUE CULTURE

Key Notes

Types of cell and tissue culture

The major types of cell and tissue culture are: organ culture, embryo culture, tissue culture (production of callus and subsequent regeneration of organs or plantlets from excised pieces of tissue) and suspension culture (single cells or cell clumps in liquid media).

Methods, media and equipment

Plant cell and tissue cultures must be initiated and maintained in a sterile environment. Growth media include a carbon source (sucrose), macro- and micronutrients, auxin and cytokinin, vitamins, water and, for solid media, a solidifying agent like agar.

Suspension cultures

Suspension cultures are started by breaking callus into liquid medium in conical flasks, which are agitated in an orbital incubator. Cells show logarithmic growth initially, then cell division ceases and growth rate declines. Subculturing to new medium is then required. Suspension cultures maintained for long periods may alter in properties due to mutation and altered gene expression.

Differentiation and embryogenesis

Differentiation to form roots, shoots and plantlets can be induced by selecting an appropriate auxin:cytokinin ratio for the medium. Suspension culture cells will also form embryoids (somatic embryos) in appropriate conditions.

Commercial and industrial applications

The major applications of plant cell and tissue culture are micropropagation, the production of high value products by cell cultures and as part of the genetic manipulation of plants.

Related topics

Biochemistry of growth regulation (J2)

Molecular action of hormones and intracellular messengers (J3)

Plant genetic engineering (K3)

Types of cell and tissue culture

Plant cell and tissue culture has been undertaken in various forms since the 1930s, although the techniques derive from the much older techniques of **plant propagation**, such as taking cuttings, which have been carried out by horticulturists for centuries. The techniques are based on the fact that plants show **plasticity of development**, the ability to change developmental path if suitably triggered. Thus, a stem section can regenerate roots and leaves to achieve a complete, functional plant and single cells can be caused to form an entire new embryo. Plant tissue culture requires sterile conditions as tissue growth is slow and material in culture vulnerable to fungal and bacterial infection. The major types of sterile plant culture are:

- **organ culture** in which an organ (flower bud, immature fruit) is grown isolated from the parent plant;
- **embryo culture** in which an isolated or immature embryo is grown;
- **tissue culture** in which cell material isolated from a parent plant is grown to form callus (an undifferentiated mass of cells) or to regenerate organs or into a whole plant;
- **suspension cultures** in which isolated cells or small clumps of cells are grown in a liquid medium.

Protoplasts, cells from which the cell wall has been removed by enzymic digestion, are frequently used in cell culture to separate single cells from cell clumps.

Methods, media and equipment

Cultures are prepared in a near-sterile environment, usually a **laminar flow cabinet,** in which a draught of sterile air blows constantly across the working surface, using equipment and media sterilized by heat or in an **autoclave.**

A wide range of **solid** and **liquid media** have been developed for different applications. The media have to contain certain key components (*Table 1*). Prior

Table 1.   Example of components of media for cell culture

Component	Function
Carbon source and osmotic balance (e.g. sucrose, 20 g l^{-1})	Energy supply for growth; in early stages at least, cultures will not photosynthesize; iso-osmotic medium required
Solid medium (agar, 7 g l^{-1})	Used in solid cultures to solidify medium

Inorganic salts, macronutrients
  Ammonium sulphate, 790 mg l^{-1}
  Calcium nitrate, 290 mg l^{-1}
  Magnesium sulfate, 730 mg l^{-1}
  Potassium chloride, 910 mg l^{-1}
  Potassium nitrate, 80 mg l^{-1}
  Sodium nitrate, 1800 mg l^{-1}
  Sodium sulfate, 450 mg l^{-1}
  Sodium dihydrogen phosphate, 320 mg l^{-1}
Inorganic salts, micronutrients
  Boric acid, 1.5 mg l^{-1}
  Copper sulfate, 0.02 mg l^{-1}
  Manganous chloride, 6.0 mg l^{-1}
  Potassium iodide, 0.75 mg l^{-1}
  Zinc sulfate, 2.6 mg l^{-1}
  Molybdic acid, 0.017 mg l^{-1}
	Supply essential micro- and macronutrients; achieve pH balance (medium brought to pH 5.5)

Vitamins, lipids and essential amino acids
  Meso-inositol, 100 mg l^{-1}
  Glycine, 3.0 mg l^{-1}
  Thiamine hydrochloride, 0.1 mg l^{-1}
  Pyridoxine hydrochloride, 0.1 mg l^{-1}
  Nicotinic acid, 0.5 mg l^{-1}
	Compounds essential for growth which cannot be synthesized by the cultures

Hormones
  2,4-D (synthetic auxin), 0.15 mg l^{-1}
  Kinetin, 0.15 mg l^{-1}
	Stimulate growth and division; altering the ratio of auxin to cytokinin influences development

to use they are autoclaved and handled under sterile conditions using sterile vessels or petri dishes.

In order to initiate a solid cell or tissue culture, a sterile fragment (**explant**) of plant material must be obtained and placed onto growth medium. Initially, an undifferentiated mass of cells termed a **callus** will be formed. The following list summarizes the stages involved:

(i)   select a healthy plant and remove a segment of tissue in a laminar flow hood;
(ii) surface sterilize it using diluted sodium hypochlorite;
(iii) excise a segment of tissue using sterile instruments and resterilize the tissue;
(iv) place the segment onto solid medium containing hormones and nutrients to induce cell proliferation in a petri dish;
(v)   seal and incubate at 22°C;
(vi) regularly observe, discard all contaminated dishes;
(vii) when callus growth sufficient, remove to fresh medium to intiate differentiation.

Frequently, a range of **auxin:cytokinin ratios** is tested to optimize growth and development. Several transfers to fresh medium may be required if the formation of plantlets is desired. The whole process may take 8–12 weeks or more to complete.

Suspension cultures

Suspension cultures are initiated by breaking cells free from a callus by gentle agitation. They consist of single cells or cell clumps suspended in aerated liquid medium. The simplest suspension culture system is a conical flask gently shaken on the platform of an orbital incubator. The swirling motion keeps the cells in suspension and oxygenates the medium. Cells in suspension culture show **logarithmic growth**, with rapid increase in fresh weight, dry weight and DNA content (indicating cell division) over the first few days. This is followed by a decline in cell division and subsequently a decrease in growth rate (*Fig. 1*). If the cells are not given fresh medium, they will then die; however, if a small aliquot is removed and **subcultured** into fresh medium the entire process can be repeated indefinitely. When the aim of the process is to harvest secondary products such as the antimicrobial dye shikonin, optimum production often occurs late in the growth cycle, when cell death is beginning to occur. The medium for maintaining cells in culture is optimized for each species and cell type and will include auxin and cytokinin to stimulate cell division.

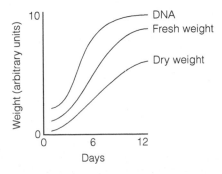

Fig. 1. Growth parameters of a suspension culture.

After prolonged culture, the cells may alter in their characteristics as a result of mutation or the activation of previously inactive genes. After repeated subculturing, for instance, genes for cytokinin biosynthesis may be activated and the cultures lose their requirement for cytokinin in the culture medium. This process is known as **habituation**.

Scaling up suspension cultures into large culture vessels is difficult as: (i) aseptic conditions must be maintained; (ii) constant aeration is required; (iii) cells are easily damaged by stirrers and changes in pressure; (iv) cells require constant agitation. These problems have been overcome in a number of designs; an example, based on using rising sterile air to agitate the cells, is shown in *Fig. 2*.

Differentiation and embryogenesis

Many experiments in cell culture are carried out with the aim of regenerating an entire plant. This is particularly important for genetic modification, where a novel gene may be inserted into a cell suspension, a protoplast or callus, and whole plants need to be produced.

Cells divide randomly to form an undifferentiated mass, known as a callus. If the auxin:cytokinin ratio of the medium is varied, the callus can be induced to differentiate to form roots and buds (Topic J1). By subculturing, intact plants can be regenerated by this method. Callus cultures may also form **embryoids** (**somatic embryos**, i.e. embryos formed in culture, as opposed to **zygotic embryos** formed sexually in a plant) on appropriate media. Cultures able to generate embryos are termed **embryogenic**.

Commercial and industrial applications

Micropropagation is the use of plant tissue culture to regenerate large numbers of plants. The technique results in genetically identical plants and is therefore **clonal propagation**. It is commonly used to produce disease-free plants, and commercially for many species including trees, potato and orchids, and as part of the procedure to genetically manipulate crops. **Somatic embryos** can be encapsulated in hydrated gel to produce a 'synthetic seed' or **propagule**.

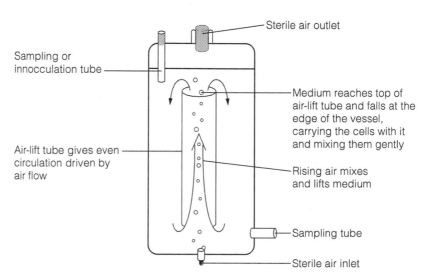

Fig. 2. An industrial-scale vessel for plant cell culture. The interior of the entire vessel is kept sterile and all the inlet and outlet ports are sealed with microparticulate filters which would prevent bacteria and fungal spores from entering.

Production of secondary products in suspension culture

Many plant secondary products (Topic F5) are difficult to produce by organic synthesis. Considerable effort has gone into using plant suspension cultures to make them. Details of important plant secondary products are presented in Topic F5; they are frequently produced by highly specialized cells. Consequently, high yields in cell culture can only be obtained if the cultures express the genes required. Frequently, this only occurs when the cells are at the end of the culture cycle. An example of a compound successfully produced in culture is the antimicrobial dye **shikonin**, produced by cultures of *Coptis japonica*.

Plant **genetic manipulation** is a widespread application of plant cell and tissue culture techniques today and without it, the commercialization of genetically modified crops would not be possible. Methods of genetic manipulation are explained in section K3.

K3 PLANT GENETIC ENGINEERING

Key Notes

The concept

Genetic manipulation involves inserting foreign genes or modifying the activity of existing genes. Methods to insert foreign genes are coupled with the methods of plant tissue culture to regenerate identical populations of plants with novel characteristics.

Basic genetic manipulation methods

Agrobacterium tumefaciens is a soil bacterium with a plasmid that inserts foreign DNA into a plant. The plasmid contains a T-DNA transferred into the plant and a VIR region that facilitates transfer of the T-DNA. Binary vectors for genetic engineering consist of one plasmid containing the VIR region and a second containing the T-DNA including the foreign DNA. Where the *Agrobacterium* system cannot be used, direct gene transfer techniques may be employed, for instance using a DNA particle gun.

Possibilities of genetic manipulation

The aims of genetic manipulation are to enhance agriculture by modifying crop plants, to minimize inputs and losses and to maximize yields and value. To date, crops have been engineered for herbicide tolerance, insect and virus resistance and post-harvest quality. In the future, a much broader spectrum of improvements of wider benefit is proposed.

Risks of genetic manipulation

Risks identified include: environmental, such as cross-pollination with native species, gene transfer; food safety (mainly the transfer of antibiotic resistance to bacteria, allergies and toxicity); and socio-economic – food supply in the hands of few multinational companies.

Related topics

Plants as food (N1)
Plant breeding (K1)

Plant cell and tissue culture (K2)

The concept

Genetic engineering (recombinant DNA technology) involves inserting foreign genes or modifying the activity of existing genes. A soil bacterium, *Agrobacterium tumefaciens*, naturally inserts its own bacterial genes into plant genomes. The result is a crown gall, a swelling of the stem at soil level caused by over-production of auxins and cytokinins produced by enzymes encoded by genes transferred from the bacterial genome. The regeneration of entire plants from single cells or explants (Topic K2) has been carried out for many decades, to produce clonal populations of plants. Together, the two provide genetic engineers with the tools for the insertion of genes from another organism into a plant and the regeneration of a clonal population of that plant. All the members of that population will express the foreign gene.

Basic manipulation methods

Agrobacterium tumefaciens contains a **plasmid**, a circular piece of DNA separate from the bacterial chromosome, known as the **Ti plasmid** (see *Fig. 1*). This plasmid contains genes which will be randomly inserted into the plant genome

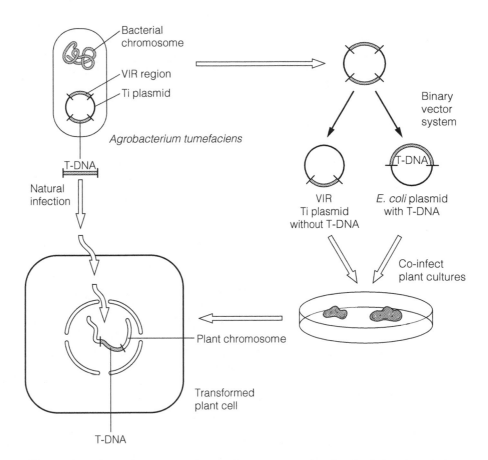

Fig. 1. Inserting a foreign gene into a plant genome using Agrobacterium tumefaciens. Agrobacterium tumefaciens *contains a Ti plasmid containing a VIR (virulence region for infection) and a T-DNA region (which is transferred to the plant genome). Vectors based on the Ti plasmid contain modifications of the DNA in the plasmid. A binary vector system is shown here, in which two plasmids have been created, one a Ti plasmid without the T-DNA region, and the second an* E. coli *plasmid with a T-DNA modified to contain genes to be inserted into the plant cell. In nature, transformation results in a swelling or gall at the site of infection.*

(**transferred** or **T-DNA**) and genes involved in the transfer of the DNA (the **VIR** or **virulence region**). Normally, the T-DNA region contains genes for auxin and cytokinin biosynthesis and for amino acid and sugar derivative production. For genetic manipulation, the Ti plasmid is modified by the removal of the genes within the T-DNA region using restriction enzymes that cut DNA at specific nucleotide sequences. The action of the restriction enzymes leaves a linear strand of DNA with 'sticky' ends where the nucleotides are unpaired and therefore able to join to a complementary nucleotide sequence in another strand of DNA.

A gene of interest identified in another organism (animal, plant or bacterium) is identified and prepared for insertion into the Ti plasmid. It is cut out of its host with the same restriction enzymes used to prepare the Ti plasmid, again leaving sticky ends. This cut DNA is then mixed with the modified plasmid DNA, the two DNA strands join at their sticky ends and are sealed together by the enzyme, DNA ligase, to form a recombinant plasmid (*Fig. 2*). The gene

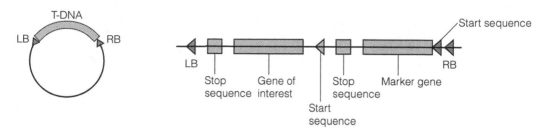

Fig. 2.   A typical modified T-DNA in a Ti plasmid. The two ends of the T-DNA, known as the left border (LB) and right border (RB) remain intact as these contain repetitive DNA sequences which are important in pasting the T-DNA into the plant genome. In between, bacterial DNA sequence has been removed and replaced with a construct of the gene of interest, transcription start and stop sequences and a marker gene (for instance antibiotic resistance).

inserted must contain a promoter (Topic B2) which will allow it to be expressed in the plant.

Frequently, a second gene will also be inserted into the plasmid, in addition to the gene of interest. This is a **selectable marker gene** that will give the plant antibiotic resistance or herbicide tolerance. This gene will also have been cloned from another organism, usually a bacterium. Any plant material now expressing these genes will show the properties the marker gene confers – herbicide tolerance or antibiotic resistance – and will grow in media containing either the herbicide or antibiotic when nontransformed material cannot. In this way they can be used to select transformed plants from nontransformed ones.

Plant material (callus; Topic K2; leaf discs, suspension cultures or organs) is then infected by incubating the cells with *Agrobacterium* containing the plasmid and grown on agar plates containing antibiotic. Only material containing the gene of interest together with the antibiotic resistance marker then grows. Clonal populations of transformed plants can then be produced by micropropagation (see *Fig. 3* and Topic K2).

Commonly, a **binary vector system** (*Fig. 1*) is used to transform plants. This system has the VIR region in a Ti plasmid modified by the removal of the T-DNA, while the T-DNA is in a second Ti plasmid. The plant is then transformed using the engineered *Agrobacterium* containing both plasmids – one including the VIR region and the other the T-DNA.

DNA constructs can be introduced directly into plant tissue directly using a DNA particle gun (*Fig. 3*). DNA is coated onto tungsten particles and fired at the specimen. The projectile is stopped by a barrier with a fine hole through which the tungsten particles carrying DNA can travel on at high speed. They then penetrate the cell and some of the DNA enters the nucleus, causing transformation.

For experimental purposes, transient (i.e. short-term) expression of a foreign gene may be studied. In this case, the foreign gene does not have to integrate stably into the genome. Cell biologists use this technique to observe the expression of genes introduced into plant organs like leaves or into protoplasts, without having to go through all the stages required to generate a transformed plant.

Possibilities of genetic manipulation

Production of genetically engineered crops has begun on a large scale. *Table 1* presents examples of uses of the technology, while *Table 2* presents some future prospects. The ability to transfer one or two genes into a genome has great

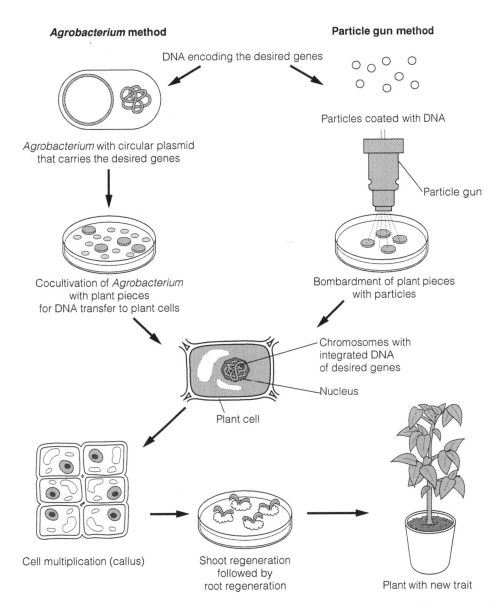

Agrobacterium method

Particle gun method

DNA encoding the desired genes

Agrobacterium with circular plasmid
that carries the desired genes

Particles coated with DNA

Particle gun

Cocultivation of Agrobacterium
with plant pieces
for DNA transfer to plant cells

Bombardment of plant pieces
with particles

Chromosomes with
integrated DNA
of desired genes

Nucleus

Plant cell

Cell multiplication (callus)

Shoot regeneration
followed by
root regeneration

Plant with new trait

Fig. 3.   *Plant material transformed either by an* Agrobacterium-*based system or by DNA from a particle gun is allowed to form callus on agar containing antibiotic, on which only cells containing the selectable marker gene can grow. Plantlets are then regenerated on agar and mature plants expressing the inserted gene are grown.*

potential in many areas of agriculture, for instance to reduce losses from pests and disease. Each situation requires a separate strategy to engineer a successful crop; *Table 1* presents some examples. Achieving the goals of *Table 2* may be more difficult, as some of the characteristics require the insertion or modification of more than one gene. As *Agrobacterium* inserts genes randomly into the genome, each transformant must be assessed separately for the consequences of transformation.

Transformed plants are being developed for the commercial production of valuable products, many of which are of clinical relevance including vaccines

Table 1. Current applications of genetically engineered crops

Goal	Use	Example method	Examples of crops transformed
Herbicide tolerance	Use of herbicides post emergence of seedlings at lower doses than required before seedling emergence	Introduce bacterial gene for enzyme which degrades the herbicide or which bypasses the point of plant metabolism inhibited	Soybean, canola (oil-seed rape), corn, cotton
Insect resistance	Reduce losses without pesticide spraying	Insertion of gene from the bacterium *Bacillus thuringiensis* gives resistance to a range of insect pests	Corn (against European corn borer) Cotton (against boll worm, tobacco budworm, etc.) Potato (against Colorado beetle)
Post-harvest quality	Increasing shelf-life and reduces losses in transport and harvest	Modified activity of polygalacturonase or other ripening enzymes	Tomato
Virus resistance	Reduced losses due to viral diseases	Insertion of viral coat protein gene into plant	Tobacco (tobacco mosaic virus) Potato (potato viruses X and Y)

for HIV/AIDS. The terms 'molecular farming' or 'Pharming' have been coined to describe the use of plants for the production of vaccines, antibodies and other pharmaceuticals for human use. The aim of this development is to grow plants expressing the recombinant protein as a field crop or in controlled and contained growth facilities. It is hoped that use of transformed plants will reduce the use of animals and expensive animal cell culture techniques.

Table 2. Targets for genetically engineered crops

Goal	Application
Salinity tolerance	Increased crop yield in areas affected by salinity (e.g. in long-term irrigation)
Drought tolerance	Increased crop yield in marginal, semi-arid zones
Waterlogging tolerance	Improved survival in temporary flooding
Enhanced flavor, storage and properties	Improved consumer acceptance; decreased losses; decreased energy inputs to processing or storage; enhanced product value or usefulness
Enhanced amino acid content	Dietary improvement and health
Antibody and pharmaceutical production	Less energy input and cost than use of animal cell culture; less use of animals
Improved disease resistance	Reduced pesticide inputs; increased yields mean population can be fed using smaller land area

Risks of genetic manipulation

The risks of plant genetic manipulation may be divided into three classes; **environmental, food safety** and **socio-economic**. Environmental risks include cross-pollination of genetically modified (GM) crops with native species, disruption of the balance of ecosystems and damage to fauna, e.g. to insect populations or natural predators. Food safety risks include the transfer of antibiotic resistance marker genes into other organisms, e.g. bacteria that may enter the human food chain. Some fear potential but as yet unproven long-term health risks from consuming GM materials. Socio-economic risks concern the fact that GM crops are patented and largely in the hands of very few multinational companies; that GM permits seed suppliers to have much greater control over the livelihood of agriculturalists, particularly in less developed nations; and that GM technology may benefit the rich at the expense of the poor. Against these arguments are the needs of a world population increasing by 1 billion approximately every 12 years, the rigorous testing of GM products and the potentially damaging effects of many current agricultural practices. Plant genetic modification is likely to remain controversial for some time as there is considerable public opinion against it, especially in Western nations.

L1 ECOLOGY OF DIFFERENT GROWTH FORMS

Key Notes

Variety of form
Plants can be mechanically independent and woody, i.e. trees or shrubs, or herbaceous. Mechanically dependent plants are climbers, epiphytes or stranglers. There are also parasites and saprophytes.

Ecology of woody plants
Trees dominate many terrestrial habitats where there is sufficient rain and a warm temperature for at least part of the year. These are mainly dicotyledons and conifers. Shrubs dominate heathlands and extend to the tundra as dwarf shrubs.

Herbaceous plants
These dominate the ground layer in many habitats, and grasslands are entirely dominated by them. Many in seasonal environments die down in the dry season or winter and some are short-lived.

Mechanically dependent plants
All these are most common in tropical rainforests and they depend mainly on trees for their support. Climbers include woody and nonwoody forms. Epiphytes include some specialist families and are much less common in seasonal environments with only bryophytes occurring as common epiphytes in temperate climates.

Related topics
Physical factors and plant distribution (L2)

Plant communities (L3)
Parasites and saprophytes (M6)

Variety of form

There is a great variety of growth forms of plants but they can be classified into a few main categories with different ecological characteristics. Most plants grow without any mechanical support and these may be **woody**, forming **trees** or **shrubs**, or **herbaceous** without any woody parts. Trees and woody shrubs range from giants of over 100 m height to many-stemmed shrubs and some undershrubs that remain prostrate on the ground but retain their woody stems. Tree ferns, cycads and most palms have a single trunk with one enormous bud and leaf rosette at its tip, and other trees have a few branches, sometimes dichotomous (Topic P1). A majority of the larger **mechanically independent** woody plants have a single main stem with side branches, though often flattening out to a domed crown in which there is no one dominant stem. Some woody plants, particularly shrubs, have underground rhizomes from which several woody stems grow, forming thickets.

Herbaceous plants are low-growing without lignified stems. Frequently the above-ground stems are ephemeral and many in seasonal regions lose their above-ground parts during the dormant season. Some are short-lived (Topic L4). Many have rhizomes or stolons and spread vegetatively.

Mechanically dependent plants include climbers which can reach the tops of tall trees and may have woody stems. Other climbers persist without much secondary thickening and some in seasonal climates grow like herbaceous plants and die back each dormant season. Epiphytes grow attached to other plants, usually trees, by their roots or stems but not extracting nutrients directly from the supporting tree. These are light-demanding plants that obtain water and nutrients directly from rain and run-off along branches. Stranglers start life as epiphytes but send some roots to the ground to become partially or totally independent. Heterotrophic plants depend on other plants or fungi for all or part of their nutrients (Topics M6, M7).

These growth forms have no direct relationship with plant classification, though some plant families are specialized to one particular form. Some growth forms include plants from widely different families and some plants are intermediate. Some individual species can adopt a different growth form in different conditions.

Ecology of woody plants

Trees dominate most ecosystems where the climate is warm and moist for at least part of the year. They cease to do so where the soil is too thin for their roots; where it is too dry; where there is permafrost in tundra regions; in permanently waterlogged places; places dominated by salt or heavy metals; and in many environments modified by grazing animals or people. They can live for a few years or decades to a few thousand years. The majority of woodland and forest ecosystems are dominated by dicotyledonous angiosperms except for the great coniferous forests of the northern hemisphere. Tree ferns, cycads and monocotyledonous trees may be common but are rarely dominant except occasionally in swamps or in colonizing situations. In many forests the trees form strata. They have a canopy layer, with taller emergent trees projecting through it in some rain forests, and understorey trees and shrubs often with more than one layer. Different species of tree are frequently involved, with the shade-tolerant species forming the understorey and faster growing canopy trees being more light-demanding, though there is overlap, with saplings of larger trees present in the lower layers. At the edges of forests and where there is a gap, pioneer, fast-growing species grow.

Shrubs and smaller woody plants occur at the edges of many forests and dominate ecosystems in the arctic regions and in some semi-deserts where the environment is too harsh for trees. Heathland is dominated by low-growing shrubs mostly less than 1 m high, and some arctic shrubs, such as willows (*Salix* spp.) rise no more than 1 cm above ground level, having all their woody parts below ground.

Herbaceous plants

Many herbaceous plants in seasonal environments die down at the end of each growing season, leaving only their roots and sometimes leaves to survive until the following season. These plants, along with small shrubs, form the ground layer in many habitats, and dominate in savannahs, temperate woodlands and grasslands throughout the world. Some herbaceous plants persist above ground, particularly in less seasonal areas, but these mostly remain less than about 3 m tall. Those that live for less than a year are the quick colonizers of newly opened areas or take advantage of the rare rains in deserts to grow quickly to set seed before unsuitable conditions return.

Grasses are all herbaceous except for the bamboo group and they are particularly well adapted to withstand grazing pressure since they have growing meristems at nodes rather than at the tip, and produce stems above these nodes. They frequently dominate in places with intense grazing pressure, and those that have meristems below the soil surface can withstand fire.

Mechanically dependent plants

Climbers, epiphytes and stranglers all require the presence of other plants, normally trees, for their growth and all are much more abundant in tropical rainforests than elsewhere. Climbers occur throughout woodland habitats with many woody climbers (**lianas**) in **tropical forests**. Many fewer occur outside tropical rainforests but most temperate woodlands have some, e.g. *Clematis* spp. Herbaceous climbers can be small and occur throughout the world, some persisting in grasslands. In tropical rainforests some of the tallest climbers do not produce woody stems, or only a little secondary thickening, e.g. members of the arum family, Araceae.

Epiphytes are dependent on a high rainfall since they only get water from it and run-off along branches. They are abundant in tropical forests. Many ferns and members of certain families, notably the orchids, Orchidaceae, and bromeliads, Bromeliaceae, are specialist epiphytes but many other families are represented. Tropical mountains have a particularly abundant epiphyte flora. Epiphytes are common in parts of the seasonal tropics but become much rarer and less diverse in drier climates. In temperate regions epiphytic flowering plants only occasionally occur, although there are a few ferns. Bryophytes are abundant epiphytes in the wet tropics, even occurring on the leaves of rainforest trees. They are the only common epiphytes in temperate woodlands, along with lichens. In tropical rainforests some epiphytes are shrubby, producing woody stems and some of these develop roots which reach the ground to become independent. The figs, *Ficus* spp., are the most prominent of these **stranglers** and can become self-supporting, killing the tree on which they started life.

The mistletoes (two related families, Loranthaceae and Viscaceae) and a few other plants are epiphytic in growth form but are partial parasites, penetrating their hosts and extracting nutrients and sugars. They are considered with other heterotrophic plants in Topic M6.

L2 PHYSICAL FACTORS AND PLANT DISTRIBUTION

Key Notes

The requirements for plant life	All plants have the same basic requirements for solar energy, water and nutrients. Their distributions are determined by adaptations to withstand environmental stress and their ability to spread, along with biotic interactions.
Temperature	The number of plant species in a community increases with increasing temperature, given sufficient water. Frost is a barrier for many plants and requires adaptations to prevent tissues from freezing. The distribution of some plants follows isotherms.
Water	In most parts of the world water is limiting at some time of year and plants must resist drought. In waterlogged conditions plants may be limited by the availability of air, though aquatic plants may derive dissolved gases from the water. High rainfall and humidity on tropical mountains leads to reduced transpiration and stunted growth. Bryophytes dominate in some wet environments. Very few flowering plants occur in the sea.
Nutrients and ions	The overall and relative quantities of elements in soils vary greatly, both with underlying geology and age and thickness of the soil. Many are essential and some are toxic and plants differ in their requirements and abilities to withstand toxicity; communities differ under different conditions. Plants tolerant to sea salt occur around all coastlines.
Disasters	Periodic or occasional disasters such as fire, hurricanes or landslides can dominate plant communities. Fire may be frequent in savannahs to less than one per century in conifer forests but still have an overriding influence.
Glaciation and plant migration	Over the last million years glaciers have expanded and contracted over the northern hemisphere making many of these areas suitable only for tundra-like vegetation. The plants have migrated differentially leading to a flexible community structure. In the tropics glacial periods were dry and rainforests were more fragmented. There is good evidence from plant fossils, mainly pollen, of post-glacial changes.

Related topics	Plants and water (E1)	Plant communities (L3)
	Uptake of mineral nutrients by plants (E4)	General features of plant evolution (Q5)

The requirements for plant life   Plants are dependent on external **temperature** and **solar radiation**, **water** supplies, and **nutrients**, normally in the soil, for their survival (Topics E1, E4, E5). The basic requirements are the same for all plants. No plant can survive when its growing temperature remains below 0°C all the year round (plants frequently have a higher internal temperature than the ambient air temperature) or in the driest of deserts where the soil is often unstable. The overall geographical distribution of plants is partly determined by their relative abilities to withstand different climates and soil conditions. These major differences across the world give rise to different **biomes**, i.e. groups of plant communities dominated by plants of similar life form, such as tropical rainforest or temperate grassland, which may have a totally different species composition in different areas but look similar and function in similar ways (*Fig. 1*).

For many plants, dispersal across unfavorable environments, such as the sea or a desert, is limited, giving rise to different plant **communities** in different regions of similar climate and frequently a different flora on islands from continental areas (Topic Q5). Interactions between plants, such as competition, and interactions with other organisms, have an overriding influence on the distribution of a species within one geographical region and can affect major distribution patterns (Section M).

Temperature

Temperature and incoming solar radiation have a profound influence on the distribution of plant communities. In general, given sufficient water, the number of plant species present increases with increasing temperature in the world. One of the major limits is **frost**; many plants cannot tolerate the presence of frost at any time of year. Frost affects the limit not just of species but of whole biomes. **Temperate** and **boreal** (northern – between temperate and arctic)

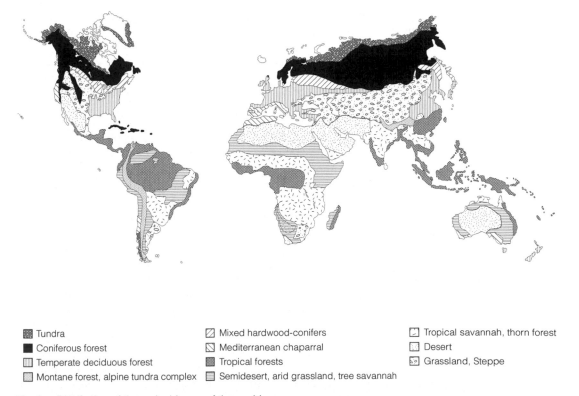

Tundra

Coniferous forest

Temperate deciduous forest

Montane forest, alpine tundra complex

Mixed hardwood-conifers

Mediterranean chaparral

Tropical forests

Semidesert, arid grassland, tree savannah

Tropical savannah, thorn forest

Desert

Grassland, Steppe

Fig. 1. Distribution of the major biomes of the world.

environments are characterized by more or less frequent frosts in one season of the year and during this period the plants become dormant, having developed mechanisms to stop their tissues from freezing. These mechanisms include becoming deciduous, dying down completely or concentrating their cell contents.

Many plants appear to be limited in their distribution by temperature at a particular time of year, following the line of a maximum or minimum **isotherm**, though the reasons for this are frequently not clear. Many plants are able to live outside their observed natural distribution in cultivation. The reasons are probably competition with other plants or other biotic interactions.

Water

On land the quantity and distribution of **rainfall** or other sources of water such as **fog** or **snowmelt** have a major influence on overall plant distribution. In most parts of the world water is limited at some season, either by a dry period in the seasonal tropics or a dry summer or frosty winter outside the tropics. In **deserts** and **steppes** it is limited most of the time. In all these environments there are specialized plants that have adaptations to limit water loss during the dry periods, and most plants become partially or totally dormant during the dry periods.

At the other extreme, in permanently waterlogged conditions and in **aquatic** environments **aeration** is often the limiting factor for plant growth and only plants specialized to withstand waterlogging can survive, although there are aquatic plants with submerged leaves that derive all their photosynthate and nutrients from the water. On many tropical mountains there is a high rainfall and frequent fog and high humidity, low solar radiation and cool temperatures. These conditions lead to a stunted forest, since transpiration, and with it nutrient supply, becomes limiting (Topic E1). In these environments **bryophytes** become particularly common since they absorb water from the atmosphere and most have no internal conduction system (Topic O2). In ever-wet cool or cold conditions on seaboards in temperate zones, in mountains and in subpolar conditions bryophytes, notably *Sphagnum* mosses, dominate and dead plant material does not decay fully, forming peat.

Where water, aeration and temperature are not limiting factors, as in parts of the tropics, **rainforests** grow. The tropical rainforests are among the most diverse terrestrial environments in plant species and life form. In these environments biotic interactions play an important role in limiting plant distribution (Topic M3).

In the sea there are very few highly specialized flowering plants, such as eelgrasses (*Zostera* spp.), the great majority of photosynthesis being done by unicellular algal plankton and, inshore, by large algae (Topic O1).

Nutrients and ions

The nutrient and ion status of soils is variable, with different elements limiting in different places and some elements being toxic (Topic E4). In many soils **nitrogen** is one of the main limiting nutrients although some plants, notably legumes, have nodules in their roots filled with nitrogen-fixing bacteria (Topic M2). In many soils more than one nutrient may be in short supply and plants compete for them. The ion status of a soil will depend in part on the underlying **geology** and certain elements, such as calcium, are most common in alkaline conditions, but in these soils iron may be limiting. Likewise in many acid soils many nutrients are in limited supply but aluminum becomes available and is a toxic element for many plants. These and other differences between soils lead to quite different plant species and communities occurring on different soils. In places rich in **heavy metals**, many of which are toxic such as lead, only certain specialized plants can grow. As a soil ages, elements will gradually leach away and it will become poorer in nutrients. In these conditions the presence of animals' dung or carcasses, ant nests or other plants can lead to great variation in nutrient status on a small scale within the community.

Areas dominated by **saline** conditions occur around all coastlines and estuaries. The great majority of flowering plants cannot grow in the presence of salt in concentrations found in sea water, with its high osmotic potential. Specialized plants with adaptations for excluding or excreting salt occur, mostly succulent and low-growing (Topic E5). Tidal estuaries which are inundated for part of each day with brackish water are often rich in other nutrients and can be highly productive for those plants that can tolerate the conditions: specialized mangrove trees in tropical estuaries; herbs and low shrubs in temperate saltmarshes.

Disasters

Some parts of the world are prone to periodic or occasional destructive forces that can dominate a plant community. **Fire** is the most common, started naturally by lightning but in many places becoming more frequent with human influence. In some savannahs the grass which dominates the ground layer is burned most years but regrows when rain returns. This maintains the savannah community. In other places, such as coniferous woodland and heathland, fires are less frequent, even as few as one per century or less, but still are one of the overriding influences on the composition of the plant community.

High winds are frequent near many coasts and on mountains, and tall-growing trees frequently cannot grow through mechanical instability. Periodic or infrequent storms or **hurricanes**, developing over the sea, can cause havoc in restricted areas, usually near coasts. Though each hurricane ploughs a different path, certain parts of the world, such as the Caribbean and the Philippines, are hurricane-prone. Some plants resist these winds better than others and in these places communities can be dominated by occasional hurricanes.

Tsunamis (tidal waves), **volcanic activity** or **landslides** can also devastate communities, sometimes maintaining them as permanent pioneer communities.

Glaciation and plant migration

Over the last million years periodic cooling and warming of the world has led to glaciers spreading and retreating in the northern hemisphere and the tropics becoming drier during glacial periods and wetter in interglacials. These fluctuations have led to major changes in the range of plant communities in the northern hemisphere with large areas of Europe and North America that now contain temperate broad-leaved forest as their natural community, dominated by Arctic-like **tundra**. In the tropics, rainforest became more restricted and fragmented during the drier glacial periods. There were many extinctions during these glacial times. The spread and retreat of the glaciers affected plant distribution throughout the north temperate region, plants with good colonizing ability surviving better than others and some tundra plants reaching their current disjunct distributions, i.e. distributions consisting of several quite separate areas, through their retreat northwards and into mountains as the glaciers retreated.

In post-glacial times (the last 10 000 years or so) we have a better record than for any previous time from analysis of pollen and other fossil fragments preserved mainly in **peat** and lake **sediments**. These show that plants have spread at different speeds north and often west as well following the ameliorating climate and that the plant communities in the northern hemisphere have changed markedly in extent and composition throughout this time.

L3 PLANT COMMUNITIES

Key Notes

Plant communities of the world	Plants define the biological communities of the world. Within one biome communities vary greatly and different communities are found in swamps, on mountains and on particular soils. Within one biome species richness and composition differs greatly and oceanic islands are frequently poor in species. In some places communities are well defined and have been classified.
Succession	When colonizing bare ground there is a primary succession of plants from pioneers to a climax community. Secondary succession happens within more mature communities after gaps are formed but end points may differ by chance. Climate changes over the last million years have meant that some communities may be in a state of continuous succession.
Dominance and diversity	Some communities are dominated by one or a few plant species but many are not. Diversity can be divided into α diversity, species living in the same habitat, β diversity, species occupying separate habitats, and γ diversity, species occupying separate geographical areas. β diversity, though subtle at times, can be explained easily. γ diversity varies from region to region. α diversity is much harder to explain but probably arises from differences in biotic interactions, particularly between germination and establishment.
Related topics	Ecology of different growth forms (L1) Physical factors and plant distribution (L2)

Plant communities of the world

Plants define the world's natural communities, providing the habitat for all its other inhabitants as well as providing the great majority of the biomass. There are many different plant communities within each **biome** (Topic L2). For instance, across the northern hemisphere in the boreal and temperate zones there is a patchwork of deciduous broad-leaved and coniferous forest, sometimes with a single species dominant in any one place, and with more broad-leaved forest as one moves south. In the Mediterranean-climate regions of the world in both hemispheres there is a rich mixture of, mainly evergreen, tree and shrub communities. In the seasonal subtropics and tropics there is a range of savannah communities, dominated by grasses with increasing numbers of trees as the climate gets wetter.

Within all these places there are patches of other communities and specialist communities in swamps, on mountains, near the sea and in response to particular soil conditions. Different continents differ in their plant communities, and species richness can vary greatly between communities within one biome. Oceanic islands are usually much poorer in species than continental areas. Herbivorous animals can have a large effect on plant communities, with large mammals frequently maintaining grassland communities. People have modified many communities and, in particular, removed trees to replace them with grassland or other open communities for pasture and agricultural communities and dwellings.

Some attempts have been made to classify plant communities, particularly in Europe where some are well defined. More often there are gradual transitions from one community to another and they are best regarded as collections of species that happen to be able to exist under certain conditions but have different limits, the composition changing continuously across its extent. Classification can be useful within one limited geographical region.

Succession

One of the best known ecological theories is that of succession, which is seen to occur in places such as sand dunes and marshes, on abandoned fields and by retreating glaciers. In a **primary** succession fast-growing but short-lived **pioneer** plant species with good seed dispersal **colonize** when the new habitat is first available. These species modify the conditions both in the soil and in the biotic environment that allow other species to colonize (Topic H5), eventually leading to a **climax** community. In a gap in a mature community, there is **secondary** succession involving pioneer species and changes in the plant community, though with less change in the soil conditions and, very often, small members of the climax community present early on. Succession will take a different form depending on the size and nature of the gap and the presence of any animals. There is a large role of chance in how any succession progresses, and the end point may not always be the same or there may not be an end point as such. Some species may act as pioneers but persist for decades or centuries.

A plant community will always be subject to change. The climate has changed markedly over the last million years with the north temperate subject to periodic glaciations of greater or lesser extent, these affecting the whole world, with the tropics becoming drier in glacial times, and the sea level rising and falling. Over a time-scale of centuries there have been more minor climatic fluctuations in any one area with dry or wet periods and cool or warm ones. As a result plants have migrated north and south and across tropical land masses, with smaller changes in community boundaries, local invasions and extinctions. If we consider the extent of climate change, the chance nature of which plants colonize any one area and the long generation time of many of the dominant plants in a community, it suggests that many communities may be in a state of continuous succession.

Diversity and dominance

Some climax plant communities are dominated by a single species within one life form. More commonly a small number of species dominate or there are no dominant species at all. Single-species dominance occurs on soils rich in nutrients, particularly in temperate zones where competition is intense and one species may become dominant through faster growth or stronger competitive ability. Dominant plant species also occur in places subjected to an environmental stress, such as waterlogging or the presence of a normally toxic element. In most places, however, there is a diversity of species within each life form, this diversity increasing towards the tropics. Diversity can be divided into three categories: α **diversity**, the diversity within one habitat or microhabitat; β **diversity**, that between habitats; and γ **diversity**, that between different geographical areas within one overall area, e.g. two mountains within one range. β diversity may include differences in subtle features such as particular nutrient-rich patches in a poor soil, e.g. in an animal dung heap, or a small change in gradient, but it is easily comprehensible and, given sensitive equipment, straightforward to explain. γ diversity varies greatly, with some areas such as the Cape region of South Africa having a particularly high γ diversity and many species with a narrowly restricted distribution, probably owing to a combination of its varied topography and stable climate history. Other places, like northern Europe, have low γ diversity, explained in part by the dramatic climate changes that have forced plants to migrate considerable distances.

α diversity is much harder to explain, beyond the differences in life form already mentioned. It probably arises from a combination of many factors. To approach an explanation we would need to see each plant's full life cycle and, in many species, we rarely see the most vulnerable stage, the period between germination and establishment. The existence of a group of individuals of one tree species may reflect conditions that applied several decades or centuries earlier. Communities are periodically disturbed, e.g. by freak weather conditions or by large animals, and germination conditions for any one species may only occur infrequently (Topic H5). Plants are vulnerable to attack from herbivores, particularly insects, and diseases and pathogens such as fungi (Topics M3, M4). This is well known in crop monocultures and will affect mostly abundant or dominant plants, and mostly at the young stages. Attack is likely to be periodic and may not be seen in a limited study period, but it will reduce the dominance by any one species. These biotic interactions are likely to account for most α diversity. In the most diverse environments in the world, such as some tropical rainforests, there is high α, β and γ diversity.

L4 POPULATIONS

Key Notes

Plant life cycles	Plants may live for a few weeks to thousands of years, reproduce once and die or reproduce many times. Semelparous plants reproduce just once, after a few weeks (ephemerals) or over a year (biennials) to many years (semelparous perennials). Iteroparous plants reproduce many times and some of these produce rhizomes and spread clonally.
Ecology of ephemeral plants	These occur everywhere but are common in temperate climates and deserts. Many have seeds that can remain dormant for many years and their numbers often fluctuate markedly between years. Many temperate species have small self-fertilizing flowers.
Biennials and semelparous perennials	Biennials are opportunist plants, often with seeds that can be dormant living in successional habitats mainly in the temperate regions. They produce showy flowers. Life cycles are flexible and they live longer if damaged. Semelparous perennials live in conditions where growth is slow or where swamping seed predators is important.
Iteroparous perennials	These range from short-lived herbaceous species similar to ephemerals to long-lived trees. They dominate most stable habitats and, in any one place, the dominants often have individuals all of similar age owing to cyclical changes in the plant community. Some spread clonally, particularly herbaceous plants on woodland floors and in wetlands.
Timing of reproduction	Early reproduction is often an advantage but iteroparous perennials are at an advantage if there is high seedling mortality.
Population dynamics	There are many hazards in a plant's life cycle, but all populations have the potential to increase exponentially, the rates depending on birth rate, death rate, immigration and emigration. If plants grow above a certain density, over time density-dependent mortality occurs, known as self-thinning. This can be described mathematically with the $-3/2$ power law which is constant for many plants. There is often great variation in size within a population and small plants die first.
Populations of clonal plants	A 'population' of a clonal plant can be one genetic individual. Clonal plants can regulate their own shoot density although sometimes shoots compete. Growth behavior differs in different environments and they can exploit rich patches effectively while crossing less favorable ones. The disadvantage of clonal growth is mainly disease susceptibility.
Related topics	Regeneration and establishment (H5) Polymorphisms and population genetics (L5)

Plant life cycles   A complete life cycle of a plant, from germination through its reproductive life to death, varies from about 3 weeks to several thousand years. Plants may be divided into those that flower and set seed once in their lives, known as **semelparous** or **monocarpic**, and those that reproduce more than once, known as **iteroparous** or **polycarpic**. Semelparous plants may live for a few weeks or for many years. Short-lived ones are known as **ephemerals** or, commonly, **annuals** although most complete their life cycle in much less than a year, and the shortest-lived, such as *Boerhavia* spp. (Nyctaginaceae) may complete their life cycle in less than 4 weeks. Those that live longer than a year are mostly known as **biennials**. These normally grow as a leaf rosette in their first year, flowering and dying in their second. Longer-lived semelparous plants are rarer but some of these have leaf rosettes of ever-increasing size over several years, such as the century plants, *Agave* spp. (Agavaceae), which usually live for 15–70 years rather than a century. Some are woody, notably some bamboos which can flower after 120 years, and the royal palm, *Roystonea*, which can live for 80 years and reach 30 m before flowering and dying. Those that live for more than about 3 years are known as **semelparous perennials**.

Iteroparous plants range from those that live for just a few years and normally reproduce less than five times to some of the longest-lived of all organisms such as the huon pine, *Dacrydium franklinii*, in Tasmania and the bristlecone pine, *Pinus aristata*, in Arizona, both of which can live for over 4000 years. Living clones of creosote bushes, *Larrea tridentata* (Zygophyllaceae), in the deserts of western USA may be even older, over 11 000 years old, radiating outwards from a central source. Nearly all woody plants are iteroparous.

All plants have meristems and if shoots remain vegetative they have indeterminate growth, i.e. continue to grow throughout their lives, remaining 'forever young'. Many have developed underground rhizomes or stolons which grow continuously and can form large patches or grow a long way from where they first germinated. These produce roots as they grow and the plant can then be split to form independent physiological units. These clones, or parts of them, can live for many centuries and the definition of an individual becomes obscured since a genetic individual and a physiological individual become different. Clonal spread is frequent among herbaceous perennials, less so among woody plants.

Ecology of ephemeral plants   These occur in most plant communities but are common in deserts, regions with a Mediterranean climate of cool wet winters and hot dry summers, agricultural and disturbed land and places with unstable soils like sand dunes. In mature plant communities such as woodlands and permanent grasslands they are mainly associated with disturbed areas, e.g. from animal digging or tree falls. Some will germinate at almost any time of year and desert ephemerals germinate in response to a substantial fall of rain. There is a group of 'winter annuals' in temperate climates that germinate in the fall and flower and set seed in the spring before the summer drought. In many ephemerals generations of adult plants do not overlap, but they may have seeds that can remain dormant for a long period so seeds from different years may germinate together. Some appear to rely on each year's seed crop and have limited dormancy. Many ephemerals in temperate zones are small plants with small flowers that can self-fertilize, but among desert ephemerals there is great variation and many have large colorful flowers. Most populations of ephemeral plants fluctuate markedly from year to year.

Biennials and semelparous perennials

Biennials are opportunist plants, like the ephemerals, and many have long-lived dormant seeds. They occur in early successional stages where a mature habitat such as a woodland or grassland has been disturbed such as by tree falls, or at the edges of grasslands, mainly in temperate climates. They do not occur on agricultural land in general. Frequently the life cycle is flexible and a normally biennial plant can perennate (i.e. live for longer and flower a second time) if its flowers or fruits are eaten or damaged. Frequently the flowers are produced in showy inflorescences attractive to flower-visiting animals, this being especially true for those plants that live for longer than a year before flowering. By the time they flower the place they are occupying is often not suitable for the next generation so the seeds must disperse or lie dormant.

Some semelparous perennials are associated with regions where growth is restricted such as deserts or tropical mountains; others such as the bamboos are thought to reproduce once, so that the huge numbers of seeds produced will swamp any potential seed predators. Frequently in these species all the plants in an area will flower synchronously.

Iteroparous perennials

These range from short-lived pioneers that behave in a similar way to ephemeral and biennial species through to long-lived species with much more stable populations. The dominant species in most communities are long-lived nonclonal species, such as most trees and many of the dominant plants of open areas. The adult plants of the dominant species in some communities are of similar age where there is some kind of cyclical change and the main germination is at a time when the area is more open, through a natural disaster. Understorey trees and other plants in a woodland, by contrast, normally have a wide range of ages including many seedlings and saplings, often with smaller numbers of adults.

Clonal species dominate certain habitats, particularly wetlands and woodland floor habitats, and one genetic individual can cover a large area, even an entire habitat. The vegetative branches, or **ramets**, can utilize the resources of the mother plant as they grow and this gives them a huge advantage in establishing themselves over nonclonal plants that rely on new seedlings.

Timing of reproduction

In general it will be selectively advantageous for a plant to reproduce as early as possible. This will often lead to maximal population growth, e.g. if a plant produces 10 seeds after 1 year and then dies, the following year each of those 10 seeds can grow and produce 10 seeds, so any plant that delays reproduction until the second year will need to produce 100 seeds to equal the potential output of one reproducing in its first year. Although this does not take into account mortality at any stage, which is likely to be different for seeds and established plants, the potential advantage of reproducing early is clear and ephemerals are thought to be among the most advanced of plants. Ephemeral plants put most of their resources into reproduction, but an iteroparous perennial needs to retain some resources for its own survival so seed production is likely to be less. To determine which is at an advantage the survival of seedlings (pre-reproductive) and of adults must be taken into account. We can say that for an ephemeral plant:

$$\lambda_e = cm_e$$

where λ_e is the rate of increase, c is the seedling survival and m_e is the mean seed production.

For an iteroparous perennial:

$$\lambda_p = cm_p + p$$

where p is the adult survival.

We can derive from this that an ephemeral plant will reproduce faster if $m_e > m_p + p/c$. This indicates that one of the important terms is the ratio between adult survival and seedling survival (p/c): if there is good seedling survival it will be advantageous to be ephemeral; if seedling survival is poor an iteroparous perennial will be at an advantage. Studies have confirmed that this is largely true.

Population dynamics

There is often high mortality of the seeds, and there may be many natural hazards before they can germinate, such as seed-eating animals, rain and wind. After germination the main competition occurs during active growth to flowering, and this is likely to involve competition with associated plants as well as members of the same species. After this an established plant faces many fewer hazards, so the greatest selection is likely to be in the seedling stage. In general, plants live in populations in the same way as animals, with birth rates, death rates, immigration and emigration. Any change in that population must take these into account and this can be expressed by this **basic equation** of population change:

$$N_{t+1} = N_t + B - D + I - E$$

where N_t is the number at time t, B is number of births, D is number of deaths, I is immigration, E is emigration. When $B + I = D + E$ the population is stable, but when $B + I > D + E$ the population will increase **exponentially**, i.e. increasing by the same factor each year, giving a logarithmic growth curve (*Fig. 1*). This has been recorded when a plant is colonizing a new area. Clearly this rate of increase can only continue for a short period, after which the population will be regulated by some finite resource such as germination sites or nutrient availability.

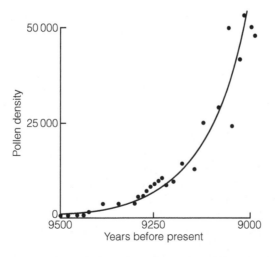

Fig. 1.   Exponential increase in the population of the scots pine, Pinus sylvestris, *in Norfolk, UK, from pollen records in peat deposits (from Bennett, K.D., Nature 1983;* **303**, *164–167).*

As density increases the plants may start to interact with each other, limiting each others' growth. This may be manifest in smaller plants and/or lower seed set, particularly in small ephemerals, but with longer-lived plants some will die before reaching maturity. This is most obviously seen in trees where seedlings can cover the ground but there is less than one mature tree per 10 m². There is a process of **self-thinning**. The idea has been best developed in relation to crop plants and their yields since farmers will want to sow a crop to reach maximal yield while minimizing seed wastage. Experimentally it was found that mortality started earlier, and at a smaller weight with denser sowings. If the relationship between sowing density and mean weight is plotted using a log scale on both axes the lines converge on a straight line (*Fig.* 2). This line has a slope of approximately –3/2 for many unrelated plants (including trees) and can be described by the equation:

$$\log w = c - 3/2 \, \log N$$

where w is mean plant weight, c is a constant, N is plant numbers. This is known as the **–3/2 power law**. In sparse populations interaction between individuals will be less and the slope of the line will flatten out, until it reaches 0 when there is no **density-dependent** interaction. Although this slope is remarkably constant across plants the c value varies hugely. This is the intercept term on the y axis, i.e. the density at which there is no interaction between individuals, clearly totally different for a small ephemeral and a tree. There are other factors that influence numbers of plants that are dependent on the density of the population, such as disease and herbivore attack. Their effects are often intermittent and they are much harder to study than competition, but they can be equally or more important in determining population density (Topics M3, M4).

If a population of a large plant such as a tree starts out at high density, over time there will arise a marked size difference between the largest and the

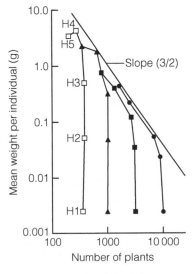

Fig. 2.   *Self-thinning in four populations of rye-grass,* Lolium perenne, *planted at different densities. There were five successive harvests to give these lines (from Kays, S. & Harper, J.L.,* Journal of Ecology *1974;* **62**, *97–105).*

smallest individuals, and the smallest will be the first to die in the self-thinning process unless some other influence intervenes. Initial size difference may arise from slightly different germination times or genetic differences between individuals in their vigor of growth but this will be accentuated by competition once they start interacting.

In general a plant population is affected by many processes that are **independent** of a plant's density such as frosts or floods at particular times and in density-dependent interactions with members of its own and other species.

Populations of clonal plants

Plants that spread **vegetatively** can regulate the density of their own shoots, although there can be competition between shoots as described for nonclonal species. An entire 'population' can be made up of just one genetic individual and this may be true of a hillside covered with bracken, *Pteridium aquilinum*, or a reed-bed dominated by *Phragmites australis*. Clonal plants may grow densely to dominate an area or as long strands invading new places, and these invaders can tap the resources of the mother plant, giving them a great advantage over seedling invaders. The invaders can change their '**behavior**' (in a plant, behavior means changes in growth form) in response to local conditions, growing densely where there are many nutrients but sparsely between rich patches.

There appears superficially to be such a great advantage to clonal spread that it is surprising that, although it is common, more plants do not spread this way. The main disadvantage lies in the fact that clones are all identical genetically. This leaves them susceptible to insect attack or disease since, if a predator or pathogen can overcome the resistance of one member of the clone, all will be susceptible. This was shown graphically in Europe in the 1970s with the outbreak of elm disease attacking all the clonal elm species, e.g. *Ulmus procera*. In agriculture, clonally produced crops require greater uses of pesticides.

L5 POLYMORPHISMS AND POPULATION GENETICS

Key Notes

Variation
Species may vary geographically or in response to ecological conditions. Some variation is continuous, some discontinuous. When there is discontinuous variation within a population it is known as a polymorphism. Polymorphisms in enzymes and DNA can be detected by electrophoresis.

Gene flow
There are two components: pollen flow and seed dispersal, both of which are restricted in all plants. Self-incompatible wind-pollinated plants with good seed dispersal have potentially long-distance gene flow compared with self-fertile plants with poor seed dispersal.

Natural selection
The importance of natural selection in the maintenance of genetic variation is much disputed. Very little of the biochemical variation detected has a known selective basis but it may be hard to detect. Heterozygous plants appear to be at an advantage in many species, indicating that cross-fertilization will be selected for.

Population genetic structure
The total amount of genetic variation differs greatly between plants and how it is distributed between individuals within each population or between populations. This depends on the breeding system and seed dispersal and if these are restricted then restricted gene flow is the dominating influence. If the plant is self-incompatible with good seed dispersal, frequently the dominant plants, natural selection appears to be more important. Isolated small populations are frequently limited by the small number of founders or random extinction of some morphs.

Related topics
Breeding systems (G3)
Ecology of flowering and
    pollination (G5)

Fruit and seed dispersal (H3)
Populations (L4)

Variation

Individual plants in any one species vary. Marked differences between individuals are most commonly seen as geographical variation or variation in response to ecological conditions. These are known as **subspecies** if the variation appears to have no obvious ecological basis, or **ecotypes** if it is a response to ecological conditions. Plants are only considered to be ecotypically distinct if the differences have a genetic basis. Plants can respond directly to environmental conditions, e.g. by longer stems and smaller leaves in dark places, but if any differences remain when the plants are grown in uniform conditions it indicates a genetic basis. Many ecotypes are morphological, such as short-growing ecotypes from exposed or heavily grazed places, and some may be physiological, such as salt-tolerant ecotypes near coasts.

Variation may occur within populations too, and some plants show a range of morphological variation. This may be continuous, e.g. in leaf shape, or discontinuous such as variation in flower color or the presence of hairs on the seeds. Some of the variants, or **morphs**, occur at a low frequency of less than 1 in 20 plants, and are maintained only by the occasional recurrent appearance from mutation or rare gene combination. Many blue or purple-colored flowers have occasional white morphs at low frequency. When morphs occur at greater frequency than 1 in 20 the feature is **polymorphic** and the plant shows a polymorphism.

The genetic basis of variation can be studied more directly by looking at protein variation, usually detected as isozymes, and DNA variation (Topic B3). A great range of polymorphic variation occurs and some of this has given valuable insights into the genetic structure of plant populations.

Gene flow

There are two phases of gene dispersal in a plant's life cycle: pollination in which only the pollen moves and which determines how the genes will mix to form the next generation; and seed dispersal in which the new generation moves. Study of plant polymorphisms has given much insight into gene flow in plants, since the distribution of individual genes can be mapped between individuals within a population and between populations.

Gene flow is restricted in all plants, with the majority of pollen and seeds dispersing only a short distance. This potentially divides many populations into part-separated **neighborhoods**, defined initially as 'an area from which about 86% of the parents of some central individual may be treated as if drawn at random'. The most extreme restriction in pollen flow is self-fertilization where there is no genetic mixing, and after a few generations all the offspring will be genetically identical. If this is coupled with poor seed dispersal then gene flow is minimal. The opposite extreme is found in self-incompatible plants with pollen and/or seeds dispersed by wind (Topic H3). These plants may have effective long-distance gene flow and neighborhood size will be large. How far genes travel will depend on the effectiveness of pollinating animals and seed dispersers and the occurrence of compatible plants and seed germination sites (Topics G5, H5).

Natural selection

There is much debate on the relative importance of natural selection and **restricted gene flow** in determining how plant populations are organized genetically. Some isozymes are known to respond differently to particular environmental conditions (Topic B3), e.g. isozymes with different sensitivities to temperature, but in most no such differences are known and the variation has no known basis in natural selection. DNA sequences are usually assumed to be neutral in their effect and to be markers for relationships or phylogeny. This does not necessarily mean that natural selection is not important, just that we have not detected the mechanism. For all conclusions on the genetic organization of plant populations, the variation is assumed to be neutral in operation. If selection is operating, favoring one or another of the isozymes in different conditions, such as a wet or dry microhabitat in any one area, then any conclusions about gene flow must take this into account. It seems likely that many of the observed polymorphisms in plant populations are maintained in part by natural selection, since their distributions are not as predicted considering only gene flow, and a relationship between certain enzyme forms and micro-environmental conditions has been found in some plants. It may involve the detected enzyme morphs working in a particular context within the plant, or a response of other linked genes. DNA variation must ultimately have a selective origin but much is likely to be conserved. Variation can sometimes be so great that it has limited use in detecting genetic structure.

There is evidence from some species that **heterozygous** plants with two different versions of a gene are at an advantage over **homozygous** plants with two the same. Heterozygosity normally results from a cross between two parents that are genetically distinct and frequently a self-incompatible plant will be heterozygous at many loci. In self-fertilizing plants each generation will lose, on average, half the heterozygous genes, so it will rapidly become homozygous at most gene loci. The advantage of cross-fertilization has been well known to plant breeders for centuries (Topic K1), known as **hybrid vigor**, and is mainly because of increased heterozygosity. Many deleterious mutations will be masked by an alternative form and, if an enzyme occurs in two different forms, it may be able to work across a wider range of conditions. Although there is evidence from some species that heterozygotes are at an advantage, it does not appear to be universally true. Many polyploid plants, particularly those deriving initially from hybridization, can retain some heterozygosity despite self-fertilization through having two different genes from different parent species on their **homeologous** genes, i.e. the equivalent genes from the two parents. Many polyploids do self-fertilize.

Population genetic structure

The total amount of genetic variation within a species is highly variable, with species that are restricted to a small area usually much less variable than widespread species, although there are many exceptions. It is likely that the distribution of genes within a population results from a balance between natural selection and restricted gene flow. The variation that occurs can be divided into that which exists within each population and that which occurs between populations.

In some plants the populations differ but individuals within each population vary little or not at all, a situation characteristic of self-fertilizing plants and others with restricted gene flow. In these plants restricted gene flow has the predominant effect. In others, individuals within each population vary widely but each population has a similar range of variation. This is true of plants with the potential for long-distance gene flow (see above) and particularly of many plants that dominate plant communities. In these, natural selection is likely to be much more important. If there is strong natural selection on plants it is likely that there will be mechanisms to avoid self-fertilization since it is nearly always disadvantageous (Topic G5). Studies that have looked at gene flow directly by examining pollen and seed dispersal have consistently underestimated the extent of genetic mixing and the effective gene flow within these species. In addition, many plants have some propensity to hybridize with related species, often producing partially fertile offspring. These can then back cross with one or both parents and some genes from the second species can be incorporated into a population. Alternatively hybridization and a chromosome doubling may lead to polyploidy and the new individual isolated from its parents to form a new species (Topic Q5).

Those plants that spread clonally may have a 'population' without any variation, i.e. consisting of a single genetic individual. In most clonal species, though, there is sexual reproduction, often with a mechanism favoring cross-fertilization and, taken overall, their genetic structure may be similar to that of nonclonal species but with fewer individuals, each covering a large area.

Small populations of many plants that are isolated from any other population often have little genetic variation. This may be because of the **founder effect**, each being founded by one or a small number of plants at one time so they have always been restricted genetically, or it may be because of **genetic drift**: some variants becoming extinct at random because they cannot be maintained in small populations.

L6 CONTRIBUTION TO CARBON BALANCE AND ATMOSPHERE

Key Notes

Global carbon dioxide

Carbon cycles between gaseous forms like CO_2 in the atmosphere and forms in which it is fixed, for instance in living organisms. Higher plants have a major role in removing carbon dioxide from the atmosphere. Global atmospheric CO_2 concentration was about 190 ppm at the end of the last ice age, but it has risen to 375 ppm today. Burning fossil fuels and widespread destruction of forests are responsible for this increase.

Global warming

Rising global atmospheric CO_2 concentration has resulted in increased mean surface temperatures on earth (of about 0.5°C over 50 years) as CO_2 absorbs infra-red radiation. Altered weather patterns are more significant to plants than small changes in average temperature.

Plants and rising carbon dioxide

C3 plants are limited by CO_2 levels and growth rates in at least some species increase with increased CO_2. In natural communities, growth is limited by other factors like nutrient availability. C4 species are not likely to show enhanced yields with rising CO_2, as they do not show photorespiration and have a much lower CO_2 compensation point. Plants reduce atmospheric CO_2 by fixing carbon; however, this only decreases global atmospheric CO_2 concentrations if the plant material is not degraded to release CO_2 to the atmosphere.

Global warming and biodiversity

Altered weather conditions will affect sensitive biomes and species at the limit of their distribution. Unless a species can adapt to a changed environment or migrate to less hostile areas, it will become extinct. This is particularly likely to be true of long-lived tree species.

Related topics

Major reactions of photosynthesis (F2)

C3 and C4 plants and CAM (F3)

Global carbon dioxide

Figure 1 illustrates the global carbon cycle. Carbon cycles between 'fixed' forms (organic carbon and carbonates) and the atmosphere (e.g. as CO_2). Significant pools of carbon are present in the atmosphere, the oceans, soils and land plants. A very large amount also exists in carbonate rocks and in buried fossil fuels. Plants play a vital role in this cycle by fixing carbon from the atmosphere. At the end of the last ice age, about 18 000 years BP, the global atmospheric CO_2 concentration was about 190 ppm (parts per million). Since then it has risen, reaching about 250 ppm in 1700s and 375 ppm today; and it continues to rise (*Fig. 2*). This rise has been largely attributed to the burning of fossil fuels and the destruction of forests. Carbon is removed from the atmosphere when living organisms die and are preserved in deposits of coal, oil, peat and other sediments. It is also deposited as calcium carbonate

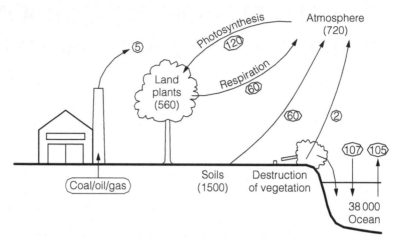

Fig. 1.   A global carbon cycle. Pools of carbon are expressed as 10^{15} g C and fluxes as 10^{15} g C year^{-1}.

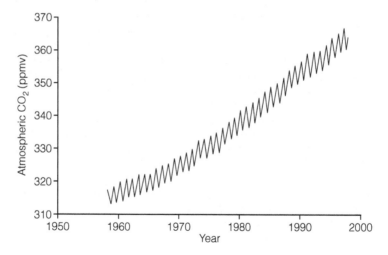

Fig. 2.   Mean monthly atmospheric CO_2 concentrations measured at Mauna Loa, Hawaii.

minerals, including the exoskeletons of living organisms. As parts of the world's carbon deposits are extracted from the ground and burned, vast amounts of CO_2 are released to the atmosphere. Large-scale destruction of forests gives further CO_2 release.

Global warming   Rising global CO_2 levels have a number of consequences, the chief of which is a net rise in global temperature. Solar radiation traveling through the earth's atmosphere is partly reflected back into space by the earth's surface and clouds. Other parts heat up the earth, this heat being re-emitted by the earth as longer wavelength radiation. This infra-red radiation is partially absorbed by the so-called greenhouse gases, CO_2 being one of them and the one that is changing most rapidly in concentration. Others, such as nitrous oxide, N_2O, and methane, CH_4, are present in small quantities in the atmosphere but the changes in CO_2

are responsible for about half the global warming that we are seeing. The degree of warming will increase with more CO_2. A rise of about 0.5°C in global mean surface temperature has occurred over the last quarter century. Altered average temperatures have been accompanied by more extreme weather changes. These include earlier and later frosts and heavier rainfall in some areas and drought in others. With drought comes the possibility of fires and these have been widespread, particularly in the Asia/Pacific region. Rising sea level accompanied by melting ice-shelves in the Arctic and Antarctic are also consequences of global warming. World sea levels have risen by more than 10 cm in the last century and their continued rise will result in losses of coastal land and island communities.

Plants and rising carbon dioxide

At present, photosynthesis in C3 plants (Topic F2) is limited by CO_2 levels and growth rates in at least some species increase if CO_2 concentrations are increased. Growth rates may increase by 30–60% if CO_2 concentrations are doubled, but this may only be temporary as growth and development are regulated in other ways (Topic J1). In natural communities, growth is likely to be limited by other factors like nutrient availability. Yields of some glasshouse crops, like tomato, have been increased by enriching the air around them with CO_2. C4 species (Topic F3) are not likely to show enhanced yields with rising CO_2, as they do not show photorespiration and have a much lower CO_2 compensation point (Topic F3). *Figure 3* illustrates net photosynthetic rates of a C3 plant (like wheat) and C4 plant (like maize) at different CO_2 concentrations but otherwise optimal environments. While the C4 plant is more efficient at current global CO_2 concentrations of around 360 ppm, this efficiency advantage will diminish and eventually C3 species will exceed C4 in efficiency if CO_2 concentrations continue to rise.

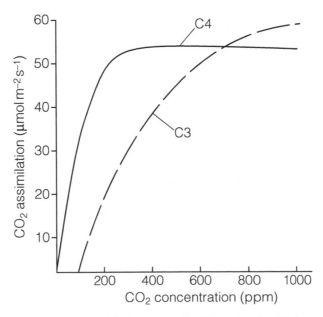

Fig. 3.   *Photosynthetic responses of a C3 plant and a C4 plant to rising CO_2 concentrations in experimental conditions. Atmospheric CO_2 concentrations are above 360 ppm and increasing by about 10 ppm every 10 years.*

While in the short term, plants reduce atmospheric CO_2 by fixing carbon, in the long term this only decreases global atmospheric CO_2 concentrations if the plant material is not burnt or rotted – processes which release CO_2 to the atmosphere (*Fig. 1*). The world has, in the geological past, gone through cycles of CO_2 concentration, and glaciations are associated with low CO_2 levels caused by vigorous plant growth over the previous few million years leading to large deposits of carbon in peat and other organic matter (Topic Q5).

Global climate and biodiversity

Overall, the impact of changing climate is likely to be greater than direct effects of CO_2 concentration. While crop yields may increase in optimal environments, losses due to drought, flooding, sea level increases and disease will also increase. Plant communities will be changed as the competitive abilities of some species increase while others decline. Biomes likely to be sensitive to climate change include tundra (loss of permafrost), savannah and deciduous tropical forests (drought and fire) and deserts (rising temperatures). Other biomes, like tropical rainforests, are at risk from both human activity and drought. If a species is to survive, it must either adapt to its new surroundings or migrate through seed dispersal at a rate equal to that of climate change. For many long-lived species this may not be possible and they may become extinct.

M1 MYCORRHIZA

Key Notes

The association	Mycorrhizal fungi are associated with the roots or rhizoids of most land plants, including bryophytes and spore-bearing plants. A few weedy plants are not infected. All three fungal groups form mycorrhizae.
Types of mycorrhiza	There are several types: arbuscular mycorrhizae, the commonest form, which penetrate the cortical cell walls in roots; ectomycorrhizae, characteristic of temperate trees, which form a sheath around the absorbing roots; and several specialized types associated with heathers and orchids. Arbuscular fungi are zygomycetes and fruit underground; others are ascomycetes and basidiomycetes.
Nature of the symbiosis	Normally the interaction has mutual benefit but it varies from the fungus being pathogenic to the plant being parasitic on the fungus. Usually sugars from the plant are absorbed by the fungus and the fungus absorbs soil nutrients and transfers them to the plant. Mycorrhizae have extensive hyphal networks and ectomycorrhizae have hyphae that fuse with neighboring fungi.
Effects of the fungi on plants	Mycorrhizae are highly efficient at absorbing nutrients from the soil, particularly nitrogen and phosphorus, and ectomycorrhizae are able to digest organic matter. Sugars from the plant are normally absorbed and fixed by the fungi, but in some interactions the plant absorbs sugars from the fungus and the interaction can differ in different species and change in the lifetime of one plant. They can inhibit other fungi, including decay fungi and plant pathogens.
Community interactions	These are not well understood but effects can be profound. Mycorrhizae may be unimportant in early succession but there may be a succession of fungi. In a mature community one species of fungus may enhance the growth of one plant but inhibit another. A diverse plant community relies on a diverse fungus community. Ectomycorrhizal networks may prevent nonmycorrhizal species from invading. Successful conifer forestry needs mycorrhizae in the soil.
Related topics	Roots (D2) Fungal pathogens and endophytes (M4) Uptake of mineral nutrients by plants (E4) Saprophytes and parasites (M6)

The association

Mycorrhizae are the associations between the roots of plants and fungi. The great majority of plants, probably over 90%, have a mycorrhizal association and it is thought to have been vital in the initial colonization of land. Mycorrhizal associations are known in the rhizoids of bryophytes and the gametophytes of spore-bearing vascular plants as well as in the roots of all groups of vascular plants, except perhaps the horsetails. Their effects are profound on many aspects of plant life and are only beginning to be studied. Mycorrhizae fall into several categories with rather different properties deriving from all three main groups of fungi. The basic structure of a fungus is the **hypha**. These are strands of cells a single cell thick, although sometimes without dividing cell walls, that grow through a medium. They secrete enzymes and absorb nutrients.

A few plant families and genera are rarely associated with mycorrhizae, and these are mainly weedy species including the cabbage family (Brassicaceae), the goosefoot family (Chenopodiaceae) and the temperate sedges (*Carex* spp.).

Types of mycorrhiza

The most common form of mycorrhiza is the **arbuscular mycorrhiza (AM)**, sometimes known as **endomycorrhiza** or vesicular-arbuscular mycorrhiza. In these the fungal hyphae penetrate through the cell walls of the root cortex, forming a mass of twisted hyphae apparently within the cells, known as arbuscules (*Fig. 1*). The hyphae do not actually penetrate the cell membrane but this invaginates many times, much increasing its surface area, around the hyphae. In some species, hyphae form oil-filled vesicles, perhaps as food stores, in invaginations in other cells. The fungus does not penetrate beyond the cortex of a root. AM are found in most herbaceous plants, most tropical trees and some temperate woody plants and are the main mycorrhizae in most habitats. Fossil evidence suggests that AM were the initial mycorrhizal associations. They are **Zygomycetes** in the order Glomales, producing subterranean fruiting bodies, but with only around 170 species known.

The other widespread type is the **ectomycorrhiza (EM)**, in which the fungi form a dense mat of hyphae known as the **Hartig net** around the outside of the short absorbing roots and may surround cortical cells but do not penetrate the cortical cell walls at all (*Fig. 2*). This type is found in conifers, especially the Pinaceae, and a small proportion of all flowering plant species, but is characteristic of many of the dominant trees of temperate and boreal environments and some tropical trees, particularly those growing on nutrient-poor acidic soils. A few thousand species of **basidiomycetes** and **ascomycetes** form these associations, although many fewer plant species are involved than with the AM.

A few other types are known. **Ericoid** mycorrhizae are ascomycetes in a specialist association with many members of the heather family group. They can be of great ecological importance since these plants dominate heathland. **Arbutoid** and **monotropoid** mycorrhizae are basidiomycetes that occur only on a few members of the heather family group, the monotropoid only on the 'saprophytic' *Monotropa* (Topic M6). All three of these groups have some

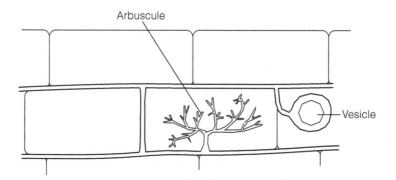

Fig. 1.   Root cells infected with an endomycorrhiza showing an arbuscule and a vesicle.

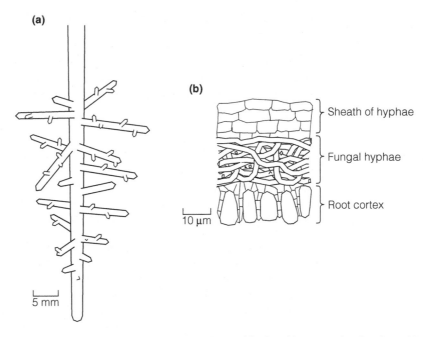

Fig. 2. *Roots infected with ectomycorrhizae: (a) absorbing root showing branching and absence of root hairs; (b) dense net of fungal hyphae around root.*

penetration of root cortical cells but not forming arbuscules. A thin sheathing is involved in some associations. **Orchid** mycorrhizae are basidiomycetes that only infect orchids. They can penetrate the cells, as in an arbuscular association, forming short-lived coils and do not form sheaths. The minute orchid seeds will not normally germinate without a mycorrhizal infection and the plants remain dependent on the fungi. A final group, the **ectendomycorrhiza**, is not well characterized but can penetrate the cortical cells and form a sheath. They are known mainly from conifers but it seems that some may form ectomycorrhizal associations as the tree grows.

Nature of the symbiosis

The symbiotic relationship between plant and fungus varies from being one in which the fungus gains from the plant but is, in effect, a **pathogen**, through a range of **mutualistic** associations where both benefit to some extent, to one in which the plant gains at the expense of the fungus. Mycorrhizae resemble pathogenic fungi in many ways and it can be difficult to distinguish the two since sites of **infection** can be the same. When working to mutual benefit, sugars generated by photosynthesis in the plant are transported to the roots and taken up by the fungi, and nutrients, most importantly nitrogen and phosphorus, are absorbed by the fungal hyphae in the soil and conducted to the plant. Many plants with AM can survive without them, particularly on nutrient-rich soils, but those with EM and the specialist groups are normally **obligately** associated at least at some stage in their life cycle.

There are usually at least 20 species of mycorrhizal fungi in a plant community. Some fungi only infect one species or genus of plants, whereas others, especially AM fungi, can infect many species. Most plants can be infected by many species of fungi, so there will be a complex web of interactions. The growth form of the fungi in the soil varies greatly, with some producing a mass of hyphae close to the roots, others spreading more widely and thinly. AM grow out from a root to form an expanding fan and normally do not fuse with other neighboring fungi though may penetrate another plant. EM start growing similarly but when they meet hyphae from other EMs these can fuse, or **anastomose**, and the mycorrhizae form a network of interconnecting hyphae in the soil.

Effects of the fungi on plants

The main effect of the fungi is absorbing nutrients from the soil. Roots infected with AM retain root hairs and are able to absorb nutrients, but plants with EM and the specialist types generally lose all their root hairs and rely solely on the fungus. All mycorrhizal fungi have an extensive hyphal network in the soil. Nitrogen and/or phosphorus are vital plant nutrients that are frequently limited in supply and the fungal hyphae absorb these more efficiently and over a much larger soil area than roots alone, leading to enhanced growth of the plant, experimentally up to 200% greater. EM and the specialist types are also able to digest organic matter making organic nitrogen and phosphorus, which is normally unavailable for plants, available to be absorbed. This is particularly important in the nutrient-poor acidic soils in which these mycorrhizae are typical since it allows the mycorrhizal plants to absorb almost all the nutrients as they become available. By contrast some mycorrhizal associations can reduce a plant's growth where the soil is rich in nutrients.

Sugars from photosynthesis are conducted around the plant and the fungi absorb these as sources of energy. Once they are in the hyphae they are normally stored as complex sugars that cannot be reabsorbed by the host plant. Plants infected with mycorrhizae frequently have higher rates of photosynthesis than uninfected plants and these fungi at the same time transport more nitrogen and phosphorus into the plant. The relationship can change through the lifetime of a plant. Some seedlings (e.g. orchids) may absorb carbohydrate as well as nutrients from the fungus but as the plant matures, growth may be reduced by the fungus, so the relationship becomes parasitic first one way then the other. The 'saprophytic' plants (Topic M6) remain parasitic on their fungi throughout their lives. Different plant species respond differently to the interaction, with some fungi enhancing the growth of one species but inhibiting another. The interaction is further complicated, e.g. in legumes, by interaction with the nitrogen-fixing bacteria in root nodules.

Mycorrhizal fungi interact with other fungi in the soil, inhibiting free-living fungi involved in the decay of plant matter, so overall decay can be slower because of the mycorrhizae. They may protect the plant by preventing pathogenic fungi and bacteria from invading. The overall effect of these interactions is that the mycorrhizal plants will be at a competitive advantage over any others.

Community interactions

The effect of mycorrhizae in community ecology is not well understood and their importance has been seriously underestimated. In many habitats they are abundant and the fungi can be responsible for 25% of the respiration in the soil; over 30 m of hyphae have been recorded from 1 cm³. Difficulties of identification and isolating the fungi are serious impediments to study.

In early succession in a rich site and in agricultural soils, many plants have no mycorrhizae and it is in these habitats that the nonmycorrhizal Brassicaceae and Chenopodiaceae are most frequent. Many facultative AM plants colonize these areas and, as the community matures, more mycorrhizae invade. AM invade first and may initially be rejected by the host plants and be in competition for nutrients, the interaction changing as the community matures. There may be a succession of fungi; some plants such as willows, *Salix* spp., and she-oaks, *Casuarina* spp., can be invaded by either AM or EM and, in succession, the AM invade first. Interactions between mycorrhizal species are not well understood but it seems that during succession the fungus community normally becomes more diverse.

In a mature plant community mycorrhizae can have an enormous influence on plant diversity. From experiments in which artificial grassland communities were created with different numbers of mycorrhizal species, the community with the greatest diversity of fungus species had greater variability within the ecosystem with productivity differing in different parts and it supported the most diverse plant communities. Many plant species only survive in a community because of the fungi. Any one fungus species can enhance the growth of one plant species but inhibit another, whereas a different fungus can have the reverse effect in the same community.

In communities dominated by EM or ericoid fungi, i.e. coniferous forests, heathlands, many temperate woodlands and some tropical and subtropical woodlands on acid soils, all the dominant plants may be connected underground via the network of anastomosing mycorrhiza and the plants may compete for the nutrients within this network. Any plant that is not connected with the mycorrhiza is likely to be unable to obtain nutrients over much of the habitat. Early attempts to plant new forest species, e.g. in Australia and the Caribbean, especially conifers, failed because of the lack of mycorrhiza which had to be imported from their native habitats before the plantations were successful.

M2 NITROGEN FIXATION

Key Notes

Nitrogen fixation	Nitrogen gas cannot be used directly by plants but is fixed to N-containing compounds by free-living or symbiotic bacteria and cyanobacteria. Legumes with N-fixing bacteria in root nodules are of enormous importance ecologically and in agriculture. Some other plants have associations with a diverse group of other N-fixers.
The infection process	The legume root hair secretes a chemo-attractant which causes the bacteria to accumulate. They cause root hair curvature and enter the root cortex by an infection thread. Cell division is stimulated to form a nodule with vascular connections to the plant.
Molecular biology of nitrogen fixation	The complex interaction of host and bacterium requires the coordinated action of *NOD* genes in the host that encode nodulation and leghemoglobin and *nod*, *nif* and *fix* genes in the bacterium that encode infection, host specificity and components of N fixation.
Biochemistry of nitrogen fixation	Nitrogen fixation requires 16 moles of ATP per mole N and almost anaerobic conditions created by the oxygen-binding protein leghemoglobin. The bacteria in the cytoplasm are surrounded by the peribacteroid membrane. N fixation is catalyzed by dinitrogenase in three stages: (1) reduction of the Fe protein; (2) reduction of the MoFe protein by the Fe protein (requires ATP); (3) reduction of N by the MoFe protein. N is exported in high-N-containing compounds like amino acids or ureides.
Related topics	Movement of nutrient ions across membranes (E3) Uptake of mineral nutrients by plants (E4)

Nitrogen fixation N_2 is abundant in the aerial and soil environment, but unlike oxygen cannot be used directly (section E4). N fixation occurs in some free-living microorganisms (for instance some cyanobacteria, bacteria and archaea), but the most complex systems are seen in symbiotic association with some species of plants (*Table 1*). Most importantly, the great majority of legumes (members of the family Fabaceae) have symbioses with the N-fixing bacteria **Rhizobium** and **Bradyrhizobium**. Legumes form one of the world's most abundant and important families, including many trees, such as *Acacia* species, and agricultural and forage crops such as peas, beans, and clover. Their ability to fix nitrogen has been one of the major factors in their success and they dominate huge areas, especially on nitrogen-poor soils, and they enhance the nitrogen status for other plants.

Other plants that have symbiotic associations with nitrogen-fixers include the water fern *Azolla* that forms a N-fixing symbiosis with *Anabaena*, a cyanobacterium. This is used to fertilize rice-paddy cultivation. Some other plants, including

Table 1. Organisms involved in nitrogen fixation

Host plant	N-fixing organism
Alfalfa (*Medicago*)	*Bradyrhizobium meliloti*
Clover (*Trifolium*) Lentil (*Lens*) Pea (*Pisum*) Bean (*Vicia*)	*Rhizobium leguminosarum*
Bird's foot trefoil (*Lotus*)	*Rhizobium loti*
Soybean (*Glycine*)	*Bradyrhizobium japonicum*
Alder (*Alnus*)	Actinomycetes
Sweet gale (bog myrtle; *Myrica gale*)	Actinomycetes
Azolla Cycads (Cycadales)	*Anabaena* Cyanobacteria

bryophytes, have cavities with N-fixing cyanobacteria and some cycads (Cycadales) form such an association in their roots with N-fixing cyanobacteria that lose their photosynthetic ability and take carbohydrate from the plant. Alder (*Alnus*) trees, sweet gale (*Myrica gale*) and mountain lilacs (*Ceanothus*) form N-fixing symbioses with actinomycetes (a different group of bacteria).

The infection process

The legume–rhizobium symbiosis has been studied in most detail and will be considered here. The legume root hair secretes a chemo-attractant which causes the bacteria to accumulate (*Fig. 1*). The bacteria secrete lipo-chitooligosaccharides (**NOD factors**, see below) that cause more root hairs to be formed and alter root metabolism. They cause root hair curvature and the bacteria then attach to the hair by sugar-binding proteins called **lectins**. An **infection thread** is then formed, by which the bacteria pass through the root hair to the root cortex, where they proliferate. Cell division is stimulated to form a nodule within which N fixation occurs. The nodule has good vascular connections through which carbohydrates are supplied to the nodule and N-containing compounds are exported to the plant.

Molecular biology of nitrogen fixation

The complex interaction of host and bacterium requires the coordinated action of a number of key genes in both organisms. *Table 2* lists these genes and their functions. The **NOD gene family** in the host encodes aspects of nodule formation

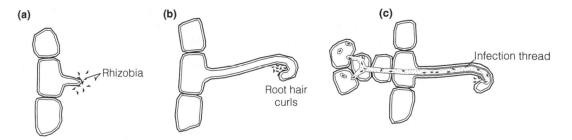

Fig. 1. *The infection process. Bacteria are attracted to a root hair (a), bind to it and cause root hair curling (b) and form an infection thread through which they penetrate the cortical cells (c).*

Table 2.   Genes involved in nitrogen fixation

Host genes and function	Bacterial genes and function
NOD genes – encode components of the N-fixing process. A gene family which are activated in order – first genes for the infection thread, then for nodule growth, and finally for leghemoglobin and enzymes involved in metabolizing fixed nitrogen	nod genes – induced by root exudates encode for enzymes producing chitooligosaccharides involved in bacterial attachment. nod genes – A, B, C and D – are located in the Sym (symbiosis) plasmid nodD activates nodA–C and induces nodE–H which give host specificity to the bacterium
	nif genes – encode for dinitrogenase components. Also found in free-living N fixers
	fix genes – encode for ferredoxin and other components of the bacteroid

and infection and the production of **leghemoglobin**, which binds oxygen, lowering the concentration of O_2 in the nodule. The **nod family** in the bacterium encodes enzymes that synthesize **NOD factors** (see above). One nod gene, nodD, is activated by root exudates and produces a product that regulates the other nod genes. Later in infection, the **nif** and **fix** genes are active in the rhizobium. These produce the enzymes and electron transport pathway required for N fixation.

Biochemistry of nitrogen fixation

Nitrogen fixation is a very energy-demanding process, requiring at least 16 moles of ATP for each mole of reduced N. It also requires almost **anaerobic conditions** as the key enzyme complex, **dinitrogenase**, is rapidly inactivated by oxygen. In the N-fixing nodule, the energy required is supplied by the plant, and low oxygen conditions by the oxygen-binding **leghemoglobin**. The bacteria are contained in the cytoplasm of cells in the nodule as **bacteroids** surrounded by a double membrane – the **peribacteroid membrane**. The bacterium supplies the plant with reduced nitrogen compounds like ammonium and amino acids. The plant benefits by increased growth and the bacterium from a food supply and enhanced environment in which to grow and replicate.

Overall equation:

$$8H^+ + 8e^- + N_2 + 16ATP \rightarrow 2NH_3 + H_2 + 16ADP + 16P_i$$

The process is catalyzed by the enzyme **dinitrogenase**, a protein complex made of two components – a large component, the **MoFe protein** and a smaller component, the **Fe protein**. Dinitrogenase functions in stages:

(i)   reduction of the Fe protein (usually by the electron donor ferredoxin);
(ii) reduction of the MoFe protein by the Fe protein (this step requires ATP);
(iii) the MoFe protein then reduces N ($N_2 + 8H^+ \rightarrow 2NH_3 + H_2$).

After fixation, nitrogen is exported to the plant in the xylem flow, not in the form of ammonium, but as high-nitrogen-containing compounds like amino acids or ureides, depending on species.

M3 INTERACTIONS BETWEEN PLANTS AND ANIMALS

Key Notes

The herbivores

Insects and mammals are important plant herbivores. Many insects are specific to one plant group. Some form galls. Mammals are not as specific, but certain plants are preferred or avoided, particularly by the smaller animals.

Insects and plants

Plant defenses against insects are mainly secondary compounds along with physical defenses such as a thick cuticle. Secondary compounds may be directly toxic, unpalatable or digestion inhibitors. When an insect evolves to be able to utilize one toxic plant group, it may spread and stimulate further evolution. Many plant families are associated with particular insect herbivores. There are more secondary compounds in tropical plants than temperate.

Vertebrates and plants

Vertebrates influence the entire plant community. Grasses have meristems at nodes and withstand grazing; many grasslands and savannahs are maintained by grazing mammals. Woodlands can be prevented from regenerating and some communities have cyclic patterns. Plants have physical defenses such as spines and stinging hairs as well as secondary compounds. A few plants provide nest sites for aggressive ants which deter other animals.

Herbivores and plant populations

Many plant populations may be controlled in numbers by herbivores, mainly insects, and this is the basis of biological control. If a plant is introduced to an area without its herbivores it may spread to places in which it would not occur in its native region. Monocultures are susceptible to herbivores and in natural communities they stimulate plant diversity. Short-lived plants may live longer if their reproductive capacity is reduced in one season.

Related topics

Plant communities (L3)
Populations (L4)
Regeneration and establishment (H5)

Fungal pathogens and endophytes (M4)

The herbivores

Plants provide a wide range of food for animals, and ultimately, all animals rely on plants as the primary producers. A few interactions between animals and their food plants are of mutual benefit, such as the pollination of flowers and dispersal of fruits and seeds (Topics G5 and H3), but most plant–herbivore relationships involve the animal eating and damaging a part of the plant. This usually involves leaves and growing shoots but can involve almost any part of the plant. The most important herbivores of living plants are insects and mammals along with other vertebrates, and molluscs in places. Some invertebrates, including insects such as bugs and grasshoppers, eat a range of plants but many insects, particularly the larvae of moths and butterflies, some beetles and others, are specific to one family, genus or even species. They may be abundant on some plants and can defoliate or kill them. Some insects form galls on plants, stimulating the plant to grow unusual structures, which enclose the insect larva, e.g. 'oak apples' on the European oak, *Quercus robur*.

Vertebrate herbivores mostly eat a wide range of plants, but in general the smaller the vertebrate the more selective it is. Certain plants are frequently preferred, such as many legumes (Fabaceae) which are rich in protein associated with their nitrogen-fixing capacity (Topic M2), or avoided owing to their toxicity, unpalatability (e.g. high silica content) or their physical defenses such as spines. Microorganisms in the gut are vital in the digestion of plant material since vertebrates cannot digest many of the structural molecules that form plants, such as cellulose and lignin, without bacteria.

Insects and plants

Plant defenses against insect herbivores include physical barriers such as a thick cuticle but the most important deterrents are secondary compounds (*Table 1*; Topic F5). These must be localized in the plant separately from sites of metabolism, e.g. in the cell wall or vacuole, if they are not to be toxic to the cell synthesizing them. They work in many different ways. Some are directly poisonous such as the monoterpene **pyrethroids**, or indirectly such as the **phytoecdysones** that mimic insect molting hormones (ecdysones). Others inhibit digestion, such as tannins or enzyme inhibitors, and others make the plant unpalatable or irritating to the herbivore. In many plants, young leaves are the most nutritious since they are softer than mature leaves, have a greater water content and have fewer secondary compounds. Many insects eat young leaves mostly or exclusively, and insects feeding on mature leaves nearly always grow more slowly.

Insects can evolve the ability to digest the toxic secondary compound in any one plant group, or can eat the leaves avoiding the toxin. Some insects can sequester the toxic substance unchanged in their bodies and use it for their own defense. If an insect group manages to utilize a toxic plant group, the insects are likely to flourish in the absence of competition and the volatile compounds given off by the plant may act as a signal for the insect to find it as a food plant. These interactions normally involve a specialized insect group, such as a family of butterflies, on one plant family. The evolution of insect herbivory in this way resembles an arms race. Plants evolve a novel toxic secondary compound and spread in the absence of much herbivory, then a group of insects evolve the ability to deal with that compound and so spread on the plants. This has happened many times, e.g. the toxic milkweed family is fed on by monarch butterflies; the passion-flowers are food plants for *Heliconius* butterflies. Many species may be involved. It is a situation in which diversity in one group stimulates diversity in the other and this may be one of the major stimuli for the production of biodiversity generally.

Certain plant families are susceptible to numerous insects, such as the cabbage family, whereas others generally have very few, such as the largely tropical Rubiaceae (which includes coffee). Tropical plants generally have a greater quantity and range of secondary compounds than temperate plants and there is a strong negative correlation between quantity of toxic compounds and latitude. This may be true even within one species, such as the white clover in which some plants are cyanogenic (i.e. producing cyanide, mainly deterring molluscs): in southern Europe almost all plants are cyanogenic, but in northern Europe very few are. This implies that there is a cost to producing secondary compounds (Topic F5).

Table 1.   *Examples of secondary compounds produced by plants*

Class of compound	Examples	Effects	Example species
Terpenes	Monoterpenes (pyrethroids)	Insecticidal	Pine (resins); *Chrysanthemum*
	Sesquiterpenes gossypol	Insecticidal	Cotton
	Diterpenes phorbol esters	Skin irritant to mammals	Euphorbias
	Triterpenes phytoecdysones	Insecticidal (induce molting)	Common polypody fern (*Polypodium vulgare*)
	Polyterpenes latex	Discourage herbivory	Rubber tree (*Hevea brasiliensis*)
Phenolics	Coumarins furanocoumarins	DNA repair damage; wide range of animals	Umbellifers (parsnip)
	Lignin	Not digested by most herbivores	Woody plants
Tannins	Various, condensed and hydrolyzable	Bind proteins and prevent digestion	Many species
Alkaloids	Various, including nicotine, morphine, strychnine, cocaine	Various, neuroactive or inhibitors of enzymes, etc.	Various
Cyanogenic glycosides	Various	Release hydrogen cyanide	Cassava (eliminated by adequate processing)
Nonprotein amino acids	Various (e.g. canavanine)	Block protein synthesis or uptake; incorporated to give nonfunctional protein	Various; Jack bean (*Canavalia ensiformis*)
Enzyme inhibitors	Proteinase inhibitors	Prevent protein digestion	Some legumes, tomato

In temperate latitudes it has been shown that those trees that are most abundant or have been in an area for the longest time have more species of insect herbivore than more recent arrivals or rarer trees.

Vertebrates and plants

Vertebrates can have major effects on the structure of a plant community. Grazing mammals maintain most of the grasslands of the world. Grasses themselves are well adapted to withstand grazing pressure since their meristems are at the nodes rather than the tips of stems (Topic L3). Without the constant presence of grazing mammals, many grasslands would be invaded by shrubs or trees. In many partially wooded areas, such as savannahs, the density of trees is kept low by grazing mammals and in some places there can be cyclic changes in the community with mammals creating open spaces which may then suffer drought. The drought may kill many of the mammals, allowing trees to regenerate. The most significant mammals in these places are the large herbivores, such as elephants, which occurred in almost all regions until their numbers were depleted by humans within the last 100 000 years or so. In all grasslands the particular herbivores present and their relative abundances have a profound effect on the composition of the plant community.

In woodlands, grazing and browsing mammals can prevent regeneration and keep the understorey open. Continuing dense populations of herbivorous mammals can open up a woodland by stopping regeneration. Some mammals, such as pigs, root in the soil and disturb the vegetation allowing gaps to appear for short-lived plants to colonize. Tree-dwelling herbivores, such as sloths and some primates, rarely appear to influence the plant community significantly.

Adaptations by plants to withstand grazing often involve physical defenses. Spines or thorns deter herbivorous vertebrates and are particularly frequent in places where growth is slow such as arid areas. Stinging hairs are a feature of certain families such as the nettle family (Urticaceae). Many plants have thick waxy cuticles, resin ducts or large deposits of silica in their leaves making them unpalatable. Toxic secondary compounds (*Table 1*) deter many vertebrates in the same way as invertebrates. Many tree leaf herbivores vary their diet to avoid too much of any one compound. Most starch-rich roots and tubers have toxic compounds in them, and for human use these must be reduced by plant breeding or, in cassava, by cooking in a particular way.

A few plants, such as the bullshorn acacia of central America, provide nest sites and oil bodies for aggressive ants which attack any herbivore, vertebrate or invertebrate, and may eat off any invading plant such as a climber as well. In many ways, these ants perform the same function as a secondary compound.

Herbivores and plant populations

Some plant populations may be controlled in their numbers by herbivores. This is the basis of biological control of plant pests. Examples include the reduction of the prickly pear cactus population introduced into Australia when the cactus moth was introduced, and a similar reduction of the introduced European St Johnswort (Klamath weed) in Canada by a *Chrysolina* beetle. These examples of introduction of a plant across continents without their herbivorous animals show how important such herbivores are; until these 'experiments' it was not clear that the herbivores were controlling the plant populations in their native lands. In addition, the St Johnswort apparently changed in its ecology and colonized open places in Canada that it does not occupy in its native Europe. It was reduced to semi-shade by the insect, which preferred open places. Many plants may be controlled in their numbers by herbivorous insects.

Herbivory, particularly of seedlings, along with fungal infection may prevent a plant regenerating or growing near an adult since the adult provides a source of herbivores. In agriculture, monocultures are always more susceptible to pests than planting many species together and pests can build up in any one place making rotational cropping more effective. In a natural plant community the result will be an increase in plant diversity in the community as a whole, and this may be one of the reasons why tropical rainforests are so diverse.

Some plants, particularly ephemerals and other short-lived plants, may live longer as a result of herbivory. If the flowers or other parts of a shoot are removed and the plant fails to reproduce in any one year or produces below a certain proportion of its potential seeds, it may live a further year or longer.

M4 FUNGAL PATHOGENS AND ENDOPHYTES

Key Notes

Fungal diseases	Fungi are important saprophytes that recycle dead plant material. They also cause a wide range of plant disease. Fungal spores germinate on a leaf surface or wound site and form a mycelium through the plant. Their life cycles may involve more than one host plant (e.g. wheat rust, which is heteroecious) or a single host plant (e.g. smut fungi, which are autoecious).
Plant defenses against fungal diseases	Defenses may be mechanical, preventing spore adherence or penetration, or chemical, e.g. saponins that disrupt fungal membranes. Other defenses are induced when a pathogen attacks, genes being activated by an elicitor binding to a plasma membrane receptor. In a hypersensitive response, cells fill with phenolic compounds and form a necrotic lesion. A lignin or callose barrier may form. In many species, antimicrobial phytoalexins are produced. Systemic acquired resistance occurs when resistance is enhanced as a result of an infection. Fungi may stimulate changes in metabolism or phytohormone levels in the plant leading to changes in growth form.
Fungal endophytes	Fungi are often found growing freely within plants which have no, or minimal symptoms. They can be particularly dangerous to humans as they can be very toxic. Best known is ergot, a fungus growing in rye that generates a toxic precursor to LSD. Some may become parasites as the plant ages.
Related topics	Amino acid, lipid, polysaccharide and secondary product metabolism (F5) Stress avoidance and adaptation (I5) Mycorrhiza (M1)

Fungal diseases   Different fungal species possess a wide range of methods capable of degrading a vast range of materials, including lignin and cellulose. Many fungi are **saprophytes** and live by degrading material from plants that have already died, and these are essential in soil formation. Others are **pathogens** that cause a wide range of plant diseases resulting in significant losses in agriculture. *Table 1* gives some examples of fungal diseases and the organisms that cause them.

Infection occurs when a **spore** germinates on a leaf surface or wound site. The fungus penetrates, often through stomata, and forms fine threads of cells (**hyphae**) which ramify to form a mat (**mycelium**) throughout the plant. Infection in cells frequently resembles the arbuscules of arbuscular mycorrhizae (Topic M1).

Fungi have complex life cycles, which may involve one or more host plant (**heteroecious** life cycle). **Wheat rust**, *Puccinia graminis* for instance, a **Basidiomycete**, invades two host plants. Commonly, the first host is the

Table 1. Examples of fungal diseases and the causal fungi

Species and plant disease	Phylum of fungus
Various species (soft rots)	Zygomycota
Corn (brown spot)	Chytridiomycota
Various (powdery mildew)	Ascomycota
Elm (Dutch elm disease)	
Cereals (smuts, rusts)	Basidiomycota

barberry (*Berberis* sp.) from which **aeciospores** (containing two nuclei, one + strain, the other – strain) are produced which infect the wheat. The wheat then produces **uredospores**, which spread the infection in the wheat field. The fungus overwinters as diploid **teliospores**, produced when the two (+ and –) nuclei fuse (**karyogamy**). Meiosis occurs in spring to release **basidiospores** (+ or – strain) which then infect the barberry host (*Fig. 1*). **Smut fungi** have a simpler life cycle involving only one host (**autoecious**) and produce masses of **teliospores** (the resting stage of the fungus). They are known to infect at least 4000 species of flowering plants, many of economic importance.

Plant defenses against fungal diseases

Some defenses are **mechanical barriers**, for instance the **waxy cuticle**, which prevents spores from adhering to the leaf or the fungus from penetrating the leaf. Pathogenic fungi can often only infect a wound site or enter the leaf through stomata. Other defenses are **endogenous chemicals**, for instance the **saponins** are **triterpenes** (Topic F5) produced by plants which disrupt the membranes of fungi.

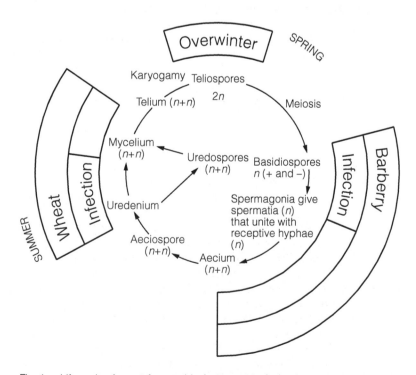

Fig. 1.   Life cycle of a rust fungus, black stem rust of wheat.

Other defenses against fungi are not induced until a pathogen attacks. This involves expression of **defense genes** which are activated by a signaling cascade (Topic J3) within the plant. This is triggered by the binding of a product of the pathogen (an **elicitor**) to a **receptor** in the plant plasma membrane (*Fig. 2*). In some instances, a **hypersensitive response** occurs, where cells around the fungus fill with **phenolic** compounds and die to form a **necrotic lesion**. This isolates the point of infection and may stop further spread of the fungus. The infection site may also be surrounded by deposits of **lignin** or **callose** (Topic F5) synthesized by surrounding cells which form a physical barrier to the spread of the infection. Another group of antimicrobial compounds, the **phytoalexins** are produced in many species as a response to infection. Phytoalexin is a generic name for a range of compounds that differ between species. After infection, expression of genes for the enzymes of their biosynthetic pathway is initiated and high levels of phytoalexins accumulate. An infection which has been successfully overcome may lead to the plant being more resistant to further infection. This **systemic acquired resistance** involves the spread of signaling molecules (possibly **salicylic acid**) within the plant (*Fig. 2*).

Some fungi stimulate changes in the metabolism of the plant. The plant's respiration rate increases and sometimes it shifts more to the pentose phosphate pathway rather than the tricarboxylic acid pathway (Topic F4). This allows for the production of more nucleic acid (which use pentoses) and NADPH, important for the fungus. The fungus may prevent much or all of the export of carbohydrate from leaves to storage organs. The fungus may also stimulate changes in phytohormone levels (Topic J2) which lead to unusual growths, one of the most common being 'witches brooms', a dense group of short stems in many trees such as birches (*Betula* spp.).

Fungal endophytes

Plants are often found with fungal hyphae growing freely within their tissues. They often produce either no, or minimal symptoms of their presence and may provide advantages to the host in deterring herbivores and other pathogens.

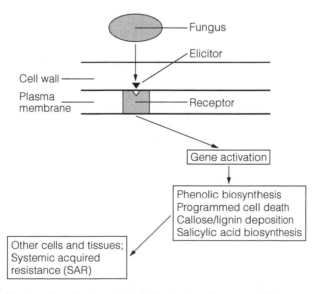

Fig. 2. Simplified model of the induction of a range of defense genes by a fungal elicitor.

These **fungal endophytes** can be particularly dangerous to humans. One of the best known, **ergot** (*Claviceps purpurea*), grows in **rye** and generates **lysergic acid amide**, the precursor of lysergic acid diethylamine (**LSD**). LSD is hallucinogenic and fatal in relatively small doses and throughout history, communities have been devastated by outbreaks undetected in the rye harvest. Other endophytes include *Sphacelia typhina*, an ascomycete which grows in the prairie grass *Festuca arundinacea* and causes cattle disease and death. Detection and elimination of endophytes in agriculture and food processing is an important area.

There is no absolute distinction between fungal endophytes and parasites and some fungi may remain as endophytes for many years before starting to damage a plant significantly. Many only reproduce sexually once the plant or the infected part is killed.

M5 BACTERIA, MYCOPLASMA, VIRUSES AND HETEROKONTS

Key Notes

Bacteria

Bacterial diseases include blights, wilts and soft rots. Most plant-infecting bacteria are Gram negative and rod-shaped, except streptomycetes which are filamentous. Some bacteria are beneficial; Rhizobia fix nitrogen and Agrobacteria have been used as the basis for plant genetic engineering.

Mycoplasma

Mycoplasma are bacteria-like organisms, lacking a cell wall, that infect sieve tubes. Commercially damaging phytoplasmas exist for many species.

Viruses

Viruses consist of a protein capsid encasing DNA or RNA and replicate using the host cells. Most plant viruses contain RNA, though a few contain DNA. They are transmitted by a vector. Infection may be limited to one site, or they may travel systemically through plasmodesmata or phloem. Control methods include: meristem tip culture; controlling vectors; and genetic modification to introduce viral capsid genes into plants.

Heterokonts

Heterokonts include oomycetes, which include some important plant pathogens. They produce biflagellate zoospores and resistant oospores that can survive many years in soil before germination. The best known is *Phytophthora infestans* which causes potato blight.

Related topics | Nitrogen fixation (M2) | Fungal pathogens and endophytes (M4)

Bacteria

Bacterial diseases can affect almost any plant and can result in great crop losses. Almost all plant-infecting bacteria are **Gram negative** and **rod-shaped**, although members of the genus *Streptomyces* will infect plants and are **filamentous**. Some bacteria are beneficial; *Rhizobium* fixes nitrogen in symbiotic association in root nodules (Topic M2). *Agrobacterium*, which cause plant galls by inserting bacterial genes into the plant genome, have been used as the basis for plant genetic engineering (Topic K3). The range of diseases caused by other bacteria include: **blights** in which necroses, areas of dead tissue, form and spread rapidly through stems, leaves and flowers; **wilts** in which xylem tissue is destroyed and the plant wilts and dies; and **soft rots** where fleshy tissue is decayed, particularly in storage tissues. *Table 1* gives examples of bacterial diseases.

Mycoplasma

Mycoplasma are bacteria-like organisms lacking a cell wall. Mycoplasma-like organisms that infect plants are known as **phytoplasmas** and many are commercially important diseases. Carrots, peaches, pears, ornamental flowers and coconuts can all be seriously affected. Phytoplasmas normally infect **sieve**

Table 1. Examples of bacterial diseases

Genera of bacteria	Symptoms
Rods	
Agrobacterium	Galls (swellings); hairy roots
Erwinia	Blights (necrosis), wilts (wilting), soft rots (decay)
Pseudomonas	Wilts, galls, blights, cankers (swellings)
Xanthomonas	Cankers, blights, rots
Clavibacter and Rhodobacter	Rots, cankers, wilts
Filamentous	
Streptomyces	Scabs

tubes. They are conveyed from plant to plant by **vector** organisms, such as aphids, that feed on sieve-tube contents.

Viruses

Viruses are noncellular structures containing DNA or RNA that use the host's synthetic capability to replicate. Many of the viral diseases known (of which there are more than 2000) are virtually symptom-less other than marked reductions in yield. Examples include tomato mosaic virus. The virus consists of a coat (**capsid**) made of protein, encasing DNA or RNA that contains the information for infection and replication of the virus. Most plant viruses contain **RNA** as their genetic information, though three (the badnaviruses, caulimoviruses and geminiviruses) contain **DNA**. Viruses are frequently transmitted from plant to plant by a **vector organism**, e.g. an insect, but seldom infect the vector. Viruses may have a limited site of infection, or may travel **systemically** through plasmodesmata from cell to cell, or through the phloem. Movement through plasmodesmata involves the synthesis of **movement proteins**, encoded by the virus, which modify the plasmodesmata. Control methods include: **meristem tip culture** (using plant tissue culture; Topic K2) in which plants are propagated from virus-free tissue; developing **resistant strains**; controlling vectors. Recently, genetic modification (Topic K3) has been used to introduce viral capsid genes into some plants, thereby preventing the virus from replicating. Diagnosis of viral disease is often difficult, with few visual symptoms. Viruses can build in plants while old individuals can be weakened by them.

Heterokonts

Heterokonts are eukaryotic organisms with one long and one short flagellum. They include diatoms and algae and the phylum Oomycota, the **oomycetes**, which include some important plant pathogens. The Oomycota produce **biflagellate zoospores**, the means of asexual reproduction of the organism, and can also reproduce sexually to produce **oospores**, which are resistant and can survive many years in soil before germination. Oomycetes form hyphae within infected tissue and are very destructive. Examples are *Phytophthora infestans*, the **potato blight**, which reached Europe in the 1840s and spread with devastating consequences of famine and emigration; *P. cynamoni*, which kills avocado trees; *Plasmopara viticola* which causes **downy mildew** of grape vines and *Pythium* species which cause **damping-off** of seedlings.

M6 PARASITES AND SAPROPHYTES

Key Notes

Parasites

There are 3000 species of parasitic plants, 20% entirely dependent on their hosts. Of plant parasites 60% are root parasites, 40% stem. Most are herbaceous ephemerals or perennials of grasslands and disturbed areas, but a few are woody and sandalwoods are trees. Many have numerous small seeds. Stem parasites have fleshy animal-dispersed fruits.

Growth of parasites

In root parasites, root exudates from a potential host stimulate germination and a radicle to grow to the host root. Once attached most parasites establish a haustorium connecting the vascular systems to absorb some of the host's nutrients. Stem parasites attach in a similar way to above-ground parts.

Economic importance of parasites

Parasites can reduce crop yields by 30%, the most important pests being *Striga* species on cereal crops in Africa and Asia, *Alectra* species on groundnuts and others and *Orobanche* species mainly on beans. These are all members of the Scrophulariaceae family group. One unrelated pest is *Cuscuta* species, stem parasites of low host specificity.

Saprophytes

Saprophytes obtain nutrients from decaying plant parts. Plants can only do this using mycorrhizal fungi and all mycorrhizal plants are partially saprophytic. A few specialized members of the two mycorrhizal family groups, the heathers and orchids, have no green parts and depend entirely on their fungi. The gametophytes of some spore-bearing plants are saprophytic.

Related topics

Roots (D2)
Ecology of different growth forms
(L1)

Mycorrhiza (M1)

Parasites

The great majority of plants use only sunlight and inorganic molecules for all their energy and nutritional needs, but there are some exceptions. There are around 3000 species of plants that are at least partially parasitic on other plants, i.e. approximately 1% of all flowering plants. Only about 20% of these are entirely dependent on their hosts, the remainder having some green parts so potentially making some of their own sugars (known as **hemiparasites**). A few of these can live without a host plant, although they rarely do. At least eight plant families have parasitic members, some of which also have nonparasitic species, so the habit has probably evolved several times within the flowering plants. Among land plants only angiosperms have become parasitic (one unusual conifer of New Caledonia may be partially parasitic on other conifers). About 60% of plant parasites are attached to the roots of their hosts, the remainder to stems. A few are specific to a single host species but most can parasitize a range of species, often within one family. They are distributed throughout the world, favoring disturbed places, though root parasites are particularly common in grassland of the Mediterranean and subtropical climates.

There is a range of life cycles among parasites. Most are herbaceous ranging from ephemeral plants to long-lived perennials, but there are several woody plants such as the **mistletoes** (Topic L1) among stem parasites, and, among root parasites, **sandalwoods**, some of which are medium-sized trees. Their reproduction is similar to nonparasitic flowering plants, and the largest known single flower, up to 1 m in diameter, is a parasite of the south-east Asian forests, *Rafflesia*. Most **root parasites** have small seeds and some, such as the broomrapes, have dust-like seeds produced in great quantity, dispersed by the wind. Almost all **stem parasites** have fleshy fruits dispersed mainly by birds. Those parasites that are specialized to one or a few host species tend to be longer-lived and produce more, smaller seeds than the generalists, many of which are short-lived.

Growth of parasites

In root parasites, the seeds germinate in response to **chemical signaling** from the roots of a host plant. A sesquiterpene chemical, **strigol**, has been identified as an important stimulant but this substance comes from nonhost plants too and may be secreted by microorganisms associated with the root rather than the root itself. Other substances, as yet unidentified, must be important. The best studied parasite is *Striga hermonthica*, a pest on several species of cereal. It seems that the parasite has several strains, each strain on one host species and responding only to exudates from that host.

Once a seed has germinated, a short radicle may grow and this must come into contact with root hairs from the host or it will die. It can then stick onto the root hairs and penetrates the root by mechanical pressure and through the action of hydrolytic enzymes. It may meet some resistance at the endodermis but once it gets through that it reaches the vascular tissue and penetrates the xylem vessels through pits or by dissolving the cell walls, making direct vascular connection with its host. Once it has penetrated, it obtains all its nutrients and energy from the host. The attachment structure is known as the **haustorium** (*Fig. 1*). In most parasite species, immediately above the site of connection the haustorium becomes swollen and full of vascular tissue and parenchyma cells. In hemiparasites, once the shoot grows out of the ground it becomes green and can photosynthesize; it reproduces by seed.

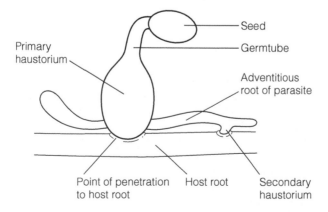

Fig. 1. A germinating seed of a root parasite such as Striga sp. with a primary and secondary haustorium where it penetrates a host root.

Some parasites do not produce well-defined haustoria and members of the Rafflesiaceae grow numerous thin strands resembling fungal hyphae through its host cortex and cambial cells. Stem parasites develop in a similar way to root parasites, forming haustoria making a connection between the vascular systems of the stems.

Economic importance of parasites

Parasites take up water and nutrients from the host and this weakens the host plants. Sometimes a parasite causes visible damage resulting in wilting, reduction in plant size, lower yield and lower quality of the crop and can cause up to **30% crop loss**. They are important agricultural **pests** in Africa, parts of Asia and the Mediterranean and their spread to other areas via contaminated seed is causing concern for food production.

The most serious pests are species of **witchweed**, *Striga*, particularly *S. hermonthica*, which seriously reduce yields of maize, rice, sorghum, millet and sugar cane in the African Sahel and Asia. They are root hemiparasites in the Scrophulariaceae, a family containing many hemiparasites. The related *Alectra* species, mainly found in West Africa, parasitize groundnut, cowpea and sunflowers. The **broomrapes**, *Orobanche* spp., in the Orobanchaceae, closely related to Scrophulariaceae, can be serious pests of beans, lentils, tobacco, tomatoes and sunflowers, *O. crenata* causing serious crop losses in the Mediterranean. A quite different group, the **dodders**, *Cuscuta* species, related to the bindweeds, Convolvulaceae, are stem parasites of dicots with low host specificity and can attack many different crops.

In natural plant communities they may reduce diversity in some places but their ecological importance is minor.

Saprophytes

A **saprophyte** is an organism that decays dead vegetation. No plant of any kind is capable of that, but many **fungi** are saprophytic, including ectomycorrhizal fungi and those associated with heathers and orchids (Topic M1). Flowering plants that are dependent on these mycorrhizal fungi for all their nutrients and energy are referred to as 'saprophytes', although any plant that has such a mycorrhizal association must be regarded as, in part, saprophytic. All orchids and some other

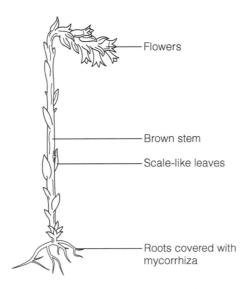

Flowers

Brown stem

Scale-like leaves

Roots covered with mycorrhiza

Fig. 2.   Yellow birdsnest, Monotropa hypopitys.

plants are saprophytic when they first germinate, being dependent on their mycorrhizae. Most later grow green stems and leaves and are partially autotrophic, but many terrestrial orchids live on woodland floors where photosynthesis is severely limited and many of these must remain mainly saprophytic.

Fully saprophytic plants are found among two of the highly specialized mycorrhizal groups, the orchids, such as the coralroot and birdsnest orchids, and the birdsnests (*Fig.* 2) or Indian pipes among the heather family group (usually in a separate family Monotropaceae). They are entirely dependent, in effect parasitic, on their mycorrhizae and have small restricted root systems, some rounded like a bird's nest or branching like a coral giving the plants their common names, and having scale-like leaves and no green parts at all. Most live in deep shade on woodland floors. The plant's mycorrhizae may be connected with trees and the genus *Armillaria* forms mycorrhizal associations with these plants and pathogenic associations with trees. Nutrients have been traced passing from tree to saprophytic plant via the mycorrhizae. In their form and color the fully saprophytic plants closely resemble root parasites.

The gametophytes of some clubmosses, horsetails and ferns (Topics P2, P3, P4) and a few bryophytes (Topic O2) are entirely subterranean and live saprophytically with their associated mycorrhizae, although in most, the sporophytes are autotrophic.

M7 CARNIVOROUS PLANTS

Key Notes

Carnivorous plants
A few plants supplement their nutrients by trapping and digesting animals. All can photosynthesize as well and have roots capable of absorbing water and nutrients but, typically, they live in acid nutrient-poor environments, often bogs.

Pitcher plants
There are three unrelated groups of pitcher plants, all of which have furled leaves sealed along a keel in the shape of a pitcher. Sugar-secreting glands occur on a flap and attract insects and other animals and these then cannot get out because of downward-pointing hairs and waxy surfaces. They are digested in collected rain water and a mix of digestive enzymes and absorbed by cells at the base of the pitcher.

Other carnivores
Sundews and butterworts have sticky secretions on their leaves that catch animals. Some other plants have sticky secretions that trap insects and carnivorous plants probably evolved from this. The rare Venus fly-trap has an elaborate hinge mechanism triggered by an electrical impulse that responds rapidly and the waterwheel plant is similar but smaller and under water. The aquatic bladderworts have an equally elaborate trap mechanism involving a partial vacuum that sucks small invertebrates into the bladder.

Related topics
Nastic responses (I3)   Saprophytes and parasites (M6)

Carnivorous plants

Carnivorous plants trap animals using **specialized leaves** or parts of leaves. All carnivorous plants are green and can photosynthesize but rely on catching animals to supplement their nutrient supply. They are characteristically plants of **nutrient-poor** acidic and often boggy soils, although a few species have colonized richer sites. Some are aquatic. These sites are especially poor in available nitrogen and supplementing this and other nutrients by carnivory is a selective advantage. The root systems of all carnivorous plants are poorly developed but can absorb some nutrients and most carnivorous plants can survive for a time without catching any animals.

The true carnivorous plants have sophisticated **traps**, sometimes involving movements by the plant, into which insects and other small invertebrates and even occasionally small vertebrates are enticed. There are about 400 species of carnivorous plants in at least five families and they appear to have arisen independently several times.

Pitcher plants

There are three separate families of pitcher plants, the **Sarraceniaceae** from North and South America, the **Nepenthaceae** of Asia, and the **Cephalotaceae** of Australia. All have a funnel-shaped pitcher (*Fig. 1*) deriving from part or all of a curled leaf fused at the edges with a keel. These families are not closely related

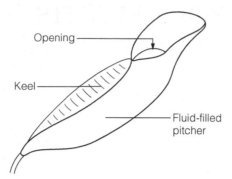

Fig. 1.   A pitcher (modified leaf) of the pitcher plant Sarracenia.

and the pitchers look alike through similar lifestyle, and are a remarkable example of convergent evolution. Details of the structure differ in the three families but all have a flap of leaf above the pitcher, mainly used as an attractant, sometimes with **sugar-secreting glands**, leading insects to the lip. On the lip of some *Nepenthes* pitchers are a group of fully wettable cells on which insects, particularly ants, slip on wet days. On dry days the ants can get out, effectively prebaiting the trap. On the insides of the pitchers are downward-pointing hairs and waxy surfaces making escape difficult or impossible. Pitchers catch rain water and the simplest drown their prey and allow bacteria to digest them. The more elaborate pitchers secrete **enzymes** such as proteases, lipase, esterase and other enzymes and acid into the pitcher. Small animals can be digested quickly, often within 2 days, leaving just the chitinous husk of an insect. Bacteria living in the pitchers may digest this too. Cells in some areas at the base of the pitcher have no cuticle and nutrients can be absorbed.

Despite the digestive enzymes in pitchers, a small community of fly larvae, crustacea and spiders live inside pitchers in some parts of their distribution, apparently resistant to the enzymes.

Other carnivores   Two unrelated plant families have leaves that act like fly-paper, catching insects in a **sticky secretion**. The **sundews** and their relatives (Droseraceae) have specialized glandular hairs on their leaves (and on stems in some) and the **butterworts** (*Pinguicula* spp.) have leaves that are sticky all over the surface. In some, the leaves roll around the animal once it is caught. They have **digestive glands** in the leaves secreting a similar range of enzymes to the pitcher plants and, with the help of bacteria secreting chitinase, insects are digested quickly. Many plants that are not regarded as carnivorous produce sticky secretions, normally on stems, buds or seeds, and insects and other small animals can become trapped and killed in these. If these animals decay while remaining attached to the plant some of these species may be able to absorb the nutrients from them and this is likely to be how carnivorous plants first evolved.

The most elaborate of carnivorous plants are the famous **Venus fly-trap**, *Dionaea muscipula*, a rather rare plant of bogs in eastern USA, and the aquatic waterwheel plant, *Aldrovanda vesiculosa*, both related to the sundews, and **bladderworts**, *Utricularia*, with about 250 species related to the butterworts. The Venus fly-trap has a modified leaf in which two lobes are joined by a hinge region (*Fig. 2*), in which **motor cells** are capable of rapid turgor changes. When

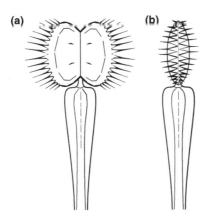

Fig. 2.   Leaf of the Venus fly-trap, Dionaea muscipula: *(a) open, (b) shut.*

an insect lands on the leaf surface and touches the two to five sensitive hairs, an electrical stimulus is sent through the leaf triggering a rapid release of pressure in the motor cells if the hairs are stimulated several times. How this trigger works is still little known. This shuts the two halves of the leaf sufficiently to trap the insect and the leaf completes its shutting more slowly through normal growth processes, crushing any animal that is caught. The waterwheel plant has a similar but much smaller aquatic trap no more than 2 mm in diameter in which it catches tiny aquatic animals.

The bladderworts have elaborate tiny bladders under water, ranging from 0.25 to 5 mm in diameter. These are spherical to conical or cylindrical and water is extracted from them until the walls become pulled inwards to develop considerable suction pressure. There is a trapdoor that is kept closed but under tension. This is released by an animal that triggers surrounding hairs and the animal is sucked rapidly in by the pressure, after which the door immediately closes again, all this taking about one hundredth of a second. The suction pressure is restored after an hour or two.

N1 PLANTS AS FOOD

Key Notes

The range of food crops

Numerous species are edible but we rely on only 20 species for 90% of our plant food and wheat, rice and maize for over half. Seeds and roots or tubers, mainly providing starch, are the staple plant foods. Leaf crops provide mainly vitamins and roughage and some seeds and fruits are rich in protein or fat.

Origin of staple crops

Wheat, oats and barley all originate in south-west Asia and were the main crops of the earliest civilizations. Rice comes from China and maize from Mexico. Ephemeral grasses have many qualities suitable for cultivation such as self-fertility, easy germination and quick growth. Certain regions of the world have been the sources of many crops. Potatoes originate in the Andes and needed much selection to be suitable for long days. Cassava is from lowland South America and other staple grain and root crops come from various tropical areas. Most of these crops have reached their present form through hybridization of related species and often polyploidy, and there are many varieties.

Other food crops

Brassicas all derive from a wild Mediterranean species and there are numerous crops from different areas. Breeding has often involved production of polyploids and, for example, cultivated bananas are all sterile triploids. Some varieties such as pink grapefruit derive from somatic mutation.

Plants for flavor

Herbs are leaves deriving particularly from Lamiaceae of the Mediterranean. Spices belong to numerous families and derive from fruits, seeds, flower buds, roots or bark. They are mainly tropical and were much sought after for flavoring food in the early days of exploration in the 15th and 16th centuries.

Alcohol

All alcoholic beverages derive from plant sugar fermented by yeasts. Grapes, native to central Asia, provide sugar-rich fruit and have been fermented in a controlled way for millennia. Germinating barley is fermented for beer and barley can tolerate short summers. Numerous other plants are used.

Crops in the twentieth century

Concentration on yield led to a reduction in the number of varieties grown in the western world. With the rise in environmental consciousness more emphasis is beginning to be placed on wild sources particularly for disease resistance. Some wild crop ancestors are rare. Destruction of the ecology of agricultural soils has resulted from monocultures with high input of fertilizer and pesticide and some areas are losing their potential for agricultural production.

Related topics

Roots (D2)	Plant cell and tissue culture (K2)
The seed (H1)	Plant genetic engineering (K3)
Fruits (H2)	Plants in medicine (N3)

The range of food crops

Plants form the bulk of our food. Many plant species are edible and prior to the advent of agriculture in the meso- to neolithic period, numerous species were eaten. One of the few remaining hunter-gatherer peoples of the world, the Kung San (bushmen) of the Kalahari, still regularly eat over 80 species of plant native to that region and use a great many more as flavors or medicines. Throughout the world over 30 000 species of plant have some edible parts, circa 10% of all flowering plants, and at least 7000 have been collected or grown as food at some time. In sharp contrast, a mere 20 species currently supply 90% of the world's food and just three grass species, **wheat**, **rice** and **maize**, supply about half of it. The potential for plant foods is being seriously underutilized.

The main importance of plants as food has been as providers of **starch**, the seeds of the three staple grasses being mostly starch. Other staple crops provide starchy **tubers** (swollen underground stems) such as **potatoes**, or **roots** such as **cassava** and **yams**. Leaves, stems and flowers have much less starch but supply vital vitamins and roughage, most importantly the **brassicas** (cabbage, cauliflower, sprouts, etc.). Many **fruits** are sugary energy sources. Proteins and fats are present in small quantities in some of these organs although traditionally animals have been our main sources. Some seeds are rich in proteins, particularly **beans** and **pulses** connected with the fact that their root nodules have bacteria that fix nitrogen, and certain fruits, e.g. **olive**, **avocado**, are rich in fats. Oil-rich seeds have become widely planted for margarines, animal feed and nonfood uses, particularly oil-seed rape (or canola), sunflowers and flax (linseed oil). Fruits and seeds come from a wide range of woody and herbaceous plants but, with occasional minor exceptions, all the vegetative parts of plants that we eat come from herbaceous plants.

Origin of staple crops

Wheat and barley are responsible for the rise of the first civilization in the Tigris/Euphrates valley, both species originating in the 'Fertile Crescent' of mountains just north of it, ranging from northern Syria through Turkey to northern Iraq. The wild relatives of wheat, oats and barley are all ephemeral grasses of this region. Rice and maize also derive from ephemeral grasses, rice originating in **China** and maize in **Mexico**, these two species being responsible for the rise of the civilizations in these regions separately from that in the Fertile Crescent. Some of the features that characterize ephemeral grasses are those that make them good crop plants: many are self-fertile, guaranteeing seed set; they produce many seeds within a year of planting; they are characteristic of disturbed ground; they have no innate seed dormancy but can germinate as soon as conditions are favorable; in comparison with other grasses their seeds are large and rich in starch favoring rapid germination. It is likely that these plants were noticed as particularly good food sources and subsequently encouraged and cultivated. Before human intervention the seeds will have dropped from the stems to be dispersed, but, for humans, those that retained ripe seeds would have been easiest to gather for subsequent sowing so there will have been unconscious selection for seed retention.

Present-day wheat is derived from at least six wild species by multiple hybridizations. Natural, then artificial, hybridization and selection for large seeds made them the staple crops and 17 000 varieties have been produced. Modern bread wheat is hexaploid and hybridizations between the species have involved polyploidy at several stages (Topic Q5). Rice and maize have fewer varieties (a few hundred), although rice derives from at least three species, maize from perhaps two.

A few particular regions of the earth, known as **Vavilov centers** (*Fig. 1*), after the Russian biologist who first described them, have supplied the most useful edible plants including the grain crops. Of our other staple crops, potatoes and cassava are both South American in origin and their cultivation contributed to the rise of civilizations there 6000–7000 years ago. Potatoes came from the Andes, cassava from the lowlands. In the wild, these and other root or tuber crops have poisonous alkaloids that are often bitter tasting; only years of selection of less bitter varieties have led to the crops of today. In cassava toxic alkaloids remain, since methods for fermenting and cooking it that remove the toxin are well known and the toxic varieties are more resistant to pests. The potato was imported into Europe in the 16th century and, later, North America, but it did not crop well or only very late in the long days of a temperate summer. It took nearly 200 years, until the late 18th century, to develop clones that produced adequate tubers under these conditions, and these probably came from a narrow genetic base. It rapidly became a staple, but it is prone to disease, the most important being the **potato blight**, the oomycete *Phytophthora infestans* (a fungus-like hetorokont; Topic M5). This appeared in Europe in the 1840s with devastating consequences for the potato crop in Ireland, which particularly depended on it, leading to the famous famine. Much potato breeding and selection has been associated with disease resistance.

Other tropical staple crops come from the various regions, the millets and sorghum from Africa, dasheen or cocoyam from Asia, and different species of yams from all three continents.

Other food crops   One species of wild **cabbage**, *Brassica oleracea*, is the ancestor of all the cabbage, kale, brussels sprout, cauliflower, broccoli and kohlrabi crops. Turnips, swede and oil-seed rape (canola) are closely related. The wild cabbage is widespread around the Mediterranean extending into Asia and the different cultivated types have originated in different parts of this range. Many other crops originated in this

Fig. 1.   Centers of origin of staple crop plants (Vavilov centers): (1) Mexican highlands, (2) Northern Andes, (3) Eastern Africa, (4) East Mediterranean, (5) Fertile Crescent of South-West Asia, (6) South-East Asia.

region including peas and lentils, both of which were first grown as animal feed with their nitrogen-rich foliage and later developed for human consumption. **Soya beans**, regarded as one of the world's most important developing crops because its seeds are rich in protein and oils, originated and was confined to the Far East, mainly China, until the early 20th century. There has been an enormous increase since then in the USA and elsewhere, and there have been extensive breeding programs for its use in the short days of the tropics. Exploration of South America in the 16th to 18th centuries led to tomatoes and capsicums appearing in Europe in addition to potatoes. Cocoa, pineapples and some other tropical crops were exported from South America to other tropical colonies. South-east Asia is the native home of pantropical crops such as bananas, sugar cane, mangoes and breadfruit, and is the center of distribution of citrus species. Apples and pears come from central Asia.

Frequently, hybridization between two or more species has resulted in varieties that are larger or more suitable for human consumption, and often polyploids. Crop **bananas** are triploid and sterile giving rise to seedless fruits that reproduce purely by clones, although all wild species are diploid. **Apples**, **pears** and **citrus fruits** are reproduced mainly by cuttings and grafts with each named variety usually being a single clone. Citrus fruits are, in addition, prone to somatic mutations, changes within the plant body that happen spontaneously without any seeds (Topic K3). One such mutation gave rise to the pink grapefruit, clones of which are now distributed in many areas.

Plants for flavor
A great variety of plants are used as flavorers in addition to those eaten as food crops. Culinary 'herbs' are the leaves, usually dried, of various herbs and shrubs of several families, most importantly the Lamiaceae, many of which are native to the Mediterranean, providing thyme, sage, mint, basil and others. The Apiaceae (umbellifers) provide parsley, caraway and aniseed. The aromatic oils and other compounds derived from these plants originate as secondary compounds, which protect the plants from herbivore attack (Topic M3) owing to their indigestibility and, frequently, bitterness. In small quantities, it is this aspect that is desirable.

Spices, prized for their strong, often 'hot' taste, derive from parts of the plant other than leaves and come from a wide range of plant families. Nearly all are tropical and spices, most importantly **black pepper**, became much sought after in the 15th century. Spices were seen as highly desirable, some as mild **preservatives**, but mainly to cover the flavor of old or decaying food and they fetched extremely high prices. Columbus's voyages to America were a by-product of the search for spices. Black pepper is native to south Asia but is now planted throughout the tropics in small quantities. Black pepper is a dried fruit, and other spicy fruits and seeds include cardamoms, allspice, nutmeg and vanilla. Spices from other plant parts include ginger deriving from a root, turmeric from a rhizome, cinnamon from bark and cloves from flower buds.

Alcohol
Alcoholic beverages are all made, at least initially, from the **anaerobic fermentation** of sugars by yeasts, unicellular ascomycete fungi. Most fruits and many nectar-rich flowers have yeasts occurring naturally, fermenting some of the sugar. Any animal may become **intoxicated** including bees feeding on alcoholic nectar which may then be unable to fly, and numerous mammals that feed on fallen fermenting fruit. The two most important crops are the **grape** with its sugar-rich fruit making wines and brandy, and **barley**, whose germinating seed

is rich in sugars, the basis of most beers and whiskies. Grape vines are naturally distributed from Afghanistan to the eastern Black Sea and were taken to the eastern Mediterranean 5000 years ago. When introduced outside its native range it has sometimes hybridized with related species of each region to produce new varieties suited to the area. It is naturally dioecious but most cultivated forms are hermaphrodite, an advantage as a crop since all will produce fruits. Seedless varieties, used for consumption as grapes, appeared first by somatic mutation.

Barley originated in the Fertile Crescent and was one of the staple crops of early civilizations. It matures quickly which makes it adaptable to short summers and is somewhat more tolerant of salt in the soil than wheat. It was formerly an important grain staple, but today it is mainly used for alcoholic drinks and animal fodder. Other grain crops such as rice and rye are used for similar drinks and many other plants are used for making alcoholic drinks including numerous fruits, potatoes (vodka) and the pith of *Agave* (tequila) and palms (arak). Some are distilled to increase alcohol content.

Crops in the twentieth century

During the last 100 years and particularly the last 50 years there has been **intensive breeding** for increased yield in many crops (Topic K1). In the western world, the advent of apparently limitless inorganic fertilizer and a great number of pesticides, along with highly selective breeding has achieved yields unimaginable in 1900. During much of this time only the highest yielding varieties were planted and, overall, there has been a serious reduction in the number of varieties. In poorer countries, including many tropical countries, breeding has been less intensive and inorganic fertilizer is not so readily available. Crops in these countries are receiving more attention since the advent of serious famines in the 1980s and 1990s, and some genetic modification is specifically targeted at tropical countries, e.g. vitamin A in rice (Topic K3)

With increasing **mechanization** of farming, monocultures, often over large uninterrupted areas, have become widespread. All monocultures are prone to pests and diseases (Topic M3) and, in farming, these crops rely on the input of large quantities of pesticide. In the 1960s, the first evidence of environmental damage came to light in relation to persistent organochlorines such as DDT and they were banned in many countries, but are still used in others. Many other pesticides, including herbicides and fungicides, are in wide use. With the rise in environmental consciousness there has been renewed interest in wild ancestors of crops where there is often genetic variation for disease resistance. These are now being used in crop breeding programs. Some of these wild relatives have a highly restricted distribution, the most extreme example being the closest relative of maize, *Zea perennis*, found for the first time during the 1980s on a range of less than 1 hectare in Mexico.

The high-yielding varieties grown utilize many soil nutrients and to grow them for greatest yield requires high input. One tonne of grain requires about 10 kg of nitrogen, 5 kg of phosphorus and other nutrients. At present, under intensive agriculture, most of this is replaced from inorganic sources. The combination of pesticide and fertilizer input damages the ecology severely, losing the soil structure and most of its microorganisms, beneficial as well as damaging, and this will continue to be a problem under existing agricultural regimes. In environments prone to drought there have been many problems of soil loss altogether and resulting loss of possibility for growing further crops or a natural vegetation cover, and these will undoubtedly become more widespread. Throughout the history of agriculture, using the soil too heavily for crops, particularly grain, has led to infertility. In addition, a covering of plants of any kind can affect the climate locally and some climates in dry areas have become drier overall in the depleted areas. The combination of lowered soil fertility and a drier climate has allowed deserts to spread.

N2 PLANTS FOR CONSTRUCTION

Key Notes

Timber	The lightness, strength and durability of the lignified xylem vessels that make wood have made it vitally important in the construction of buildings, furniture, ships and ornaments. It can be preserved for centuries, but fungi and bacteria can digest it in damp conditions.
Other uses of wood	Wood can be broken down into fiber to be used for paper, one of its most important current uses. Refined fibers are used as clothing fabrics, notably viscose. It was the most important fuel before fossil fuels were exploited commercially.
Wood use and the environment	Throughout history woods have been felled for pasture and agriculture and for their products. Most have not been replaced, but some regeneration of desirable species has been encouraged at the expense of less useful ones. Fuel wood has in some places been harvested by coppicing, a sustainable practice though leading eventually to reduced soil fertility. Plantations are usually conifers or eucalyptus for paper. These are usually poor in wildlife and, though the soil is retained, it is often acidified through slow leaf decay.
Fiber	Useful plant fibers are strings of sclerenchyma or collenchyma cells either from the leaves or stems of herbaceous plants or seed appendages. Many species have been used such as flax, hemp and sisal. The most important plant fiber is cotton, long unicellular seed appendages spun to fine thread. Demand for this had major historical repercussions. Nylon and 'artificial' fibers are made from oil derived from fossil plants.
Related topics	Plant communities (L3) Plants as food (N1) Mycorrhiza (M1)

Timber

Wood is made of elongated hollow **xylem** cells thickened with **lignin** and **cellulose** (Topic D4). Lignin in particular is strong and durable and only some fungi and bacteria are capable of digesting it. In addition, many trees secrete tannins or resinous compounds into the xylem in the **heartwood** of a tree trunk that has lost its function as water conducting cells; this acts as a preservative. Many of the vessels remain hollow and wood is light in weight considering its strength. These properties make wood excellent material for the **construction** of buildings, furniture, tools, etc., and it has been used for numerous purposes since paleolithic times. It is (or was) abundantly available across much of the habitable world. Most woods float in water and its use in building boats and ships gradually increased in importance until the early 19th century when metal ships were introduced.

The '**grain**' of some woods, formed mainly by annual growth rings (Topic D4) and its ability to be smoothed and stained have made it an important medium for sculptures and other ornaments and qualities of particular species are used to make many musical instruments, e.g. maple and pine or spruce for violins, African blackwood for clarinets. If kept dry and adequately treated, wood will last well and can last for centuries in the right conditions, although if kept permanently damp it will decay within a few months to a few decades depending on the type of wood, its size and the environment.

There is a minor but significant and constant demand for **corks** made from the spongy bark of the Mediterranean cork oak, *Quercus suber*.

Fungi and bacteria can digest the lignin and other compounds that form wood and are the organisms responsible for wood decay. All decay organisms require moisture (including the so-called 'dry-rot' fungus) and wood can be kept for centuries if it is kept dry, or treated with preservative. In the tropics termites are specialist feeders on dead wood and digest it by means of symbiotic microorganisms making it difficult to preserve wood in the wet tropics.

Other uses of wood

Wood may be broken down into fibers to make **paper** and this is now one of the largest uses for wood. Paper can be made from almost any tree species. The wood from conifers and broad-leaved trees is pulped mechanically in a water and chemical solution to extract the fibers which are then glued, pressed and bleached, the whole operation being highly commercialized. Many other fibers are suitable for paper-making and the earliest paper was made from the papyrus, *Cyperus papyrus*, from African swamps. Wood pulp is also used to make fine clothing material after chemical treatment that modifies the cellulose. One such product, sodium cellulose xanthate, is the best known fiber from wood pulp, known commercially as **viscose**.

Wood has been the most important **fuel** for burning throughout most of human history, either burned directly or first charred to **charcoal** to be used as a hotter and more efficient fuel later. Only with the mining of large quantities of coal, oil and gas since the 19th century has its importance diminished. It is well to remember that these mined fuels are all partially decayed fossil plants, including some wood in the coal, preserved by crushing and, in coal, petrifying over millions of years.

Wood use and the environment

The use of wood through history has had a profound effect on the world's plant communities. Many **forests** were destroyed to open land for pasture and agriculture, and forests had disappeared from large areas of temperate Eurasia by 0 AD. Wood used for building was usually cut without any replanting and the large quantity of wood required for ships in western Europe, mainly during the 1700s, destroyed many more. Throughout the world, forests diminished in extent through human intervention and many other wooded areas changed in character; for example, in many areas, such as much of North America, the understorey was burned to make a more open environment. Some areas of forest remained in most places for use as a source of wood. Trees were allowed to regenerate naturally and often desirable tree species were encouraged and other species removed. Useful species were increasingly planted to replace felled woods, particularly in the 18th and 19th centuries. The domination of many western European woods by the oak, *Quercus robur*, and in places beech, *Fagus sylvatica*, and the absence of the lime, *Tilia cordata*, is mainly due to human encouragement of the good timber crops. Many forests now dominated by a single species would naturally be mixed.

Fuel wood is frequently harvested on a partially sustainable basis. Many trees will regenerate new stems from a cut base and this ability has been used in **coppices** for many centuries. The understorey trees are cut on a rotational basis with a cycle of a few years or decades depending on the fertility of the site and species involved. This will periodically open the coppice to light after which it will gradually become shadier until the next cut; many plants at a woodland edge are adapted to this and have colonized coppice woodlands. Numerous small and some large wooded areas remain, particularly in northern Europe, owing to the need for coppice wood as fuel and for some construction purposes. Over centuries, soil fertility declines and it takes increasingly longer for the trees to regrow in a coppice woodland. The cork oak woods of southwestern Europe have been kept for their valuable product and remain ecologically rich.

Plantations of trees have increased in the 20th century with the decline of ancient woods and with the huge demand for paper. These have largely been plantations of **conifers**, most originating in North America, in the cool temperate zones, often on acidic soils not suitable for other crops, and **eucalyptus**, originally from Australia, and other conifers in the warmer parts of the world. These plantations suffer from the problems of monoculture described in Topic N1 and can deplete the soil of nutrients. With the dense plantings, few other plants can live except in the young early stages. Many of these trees are planted in continents different from their place of origin. This has sometimes led to growth problems because of a lack of mycorrhizal infection (Topic M1) and, frequently, a limited wildlife community compared with plantations of native species. Despite this, wildlife is often more common than on open agricultural land and the soil structure and depth may be retained, though the soil may become increasingly acidic since the leaves of many trees are resistant to decay.

Fiber

Useful plant fibers come from many species, the fibers themselves consisting mainly of **sclerenchyma** cells (Topic D1), sometimes with **collenchyma** or conducting cells. Some of these have lignified walls but, for flexibility, the lignin is usually digested out to leave **cellulose** walls for the fiber. A few derive from seed appendages but most others are strands of cells within the vegetative tissue that provide support and usually have no cell contents at maturity. In a few plants the fibers are long and at least partially separate and can be used directly for clothing (e.g. the underbark of a few trees), but more frequently they must be treated before use. This usually involves soaking and beating to remove surrounding cells and sometimes treatment with chemicals. The finer fibers must then be **spun** to make long strands.

Most fibers are extracted from the **stems** of dicots or the leaves of monocots. The most important among the dicots are **flax** (*Linum* spp.) that gives us linen, used widely before cotton, and **hemp** (*Cannabis* spp.) and **jute** (*Corchorus* spp.), both of which are used mainly for rope. Among monocots, **sisal** (*Agave* spp.) and **Manila hemp** from a banana, *Musa textilis*, are still widely used. Many other species have been used and still have local uses, such as nettle species, *Urtica*, pineapple fiber, date palms, etc. With the large demand for paper some of these crops are being tested for suitability in paper manufacture.

The world's most important fiber is **cotton**, the feathery appendage of seeds of the genus *Gossypium* (Malvaceae, the mallow and hibiscus family), native to Asia and the Americas. Cotton has been used for thousands of years in India and South America. These appendages are each a single elongated cell, the longest making the finest cotton, and now most commercial cotton comes from a South American species, *Gossypium hirsutum*. One kilogram of cotton consists of approximately 200 million seed hairs. Exploration of Asia, and later South America, introduced the fine qualities of cotton to Europe during the 15th century after which it became much sought after in Europe where there were only coarser fabrics. The huge plantations of cotton, particularly those involving slavery in the southern USA, made Liverpool the richest port of Europe and most northern English towns rich through manufacture of cotton clothing. The serious inequalities that developed between northern American states that marketed and exported cotton and southern states that grew it, was one of the major factors leading to the American civil war.

Kapok fiber derives, like cotton, from seed appendages and was used as light stuffing for mattresses and life jackets. It comes from a giant forest tree, *Ceiba pentandra*, of the Bombacaceae, related to the cotton family. **Nylon** and other 'artificial' fabrics derive from oil, itself derived from fossil plants.

N3 PLANTS IN MEDICINE

Key Notes

Ethnobotany

Ethnobotany is the science of the relationship between the chemical constituents and properties of native species of plants and their uses by indigenous peoples. As plants contain a wide range of different, biologically active secondary metabolites, the discovery of compounds new to western medicine is an important aspect of this science.

Glycosides

The cardiac glycosides or carnolides based on digitoxin and digoxin inhibit the heart Na^+/K^+-pump. Applied at appropriate dose, they stabilize the heart rhythm and are used to treat heart failure and similar conditions.

Alkaloids

Alkaloids are a wide range of compounds with many applications. The quinoline alkaloid quinine, the isoquinolines morphine and codeine, and the indole alkaloids vinblastine and vincristine, are all widely used medicinals.

Terpenes

The terpene taxol, from the bark of the Pacific yew, stabilizes microtubules and thereby inhibits cell division in tumor cells. It is therefore a potent chemotherapy agent in cancer treatment.

Related topics

Amino acid, lipid, polysaccharide and secondary product metabolism (F5)

Plant cell and tissue culture (K2)

Ethnobotany

Historically, botany and medicine were closely allied subjects as most medicines were herbal, using plant materials to supply medicinally active drugs. The earliest botanic gardens were dedicated to the production of plants with medicinal properties. Human beings have used plants as medicines for millennia and there is a rapidly growing interest in identifying medicinally active plant products. This involves everything from learning about traditional medicines to the assessment of crops and native floras for the production of new medicines. The study of **ethnobotany**, the study of plant species and their uses in indigenous societies, has been inspired by awareness of rapid loss of species diversity and of the importance of plant-derived drugs. Ethnobotany seeks to record the uses of plants by societies and wherever possible, to conserve those species for future use.

Medicines based on plant products include many different compounds with many different uses. The **cardiac glycosides**, based on the compound **digitonin** from foxglove (*Digitalis purpurea*) are widely used to treat heart disease. Drugs based on the plant **alkaloids**, the **opiates**, derived from the opium poppy (*Papaver somniferum*) are amongst the most effective pain-relievers known. Many **anti-cancer agents** such as **taxol** (derived from the pacific yew, *Taxus brevifolia*) and **vincristine** and **vinblastine** (derived from the periwinkle, *Catharanthus roseus*) are also plant products.

Plant-derived medicinal compounds are plant secondary products. The general biochemistry of the production of plant secondary products is described in Topic F5.

Glycosides

Glycosides are a very diverse group of compounds. The **cardiac glycosides** (or **cardenolides**) contain sugar residues bonded to sterols; in one example (**digitoxin**, from the foxglove *Digitalis purpurea*), the sugar residues are one glucose molecule, two digitoxose molecules and one molecule of 1-acetyl digitoxose. Digitoxose is a rare 6-carbon sugar. A second cardiac glycoside, **digoxin** is also present in the foxglove.

Cardiac glycosides are extremely toxic compounds that inhibit the heart Na^+/K^+-pump. At a suitable dose, they slow and strengthen the rhythm of the heart and are very effective in the treatment of heart failure and other heart conditions.

Alkaloids

Alkaloids are also a diverse group of compounds, all of which contain nitrogen, usually as part of a heterocyclic ring. They are synthesized from amino acids. Many alkaloids are extremely poisonous; others, including **codeine** and **morphine** are very effective **pain killers**, while others are addictive drugs (nicotine, cocaine). *Table 1* presents some of the alkaloids and their effects.

Terpenes

Terpenes are synthesized either from acetyl coenzyme A (CoA) or from 3-phosphoglycerate and pyruvate and are based on the 5-carbon isoprene unit (Topic J2). Complex terpenes are generally believed to be anti-herbivory defenses in plants; however, amongst the very diverse range of terpenes are **essential oils** such as the **menthols** (used as insect repellants and in proprietary remedies for colds and coughs). The most significant medicinal terpene of recent years has been the diterpene **taxol**, isolated initially from the bark of the **Pacific yew**, *Taxus brevifiolia*. Taxol causes a blockage of cell division in tumor cells by stabilizing and polymerizing microtubules. Taxol is effective against solid tumors. Its initial discovery led to a considerable effort to secure sufficient supplies of the compound, as its production (like many plant secondary products) is limited to one tissue (the bark) of a single species. Application of plant tissue culture was less successful than attempts at organic (chemical) synthesis and taxol-based drugs are now produced by this means.

Table 1. Major types of medicinal alkaloids and their precursors

Group	Examples	Uses
Quinoline	Quinine	Anti-malarial
Isoquinoline	Morphine, codeine	Pain relief
Indole	Vinblastine, vincristine	Anti-cancer agents used in chemotherapy

N4 PLANTS FOR OTHER USES

Key Notes

Uses of secondary compounds	Rubber, derived from a Brazilian tree, has been highly valued and is used today in tyres. Resins and other aromatic plant products are much used in ceremonies, e.g. as incense, and many dyes come from plants. Plant oils may be used in soap and tannins from trees are used to treat leather.
Drugs	The two most important commercial drugs from plants are caffeine from coffee and tea, and nicotine from tobacco. Coffee and tea are both large cash crops from tropical countries, much prized in Europe and North America. Tobacco depletes soil nutrients quickly and traditionally was grown in new soil as southern USA was colonized. Other drug plants are grown in scattered plantations and can fetch high prices.
Plants as symbols	Many plants have symbolic associations, often those with other useful properties. Trees that are worshipped usually have an unusual feature. Plants accompany all festivals, and some symbols, such as the olive for peace, are universal. Painting and writing about plants has influenced the conservation ethic.
Horticulture	Plants have been grown in gardens for millennia, for medicinal use, shade or recreation. Some plants have become rare from over-collecting. Ancient garden plants such as roses are much modified, with petals replacing stamens and multiple hybridizations producing a huge range of cultivars. They are propagated by cuttings and grafting onto roots of a different rose. Most plants can hybridize with related species and this has been widely used. Much propagation is by cloning from cuttings or root division. The growth form can be modified, e.g. bonsai.
Related topics	Amino acid, lipid, polysaccharide and secondary product metabolism (F5) / Plant breeding (K1) · Interactions between plants and animals (M3) / Plants in medicine (N3)

Uses of secondary compounds

Secondary compounds within certain groups of plants (Topics F5, M3) are among the most commercially valuable of plant products. **Rubber**, the solidified **latex** of the Brazilian tree *Hevea brasiliensis* (and occasionally other species), makes a very strong flexible solid material, particularly after treatment, and remains the main component of tyres. For many years it was the exclusive preserve of Brazil despite attempts to plant it elsewhere and fetched a high price, until in the 1870s about 70 000 seeds were, in effect, smuggled to Kew in London under the guise of botanical specimens. Just 22 seedlings from these reached Malaysia to form the basis of the millions of hectares of rubber plantation. Rubber is tapped from the tree without doing any permanent damage and a tree can produce latex for over 20 years. Ecologically this has the great advantage that an area remains forested, with associated benefits of soil stabilization, and it has helped to stop some Amazonian forests from being exploited.

Resins and similar **aromatic** compounds come from many flowering plants, especially trees and shrubs, and conifers. Traditionally these have been used extensively in perfumes and were believed to have healing and magical properties. They became used for religious and other ceremonies, e.g. frankincense, *Boswellia* spp., myrrh, *Commiphora* spp., and other members of the Burseraceae all used as incense. Other resins made fine **varnish** and lacquer and are the base for paints, glues and some cosmetics and organic solvents such as turpentine. Their value has sometimes led to their preservation along with associated flora and fauna. Similarly, many dyes come from plants. These may derive from leaves, roots, flowers, fruits or seeds and include our most ancient dyes such as henna, derived from the leaves of the Asian *Lawsonia inermis*, and woad from *Isatis tinctoria*. Many of these became associated with ceremonies. Many soaps and detergents are made partially from plant oils, particularly the oil palm, *Elaeis guineensis*, and coconut, mixed with animal tallow.

Tannins are dark phenolic compounds deriving from bark, leaves or other parts of many trees such as oaks and are traditionally used to treat leather, rendering it more waterproof and keeping it flexible. Tannins in tea leaves give tea its flavor.

Drugs

Some plants contain substances that affect the nervous system, either as stimulants or depressants and these have been used for centuries, often in traditional ceremonies. As they were introduced to temperate regions, by their nature they became enormously desirable among wealthy people and at times commanded high prices. The most important of these in commercial production today are **caffeine** from **tea** and **coffee**, **nicotine** from **tobacco**, **chocolate** that contains a range of drugs in small quantity, and various medicinal drugs (Topic N3) as well as alcohol (Topic N1). One of the most commercially valuable is coffee deriving from species of *Coffea*, mainly *C. arabica*, which comes from Ethiopia originally, but is now planted throughout the tropics. It requires adequate rainfall and is often grown in hilly districts. The markets were controlled by Arab traders until the 17th century when it was planted elsewhere and it is one of the most important export crops from South America. Tea, *Camellia sinensis*, originally from China, was also much desired. Tea plantations in India became a foundation of British occupation and high taxes on tea set by Britain on exports to North America were one of the triggers for the American war of independence. Chocolate, deriving from the understorey tree *Theobroma cacao* originally from South America, is another tropical plant of great commercial value. It contains small quantities of caffeine, tryptophan that stimulates serotonin, phenylethylamine and perhaps other substances that stimulate excitement, etc. It was used extensively in ceremonies by the Aztecs and other native Americans when Europeans first reached central America. Much commercial production now comes from West Africa and it is a crop that requires partial shade so at least some forest trees are left standing in cocoa plantations.

Tobacco, *Nicotiana tabacum*, originates in the American tropics and is planted mainly in America. It is a plant requiring a fertile soil which is depleted rapidly, and tobacco production requires high inputs. It thrived on the newly cultivated soil in southern North America. Medicinal drugs such as opiates from poppies and some, mainly illegal, psychoactive drugs such as cannabis and coca derive from plants, and small-scale plantations can raise considerable sums.

Plants as symbols

Plants have been used **ceremonially** since paleolithic times. Although they may have had uses as food, timber, medicine or narcotic drugs, certain individuals were frequently singled out for different uses. This has had an effect on the ecology of many regions of the world since certain plants have been preserved at the expense of others and some owe their survival to such a role. The ginkgo tree (Topic Q3) has an edible seed but may owe its survival to its ceremonial role as it is unknown outside cultivation. On every continent certain plants, often trees, are traditionally worshipped, leading to the preservation of certain places as sacred groves or certain strategically positioned individual trees as sacred. Often trees with an unusual shape are revered, such as the baobab with its swollen trunk in Africa, or the sacred banyan of India with its numerous aerial roots spreading the tree (a fig) over large areas. The properties or associations of some plants have led to their symbolic uses, e.g. the soothing action and antiseptic properties of rose oil probably led to its importance as a symbol of all aspects of love. All festivals involve plants, and flowers are strongly associated with birth, marriage and death. The olive branch is a global symbol of peace, coming from early Mediterranean civilizations when olive oil was one of the most valuable of commodities and used as a kind of currency.

Plants feature widely in numerous paintings and writings, these having a marked influence on the way in which we perceive our landscape and to some extent have shaped the environmental conservation ethic of today.

Horticulture

Growing plants for no use other than ornament is almost universal in human society, being known from hunter-gatherer peoples such as the San people of southern Africa and throughout settled societies. In Roman, Chinese, Arabic and other civilizations, people planted pleasure gardens, the hanging gardens of Babylon being considered a wonder of the world. Some of these plants had herbal and medicinal use or were planted for shade but most were planted purely for aesthetic reasons and certain species gained a kind of 'cult' significance, such as plum blossom in China. During the nineteenth century with the huge increase in knowledge and travel the demand for exotic plants in Europe and, to an extent, North America, became intense, leading to many plant-collecting expeditions. Some plants were over-collected, even in Europe, and became rare as a result. The interest in the rarer **ferns** of Britain at this time resulted in local extinction and the continuing rarity of a few species. Many **orchid** species are prized for their exotic blooms, but some have restricted natural ranges and are often scarce. There has grown up a black market with high prices for spectacular rarities, making them rarer and some in danger of extinction.

Enormous numbers of plant species are now cultivated purely for their ornamental value. Most plants are capable of **hybridizing** with related species, sometimes leading to sterility but more frequently partial sterility so further hybridization is possible. In some groups, such as the orchids, species from several genera can be fully interfertile, making hybrids particularly easy to generate. In nature hybrids are usually at a selective disadvantage. Most **roses** and many other garden plants have **double** flowers, with many of the stamens replaced by petals; petals probably derived initially from sterile stamens (Topic G1). Cultivated roses come from a wide range of species distributed across Eurasia from extensive breeding and hybridization through the centuries, and **cultivars**, those cultivated varieties which are marketed commercially, are mostly sterile or partially sterile multiple hybrids.

Many woody plants are reproduced by **cuttings**, small pieces of a side stem or more rarely a leaf that is separated and rooted in soil, making all plants of one cultivar genetically identical and part of one clone. Some, such as roses and fruit trees, are **grafted** onto the **rootstock** of a wild or vigorous related species because these roots are stronger and the specimen grows more vigorously. Grafting involves detaching the stem, or **scion**, around ground level and fixing it to the rootstock that has been similarly detached, following which the plants establish vascular connections. These roots may produce other stems directly, known as **suckers**, which are genetically the root not the cultivar. Occasionally cells from the rootstock of a graft occur above ground, mixed with cells from the scion, making a **chimera**, e.g. *Laburno-cytisus*, which has brownish flowers deriving from a mix of yellow flower cells from the *Laburnum* scion and purple flower cells from the *Cytisus* stock. Some herbaceous plants may be cloned by dividing the roots.

A plant is flexible in its growth form and can be trained into shapes quite unlike the wild plant. Hedge trimming stimulates axillary side shoots to grow and this can be refined further to **topiary**, making a tree a recognizable shape, or training fruit trees to espaliers or fans. The most extreme example is the miniaturizing of trees to **bonsai**, developed in Japan, through elaborate root trimming and other techniques applied over a long period.

N5 BIOREMEDIATION

Key Notes

Introduction	Bioremediation is the use of plants to extract heavy metals from contaminated soils and water. Success depends on the identification of species that can tolerate and accumulate toxins into shoots and leaves, which can then be removed and disposed of appropriately.
Hyperaccumulator species	Some species of plants can accumulate high levels of toxins without death. Usually, they can only tolerate a single toxic ion and grow slowly. Hyperaccumulation is thought to confer disease and pathogen resistance.
Bioremediation	Bioremediation of soils may be through the use of hyperaccumulators, genetic modification of crop species or by the use of chemical chelators. Decontamination of water can be carried out by rhizofiltration using species with high transpiration rates and by use of aquatic plants. In some instances (e.g. selenium, mercury) volatilization of the toxin to the atmosphere will also contribute to decontamination.
Related topics	Stress avoidance and adaptation (I5) Movement of nutrient ions across membranes (E3)	Uptake of mineral nutrients by plants (E4) Plant genetic engineering (K3)

Introduction

Many soils are contaminated with toxic pollutants. These may be from atmospheric deposition, mine spoil, sewage sludge or contaminated ground water, or may be natural deposits of toxic ions. Some species of plant tolerate high levels of soil toxins and **phytoremediation** and **bio-mining** (the use of plants to extract minerals) has developed from the additional observation that some species accumulate toxic elements to a high level. Such species, which may accumulate more than 100 times the amounts of a toxin than other plants, are called **hyperaccumulators**. However, bioremediation not only depends on the use of natural hyperaccumulators, as other species may be induced to take up high levels of soil toxins either by adding **chelating agents** (chemicals which bind to the toxin) to the soil or by genetic modification.

Toxic ions which are found in soils and may be suitable for phytoremediation are cadmium, cobalt, copper, lead, manganese, nickel, selenium and zinc.

Hyperaccumulator species

Hyperaccumulator species have been identified for most of the toxic ions found in soils, including radionuclides (*Table 1*). To be used successfully for bioremediation, hyperaccumulators must accumulate the toxic ion in their leaves and shoots, as removal of roots from soil is likely to be impracticable. In most instances, hyperaccumulators only grow slowly; it is assumed that the main selective advantage gained by hyperaccumulation is in deterring predators and pathogens, as tissues containing heavy metals are unpalatable and poisonous.

Table 1.   Examples of hyperaccumulator species for various toxic ions

Metal	Number of hyperaccumulators known	Key example species	Amounts accumulated
Cadmium (Cd)	1	*Thlaspi caerulescens*	>100 µg g^{-1} of dry weight
Cobalt (Co)	26	*Haumaniastrum katangense*	>1% dry weight
Copper (Cu)	24	*Aeollanthus biformifolius*	>1% dry weight
Manganese (Mn)	11	*Alyxia rubricaulis*	>1% dry weight
Nickel (Ni)	290	*Streptanthus polygaloides*	3–8 mg g^{-1} of dry weight
Selenium (Se)	19	*Brassica juncea*	1–10 mg g^{-1} of dry weight
Thallium (Tl)	1	*Iberis intermedia*	2 mg g^{-1} of dry weight
Zinc (Zn)	16	*Thlaspi calaminare*	>1% dry weight
Zn+Cd	1	*Thlaspi caerulescens*	>100 µg g^{-1} of dry weight
Cu+Co	1	*Haumaniastrum katangense*	>1% dry weight

The fact that they only accumulate small amounts of biomass means that the total amount of an ion extracted from the soil may be small. As most hyperaccumulators can only tolerate one toxic ion, this means that they may not be effective where a soil contains several contaminants. Examples of hyperaccumulator species are shown in *Table 1*.

Bioremediation

Bioremediation involves the use of plants to remove toxins from soils or water. Bioremediation can be used to decontaminate soil by growing plants which accumulate the toxin and are then harvested and removed. Water can be decontaminated by **rhizofiltration**, in which contaminated water is passed through the roots of plants that extract the toxins and by the use of aquatic plants which are harvested and destroyed.

The species listed in *Table 1* have potential for bioremediation of soils; however, because of the slow growth rates and poor yields, other alternatives are being explored. These include **genetically modifying plants** (e.g. arabidopsis; Topic B1) and the use of **chelating agents** to mobilize soil toxins and reduce their toxicity to the accumulator plant. Arabidopsis has been successfully modified to express mercuric ion reductase (which converts toxic Hg^{2+} to Hg^{0} which is volatilized to the atmosphere). Experiments using a chelating agent have shown that a crop of maize can accumulate >200 µg of mercury per gram of shoot dry weight from contaminated soil. Volatilization to the atmosphere as a result of uptake by plants is very significant for selenium. *Brassica juncea* (a wild mustard) accumulates up to 10 mg g^{-1} (dry weight) of selenate; it also releases large quantities of dimethyl selenate to the atmosphere resulting in 25–40% loss to the air.

Several species have been indicated as useful for **rhizofiltration**. The best are plants with extensive root systems and which have high transpiration rates.

Hybrid poplars achieved a complete removal of zinc from a solution of 800 µg ml^{-1} in 4 h and willows (*Salix* spp.), *Brassica juncea*, sunflower (*Helianthus annuus,*) and reeds (*Phragmites*) all show potential. Overall, plants like this offer a potentially effective method for the decontamination of effluent, one with negligible energy input and at low cost.

Experiments with water hyacinth (*Eichhornia crassipes*) show uptake of Cd, Co, Pb, Hg, Ni and Au from contaminated fresh water, with a total biomass production of 600 kg Ha^{-1} day^{-1}. Seaweeds can be used to accumulate iodine from sea water and as indicators of local metal contamination.

O1 THE ALGAE

Key Notes

The variety of algae	There is a wide variety of algae. Unicellular planktonic diatoms and dinoflagellates are responsible for most photosynthesis in the seas; most are haploid and reproduce mainly asexually.
Rhodophyta, the red algae	They are mainly multicellular and characteristic of deep inshore waters with low light. Unicellular and filamentous forms occur. Reproduction involves a gametophyte and two different sporophyte generations. One or more of these generations can be reduced. The gametes are nonmotile.
Phaeophyta, the brown algae	These are the largest and most complex seaweeds, dominating many intertidal regions. All are multicellular, some with holdfast, stipe and lamina. Reproduction involves a gametophyte and sporophyte, with both generations looking similar or the gametophyte reduced, in some to a gamete. Vegetative fragmentation is important in *Sargassum* and others.
Chlorophyta, the green algae	This is a highly diverse group, sharing features with the land plants. There are unicellular and multicellular forms. Sexual reproduction normally involves an alternation of similar generations or with either the sporophyte or the gametophyte much reduced. Members of the most complex group, the stoneworts, have a well-differentiated body and elaborate sex organs.
Chlorophyta and plants	The Charophyceae, one of five sections of the Chlorophyta, share many features with land plants and have colonized rocks and bark. The Chlorophyta are sometimes classified as plants.
Related topics	Introduction (A1) The bryophytes (O2)

The variety of algae

There is a large variety of unicellular photosynthetic algae and many groups are equivalent to kingdoms undergoing independent evolution (*Table 1*). Some nonphotosynthetic groups may be more closely related to true plants. All are capable of asexual reproduction either by binary fission or producing spores by mitosis.

Unicellular diatoms and dinoflagellates (*Fig. 1*) are the main members of the plankton in seas and fresh water and responsible for much of the photosynthesis of the oceans, and therefore of the world. Diatoms are enclosed in a silica cell wall consisting of two halves fitting together like a petri dish. These walls have the most intricately sculpted patterns. They form numerous fossils. Dinoflagellate species are responsible for toxic 'red tides' and for phosphorescence in the sea. Euglenas (*Fig. 1*) occur mainly in high nutrient environments in fresh water and are commonly demonstrated in biological laboratories. Some can occur as heterotrophic organisms without chloroplasts and it is likely

Table 1.   Some major divisions of algae

Division	Vegetative structure	Reproduction	Pigments
Euglenophyta (euglenas)	Unicellular or a few colonial; flagellate, no cell wall	Asexual only	Chlorophyll *a*, *b*
Bacillariophyta (diatoms)	Unicellular or colonial; silica cell wall forming a box	Asexual; sexual by cells dividing by meiosis to form gametes	Chlorophyll *a*, *c*; brown pigments
Pyrrophyta or Dinophyta (dinoflagellates)	Unicellular or filamentous; cellulose cell wall	Asexual; sexual occasional	Chlorophyll *a*, *c* or none (heterotrophic)
Rhodophyta (red algae)	Multicellular (a few unicellular); cellulose cell wall	Asexual and sexual, some involving three generations	Chlorophyll *a*; red and purple pigments. Unusual storage products
Phaeophyta (brown algae)	Multicellular; cellulose cell wall	Asexual and sexual, involving alternation of generations	Chlorophyll *a*, *c*; brown pigments. Unusual storage products
Chlorophyta (green algae)	Unicellular and multicellular; some large and complex. Cellulose cell wall. A few flagellate. Huge and variable group	Asexual and sexual; zygote usually dividing by meiosis after fertilization	Chlorophyll *a*, *b*; many other pigments and storage products like those of land plants

that euglenas and dinoflagellates have gained chloroplasts independently through engulfing a photosynthetic cell and using the chloroplast.

Most unicellular algae are haploid but the diatoms and a few dinoflagellates are diploid. Sexual reproduction involves flagellated sperm cells and the fertilized egg often lies dormant, meiosis taking place when it germinates.

Rhodophyta, the red algae

Most red algae are multicellular seaweeds occurring in deep offshore waters in dim light and their red pigment can absorb most available light. A few are used either as foodstuffs (*Porphyra*, laverbread) or as the source of agar (mainly *Gelidium*), a food additive and laboratory gelling agent, made from mucilage in the cell walls. A few secrete a calcium-rich exoskeleton which contributes to the formation of coral reefs. Red algae typically have one gametophyte and two

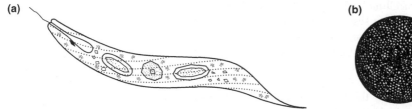
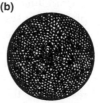
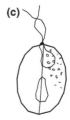

Fig. 1.   Unicellular algae: (a) Euglena, *(b)* Coscinodiscus *(a diatom), (c)* Exuviaella *(a dinoflagellate).*

sporophyte generations (*Fig. 2*), the gametophyte or the second sporophyte or both forming the main plant. In some species one sporophyte is missing from the life cycle. The sperms are nonmotile relying entirely on passive drift for fertilization.

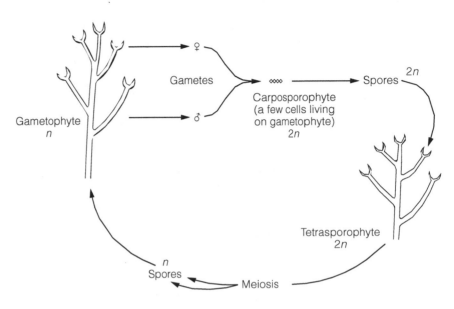

Fig. 2.   Life cycle of a red alga (Ceramium sp.).

Phaeophyta, the brown algae

These are the largest, most complex and usually most abundant seaweeds. The great kelps, *Macrocystis* and *Nereocystis*, of the Pacific can grow up to 1 m per day and reach 70 m in length; *Sargassum* forms the basis of life in the mid-Atlantic Sargasso Sea; *Laminaria* may be hung up by children to detect humidity; alginates from some can be extracted, to be used as emulsifiers in ice creams and other food stuffs. They are used occasionally as fodder for domestic animals (e.g. on Irish islands) and as fertilizer. Some are filamentous but most are differentiated into a flexible but very strong holdfast with no absorptive function, a stipe and a lamina. The lamina of some wrack species, *Fucus*, includes air bladders, though at times with a rather different gas combination from normal air (high in carbon monoxide). They may secrete toxic polyphenols inhibiting bacterial growth.

Sexual reproduction involves the typical alternation of diploid and haploid generations. The two generations may be alike in morphology, or unlike but both large and complex, or the gametophyte much reduced. In the wracks the gametophyte is reduced to gametes only, so there appears to be no alternation of generations and reproduction resembles that of vertebrates. Some brown algae can spread vegetatively by fragmentation, notably in *Sargassum*, which only reproduces in other ways when attached to the sea bed around coasts.

Chlorophyta, the green algae

This large and highly variable group has many features in common with true plants. The group includes unicellular and colonial flagellate forms, such as *Volvox*, flattened thalli, filamentous forms and complex multicellular organisms (*Fig. 3*). They occur in a wide range of mainly aquatic environments and some

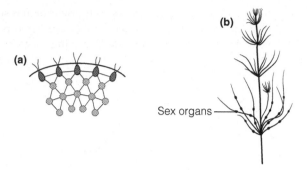

Fig. 3.   Green algae: (a) Volvox *(colonial, flagellate), (b)* Chara *(multicellular).*

are planktonic. In multicellular Chlorophyta, alternation of generations may involve both generations looking similar, a reduction of the sporophyte to a single cell or the reduction of the gametophyte to a single cell. The gametes (and sometimes the zygotes) are motile and flagellate; all require an aquatic medium for reproduction and their colonization of land is confined to damp places. Many can also reproduce asexually by flagellate diploid spores, sometimes with multiple flagella.

The most complex green algae are the stoneworts, Charales, characteristic of lime-rich fresh water. They can reach 10 cm or more in length and have rhizoid-like cells (Topic O2) that attach them to soft substrates, making them strongly resemble some aquatic flowering plants. Specialized male and female reproductive organs are produced by the branch whorls and are large enough to be visible to the naked eye. The fertilized zygote may undergo meiosis before regenerating the plant body but this is not confirmed and it is not clear whether the main plant is diploid or haploid. The name stonewort derives from secretions of calcium carbonate on the outside of the branches.

Chlorophyta and plants
There are five sections within the Chlorophyta and one of these, the Charophyceae, which includes filamentous algae like *Spirogyra*, some thallose forms and the stoneworts, are thought to be the closest living relatives to the land plants. In their cellular structure, mitosis, details of enzyme structure and in their DNA they resemble land plants. They show a variation in form and a wide ecological tolerance with species colonizing bark and rock on land. The protonema of a moss strongly resembles a green algal mat (Topic O2). For these reasons the Charophyceae or the whole of the Chlorophyta, though not the other algae, are sometimes classified as plants.

O2 THE BRYOPHYTES

Key Notes

Description	Three of the four major plant divisions, liverworts, hornworts and mosses, are known as 'bryophytes'. These divisions share some features. The gametophyte is the dominant plant with the sporophyte growing attached to it.
Early evolution	Fossil liverwort spores from the Silurian are the earliest evidence of land plants. Liverworts share features with green algae, and hornworts and mosses diverged from liverworts early, with vascular plants branching off from the moss line.
Vegetative structure of liverworts	Liverworts are highly variable with leafy and thallose forms. Leafy liverworts usually have three ranks of leaves, all one cell thick, one rank usually reduced in size. There are many different shapes and arrangements. Thallose forms have thicker structures which may have cavities and pores.
Vegetative structure of hornworts	Hornworts produce a simple thallus similar to some thallose liverworts but, in most, cells have a single chloroplast with a pyrenoid.
Vegetative structure of mosses	The first growth from the germinating spore is a filamentous protonema from which the main plant grows. Mosses have spirally inserted tapering leaves, mostly one cell thick, but are highly variable in growth form and in details of leaf structure and cell shape. Many have a midrib and *Polytrichum* has lamellae. *Sphagnum* has large dead hyaline cells that retain water.
Water relations	Bryophytes rely on surface water and capillary action, often enhanced by leaf arrangement, and many have rudimentary water-conducting hydroids in their stems, resembling tracheids. *Polytrichum* also has solute-conducting cells resembling phloem.
Ecology	They occur in all habitats, particularly in wet places and deep shade. Mosses dominate some subpolar regions and bogs where they form peat. They are abundant on the floor of wet woods, by streams and as epiphytes. Some can tolerate desiccation and occur on rocks where no other plants can grow.
Interactions of bryophytes	Pioneer communities may lead to succession. Different growth forms occur and interactions between species may be competitive or beneficial. A few are heterotrophic associated with fungi. A few specialist invertebrates eat bryophytes, and a few fungi infect them. Nitrogen-fixing cyanobacteria can colonize.
Human uses	They have been used for bedding and as filling in buildings and have minor interest as ornamentals. *Sphagnum* has been an important wound dressing. Peat, based on *Sphagnum*, is an important fuel. Scientifically they have been useful as genetic tools and in monitoring pollution. Peat has preserved numerous biological and archaeological remains used for reconstructing history.
Related topics	Introduction (A1) The algae (O1)	Reproduction in bryophytes (O3)

Description

'Bryophytes' is a general term covering three of the four major plant divisions, the **liverworts**, Marchantiophyta or Hepatophyta, the small group of **hornworts**, Anthocerophyta and the **mosses**, Bryophyta. They are almost entirely land plants, with a few in fresh water. All are small, often less than 2 cm high, with the largest reaching up to 1 m above the substrate. The three divisions share many features but also differ markedly.

All three groups show a typical alternation of generations (Topic A1), the main plant body being the haploid **gametophyte**. The **sporophyte** consists solely of a basal foot with a stalk and capsule and remains attached to the gametophyte throughout its life. The sporophyte lives only for a few days in most liverworts, up to a few months in mosses and hornworts, in contrast to the perennial gametophyte. Meiosis in the capsule leads to spore production and dispersal. Vegetative fragmentation is an important mode of propagation in many species and some produce specialized asexual **gemmae** (rounded groups of cells that disperse to form new plants).

Early evolution

Fossil spores resembling those of liverworts go back to the Silurian era, about 450 million years ago, making them the earliest land plants. Liverworts and green algae share some DNA structures that are different in hornworts, mosses and vascular plants. Fossils identifiable as mosses appear later in the record with definite fossil mosses appearing in Carboniferous deposits. There are no clear hornwort fossils until the Tertiary.

Putting evidence from morphology, fossils, proteins and DNA together, there appears to have been a single origin of land plants; liverworts were the first, or basal, group that retain features otherwise found only in the Chlorophyta (Topic O1). Hornworts and subsequently mosses diverged off the liverwort line early in their evolution. The vascular plants branched from the moss line in the late Silurian or early Devonian eras, 400–420 million years ago.

Vegetative structure of liverworts

Liverwort gametophytes are highly variable in structure, ranging from leafy shoots with leaves one cell thick to flat thalli many cells thick and of indeterminate growth (*Fig. 1*). The **leafy liverworts** usually lie prostrate or nearly so in damp places or as **epiphytes** and normally grow up to a few centimeters in

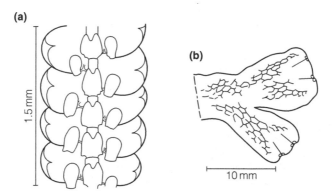

Fig. 1. Vegetative shoots of liverworts: (a) leafy liverwort, Frullania, from the underside, showing small notched underleaves and helmet-shaped lobes of main leaves, (b) thallose liverwort, Conocephalum.

length. They consist of a simple stem, leaves and often **rhizoids**, single cells forming a hair-like projection into the substrate. Most have three lines of simple, toothed or lobed leaves but, in many, one of these is much reduced in size to form '**underleaves**' on the ventral side, or is lost. All the leaves are a single cell thick. The classification of leafy liverworts is mainly based on the shape and arrangement of the leaves.

Thallose liverworts have a thallus (*Fig. 1*) several cells thick that branches dichotomously. Some consist of little more than this but, in the more complex forms, there is a thickening of the central part into a midrib, the presence of cavities in the upper surface with a pore for gas exchange and a largely nonphotosynthetic body of cells beneath. Unicellular rhizoids and multicellular scales anchor them to the substrate. Some species produce gemmae. Most liverworts contain oil bodies in their cells and sometimes these are fragrant when the plant is crushed, e.g. *Conocephalum*. They also produce antibacterial products but the potential of these for human use has not been examined.

Vegetative structure of hornworts

The gametophyte of the hornworts resembles that of a simple thallose liverwort, with no midrib and with no dichotomous branching. They differ in having cavities, often with a symbiotic nitrogen-fixing cyanobacterium in them. In most the cells contain a single chloroplast with a pyrenoid, a separate body in which the carbon-fixation reactions of photosynthesis (Topic F2) take place. This feature occurs in the green algae but no other land plant has a pyrenoid, carbon-fixation occurring throughout the chloroplast, and most land plants have many chloroplasts per cell.

Vegetative structure of mosses

The mosses are the most abundant and ecologically important of the three groups and a few grow to 1 m above the ground. After germination a spore produces filaments of undifferentiated cells, known as the **protonema**, resembling filamentous algae, which spread over the substrate. These normally grow for only a few days in this way before growing into the main gametophyte, though they persist in a few species. Some mosses have simple stems with no specialized cells but some have elongated **hydroid** cells that conduct water. Most are branched and the shoots may grow upright, often forming a tight clump or mat, or trail along the substrate. Rhizoids, sometimes multicellular, grow into the substrate, anchoring the plant but with no special adaptations for absorption, and some mosses grow similar hair-like structures along their stems. Mosses normally have spirally arranged leaves, tapered, often to a point, so in general not closely resembling those of leafy liverworts (*Fig. 2*). One genus, *Fissidens*, has two ranks of leaves. A few mosses produce gemmae.

The majority of mosses have leaves one cell thick (*Fig. 2*) but there is frequently a line of narrow thick-walled cells forming a midrib or nerve and sometimes around the edge. The nerve can extend beyond the end of the lamina as a hair point. *Polytrichum* and its relatives have leaves several cells thick and lamellae of cells along the leaf (*Fig. 2*), making the leaves fully opaque and much tougher than those of other mosses. The leaves of the bog mosses (*Sphagnum*) have normal photosynthetic cells interspersed with much larger dead cells, known as **hyaline** cells, with prominent spiral thickening and pores. The hyaline cells act as a water reservoir giving the mosses their sponge-like character and allowing them to be particularly effective as bog builders. The details of leaf and leaf cell structure and their arrangement are some of the main features distinguishing moss species.

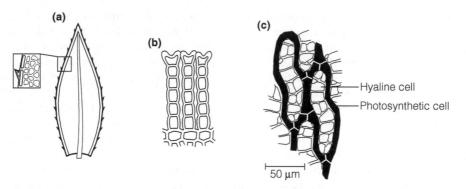

Fig. 2.   Leaves of mosses: (a) Mnium, *showing midrib and border, (b) cross-section of* Polytrichum *leaf showing lamellae, (c) leaf cells of* Sphagnum, *showing photosynthetic cells and hyaline cells.*

Water relations

Most bryophytes are dependent on surface rain water and capillary action across their surfaces. The leaves frequently form sheaths along the stem enhancing water flow. Many are able to dry out and remain dormant in a place free from competition such as a wall top. The leaves usually distort in the dry state (often twisted) and some can remain dormant for months, but they absorb water and resume growth within minutes of rain starting. The cushion form of many mosses will slow water loss.

Many mosses have rudimentary conducting cells in their stems, especially elongated water-conducting hydroids that lose their contents and connect by pores, like tracheids of xylem tissue (but without lignin). In the tallest and most complex of bryophytes, *Polytrichum* and its relatives, hydroids are well developed and these mosses are the only ones that also possess cells that conduct solutes effectively around the plant. Solute-conducting cells have pores and oblique walls and the nuclei degenerate. They are associated with cells with high metabolic rate and resemble phloem sieve elements and companion cells (Topic D3).

Ecology

Bryophytes tend to occupy places that other plants cannot grow in and frequently fit in spaces between other plants. They occur throughout the world and in almost every habitat, living mainly in wet places and places with dim light, e.g. under dense woodland, or in places with little or no soil, as epiphytes or growing on rocks. They dominate some plant communities in polar and subpolar regions, where extensive areas are covered with moss carpets, sometimes with very few or no other plants (though lichens, a symbiosis of fungus and alga, frequently occur too). In acid cold conditions, such as peat bogs, bog mosses (*Sphagnum*) dominate. They are able to store water and secrete acid from their cells inhibiting other plants and creating conditions of limited decay. The result is that few other plants can grow on the *Sphagnum* cushions. *Sphagnum* moss is the basis of much of the world's **peat**, partially decayed, waterlogged, compressed plant matter that gradually accumulates in boggy ground.

Extensive bryophyte communities occur on the floor of wet woods and by streams. Numerous bryophytes occur as epiphytes throughout the world and they can smother tree branches, particularly on tropical mountains. In temperate regions they are often the only epiphytic plants. In tropical rainforests some species grow on leaves, when they are known as **epiphylls**. Smaller bryophyte species may be some of the only plants growing on walls and as epiphytes on tree trunks in drier woods; growth is slow and interspersed with frequent dormant periods, but they are often the only visible life form growing there except for the even more resistant lichens. In almost all situations mosses are more common than liverworts, which tend to fit around the mosses. A few mosses have specialized ecology such as species in flowing fresh water and those that grow on mobile sand dunes. A few species have colonized deserts relying on dew for water, and some live by hot springs.

Interactions of bryophytes

Bryophytes (mainly mosses) may follow an ecological succession similar to that of flowering plants (Topic L3), e.g. colonizing bare rock, although change is often slow. In cold and waterlogged climates where mosses predominate, moss hummocks persist for decades with little change. Pioneer species are usually low-growing but retain some water and allow other species to colonize. There may be competition between species, but frequently the presence of one bryophyte stimulates others to grow and the role of competition and beneficial interactions is not well understood.

One liverwort, *Cryptothallus*, is entirely subterranean and heterotrophic and its thallus is intimately associated with fungi, analogous to mycorrhizae (Topic M1). A few mosses, such as *Buxbaumia*, are similar but have a few chloroplasts.

A few invertebrates, mainly fly larvae, are specialist moss feeders but bryophytes are nutrient-poor and are not eaten by animals to a great extent. Many microscopic animals live in moss cushions and birds use them as nesting material. Fungi can infect mosses, particularly in the high latitudes and other fungi are responsible for decay. Nitrogen-fixing cyanobacteria (Topic M2) can associate with certain species and enhance their growth.

Human uses

Mosses have been used for centuries as bedding material owing to their soft quality. They were also used as padding for building, e.g. between timbers, and blocking air vents in chimneys, etc. The wiry stems of *Polytrichum* have been used for baskets. *Sphagnum* can be used as an absorbent anti-bacterial wound dressing and to retain water in window boxes or plant nurseries. Mosses have occasionally been used as ornamentals, particularly in Japan. Peat derived from *Sphagnum* has been an important domestic fuel when dried out, e.g. in Ireland where there are extensive bogs and large areas without trees, and some power-generating stations use peat. Peat is also used in the horticultural trade as a soil conditioner, but many peat bogs have been destroyed through over-extraction and its use is not favored by conservationists.

In science bryophytes have several uses. They are haploid and have been useful in genetic studies since geneticists can look directly at gene expression (although many are polyploid). Some species are sensitive to pollution, particularly by sulfur dioxide, and their presence, along with that of lichens, has been used to monitor pollution levels. Peat accumulation has led to the preservation of numerous remains of plants, animals and human artifacts. Reconstruction of vegetation and human history over the past tens of thousands of years has been possible through examination of these remains.

03 REPRODUCTION IN BRYOPHYTES

Key Notes

Sexual reproduction	All bryophytes have antheridia with sperm and archegonia with eggs. Antheridia are spherical to oblong bags of sperm and archegonia are cylindrical with a bulbous base containing the egg. They may be scattered along stems or clumped at the stem tips. Sperms are motile and require water. Some have sterile hairs interspersed with the sex organs and a few have explosive dispersal of the sperms.
The sporophyte of liverworts	This is short-lived and consists of a foot embedded in the gametophyte, a colorless stalk and a roundish capsule, usually black. Almost all nutrients come from the gametophyte. The capsule splits into four valves at maturity and usually has sterile hairs (elaters) that aid spore dispersal. Some have reduced sporophytes.
The sporophyte of hornworts	This consists of a long-lived foot and capsule only. The capsule is green and photosynthetic, cylindrical to 4 cm, with a central column and elaters. Spores disperse when it splits from the tip.
The sporophyte of mosses	Typical mosses have a foot, a strong stalk and capsule. A protective calyptra from the gametophyte may persist until maturity. The tip of the capsule has peristome teeth that open and close with different humidity dispersing the spores. Details of capsule structure distinguish species. Bog mosses and rock mosses have a capsule borne on an outgrowth from the gametophyte. In bog mosses spores disperse explosively; in rock mosses, the capsule splits.
Related topics	Introduction (A1) The bryophytes (O2)

Sexual reproduction

All three bryophyte divisions have **antheridia** which produce motile sperm and **archegonia** containing eggs. The structure of these organs is fairly uniform across the three divisions and more elaborate than in other plant groups. The antheridia (*Fig. 1*) are near-spherical to ovoid sacs on short stalks with a wall one cell thick enclosing sperm mother cells and eventually the sperms. The sperms themselves have two flagella and can only swim a very short distance, but in some species the sperm mother cells are dispersed passively in water first. The archegonia (*Fig. 2*) are cylindrical with an inflated base. A jacket of sterile cells encloses the egg in the inflated basal part and 10 or so **canal** cells in the neck above that degenerate at maturity. Archegonia are usually on short stalks. These sex organs range in size from about 0.1 mm long in most species up to about 1.5 mm for a few archegonia.

In most liverworts and some mosses the sex organs are scattered along the stems but in others they are clustered at the branch tips in 'inflorescences', which may be surrounded by a cup-like cluster of leaves making them conspicuous. Sterile hairs may be associated with clusters of either sex organ and these

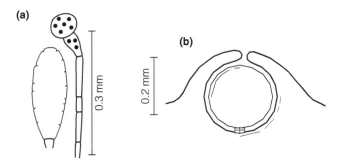

Fig. 1. *Antheridium of (a) a moss, with sterile hair, (b) a thallose liverwort sunk in the thallus.*

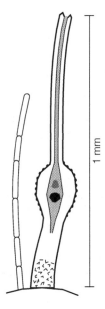

Fig. 2. *Archegonium of a typical moss, with sterile hairs.*

help retain water and perhaps aid dispersal. In some thallose liverworts and hornworts antheridia are sunk in pits and sperms are discharged explosively. Hornworts have small sunken archegonia too. The most striking sexual reproductive structure is the archegonial stem in some thallose liverworts, e.g. *Marchantia*. This is 2 cm or so long and looks like a small parasol with the archegonia on the underside; antheridia are produced on the upper side of a similar but smaller structure.

Some bryophytes are monoecious, bearing male and female organs and, in these, capsules are often abundant. In dioecious bryophytes, each plant bearing only male or female organs, sporophyte formation is often rare.

The sporophyte of liverworts

There are three parts to the typical liverwort sporophyte, a **foot** embedded in the gametophyte, a colorless **stalk** up to about 2 cm long and a **capsule** at the tip, 0.5–2 mm across and usually glossy black (*Fig. 3*). It is short-lived, normally lasting only a few days. The foot is embedded in the gametophyte and draws

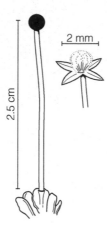

Fig. 3.   Sporophyte of a liverwort.

nutrients from it throughout the life of the sporophyte. When the spores are mature the capsule wall splits into four valves. Spores are mixed with sterile hairs, known as **elaters**, that assist in spore dispersal as they have spiral thickening and respond to drying out by jerky contracting movements. In humid conditions they lengthen and fewer spores are dispersed. Details of the form of the sporophyte can be used as reliable, if microscopic, characters to distinguish liverwort groups and particular species.

In a few liverworts the sporophyte is smaller and in the small thallose liverwort *Riccia*, there is just a sac of spores which remains embedded in the gametophyte until the gametophyte itself decays around it. The sporophyte generation has all but disappeared in this liverwort.

The sporophyte of hornworts

This has a large foot, no stalk but a cylindrical capsule 2–4 cm long directly on the foot (*Fig. 4*). They last for several weeks, unlike those of liverworts, growing from the base and can outlive the gametophyte. The capsule is green and photosynthetic with stomata on the outside, so does not rely solely on the gametophyte for its growth. There is a column of nonreproductive cells with attached elaters in the middle. Spores mature in the hollow part around the column and are dispersed when the capsule wall splits in two starting at the tip. The capsule wall, the central column and the elaters all twist and aid

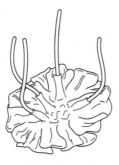

Fig. 4.   Sporophyte of a hornwort.

dispersal. Frequently the capsule continues growing from the base when spores are being dispersed from the tip.

The sporophyte of mosses

There are three different groups of mosses, typical mosses, bog mosses, and rock mosses, and they differ mainly in the structure of the sporophyte. The typical mosses (the great majority) have a sporophyte that lasts for several weeks (*Fig. 5*). There is an anchoring foot embedded in the gametophyte, a tough stalk that elongates early in the sporophyte's life and persists, and a capsule. The stalk contains water-conducting hydroid cells in many mosses (Topic O2). As the sporophyte grows there is often a cup-like piece of the gametophyte derived from the archegonium, known as the **calyptra**, attached to the capsule. This protects the developing sporophyte and may fall only when the capsule is fully mature. The capsule is usually photosynthetic and has stomata and a central column of sterile tissue. There is a lid at the tip of the capsule that is thrown off when the capsule is mature. Inside this, most have one or two layers of teeth, the **peristome**, the number and arrangement distinguishing different genera. They respond to humidity with movements that open the capsule for spore dispersal in dry weather and close it in wet or involve twisting movements or active ejection of the spores by a combination of outer and inner teeth sticking together.

One unusual family of typical mosses, the Splachnaceae, lives mainly on dung and its spores are dispersed by flies. Their extraordinary sporophytes have long stalks up to 10 cm and a swollen base of the capsule which can be up to 2 cm across and brightly colored resembling a toadstool. These produce an odor that attracts flies which carry off the sticky spores to another dung heap.

In bog mosses (*Sphagnum*) and rock mosses (*Andreaea*) (a group of small almost black mosses that grow encrusted on mountain rocks) the capsule has no stalk but is raised on an outgrowth of the gametophyte (*Fig. 5*). In *Sphagnum* the capsule is a brown spherical structure about 2 mm across, in which air pressure builds up as it dries until the tip is explosively blown off and the spores are discharged. In *Andreaea* the capsules are tiny, some only about 0.5 mm across, and they split into four sections except at the base and the tip. The cracks gape in dry weather to disperse the spores.

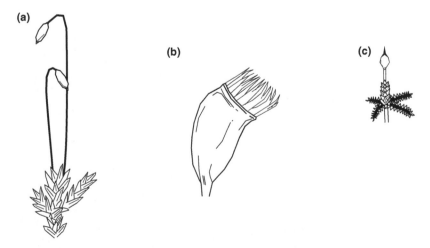

(a)

(b)

(c)

Fig. 5.   Sporophytes of mosses: (a) typical moss, (b) capsule of typical moss with peristome teeth, (c) Sphagnum.

P1 EARLY EVOLUTION OF VASCULAR PLANTS

Key Notes

The earliest vascular plants	Earliest fossils of land plants, *Cooksonia*, occur in late Silurian rocks. It had photosynthetic stems but no leaves or roots and no stomata. By early Devonian several genera occur. They were low growing plants less than 50 cm high bearing sporangia at the tips (Rhyniopsida), laterally (Zosterophyllopsida) or in bunches (*Psilophyton*). *Aglaophyton* may provide a link with bryophytes.
Later developments	There was rapid diversification through the Devonian era with developments of monopodial branching and trees belonging to the lycopsids, ferns and other living groups. Their greatest abundance was in the Carboniferous during which they reduced the CO_2 levels by 10 times, cooling and drying the climate.
Origins and evolution	Compared with an aquatic environment, land plants require structures to withstand changes in temperature and humidity, wind, rain and desiccation. They require a conducting system for water and nutrients and mechanical strength. Spores are more resistant to desiccation than gametes so sporophytes become the main plant.
Life cycle	Fossils are of sporophytes with sporangia having no or limited dehiscence. Fossil gametophytes are little known, but some probable gametophyte fossils have cup-like structures at the stem tips bearing archegonia and antheridia.
Homospory and heterospory	Homosporous plants produce one type of spore that germinates to produce a hermaphrodite gametophyte; heterosporous plants have two types of spore, one producing only male gametophytes, the other female. Heterospory has evolved several times. Most early plants were homosporous but heterospory probably appeared early and increased during the Devonian.
Related topics	The bryophytes (O2) Horsetails (P3)
	Clubmosses and quillworts (P2) Ferns (P4)

The earliest vascular plants

Vascular plants first appeared probably in the Silurian era (*Table 1*). The oldest fossils are those of *Cooksonia* (*Fig. 1*) in the Rhyniopsida from late Silurian rocks, a little over 400 million years BP. Fossils of *Cooksonia* have been found in several places in Europe and North America. These plants had photosynthetic stems 5–8 cm high that branched **dichotomously**, i.e. into two even branches at each point, but no leaves or roots. Some had **rhizomes**, horizontal underground stems, and subterranean rhizoids, one cell thick, growing out from the rhizomes or stems that may have absorbed water and anchored the plant. The earliest fossil *Cooksonia* species had no stomata and had a simple vascular system with

Table 1. Approximate time-scale of first appearance of major vascular plant groups in the fossil record

Period	Millions of years BP (start)	Spore-bearing plants other than ferns	Ferns	Seed plants
Present day	1			
Tertiary	65			
Cretaceous	145		Leptosporangiate ferns	Gnetopsida; dicotyledons, monocotyledons
Jurassic	205			Earliest flowering plants?
Triassic	250			*Caytoniales, *Bennettitales
Permian	290			Ginkgoopsida
Carboniferous	360	Tree and heterosporous Lycopsida; tree Equisetopsida	Heterosporous ferns	Cycads Conifers *Cordaitales (perhaps earlier)
Devonian	410	Homosporous Lycopsida and Equisetopsida; *progymnosperms *Other Rhyniopsida, etc	Eusporangiate ferns (including tree ferns)	*Pteridosperms
Silurian	440	*Cooksonia (end Silurian) Earliest land plants?		
Ordovician	500			
Cambrian	570			

*Extinct groups.

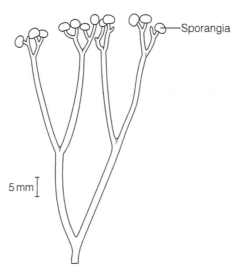

Fig. 1. The earliest known fossil land plant, Cooksonia.

tracheids. The carbon dioxide concentration in the atmosphere was considerably higher, probably more than 10 times as high, than at present and these plants may have obtained sufficient through diffusion into their stems, or may have absorbed carbon dioxide from the ground through their rhizoids, like living quillworts (Topic P2). They had sporangia at the branch tips.

At the beginning of the Devonian, around 400 million years ago, several genera are known from many sites, mainly in Europe. The most famous are the beautifully preserved fossils in siliceous cherts from Rhynie in Scotland, from which the group name Rhyniopsida derives (*Table 1*). These plants were more varied and complex, though *Rhynia* itself resembled *Cooksonia* in many ways. All the later plants had stomata and the potential for gas exchange in the stems. The attachment of these plants to the ground was by means of horizontal or arching rhizomes or a swollen corm-like section of the stem; thin, thread-like rhizoids grew out from these to penetrate the substrate and absorb water and nutrients.

Zosterophyllum (*Fig. 2*) had lateral sporangia and some members of the Zosterophyllopsida had spine-like outgrowths of the stem resembling tiny leaves, but without any vascular connection with the stem and they are classified as a sister group to the lycopsids (Topics A1, P2). A third genus, *Aglaophyton*, so far found only in the Rhynie chert, resembled *Rhynia* morphologically but the vascular system had no thickening (lignin in other vascular plants) and was like that in moss sporophytes so may provide a link with bryophytes. A fourth genus, *Psilophyton* (*Fig. 3*), had a single main branch with side branches, i.e. were **monopodial**, with sporangia in terminal bunches and is placed among the Euphyllophytina (Topic A1).

Together these fossil plants suggest that the major division among vascular plants between Lycophytina and Euphyllophytina was established early in the Devonian, and *Aglaophyton* may provide a link with the bryophytes.

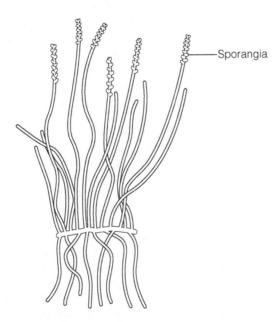

————Sporangia

Fig. 2.   Zosterophyllum *(Zosterophyllopsida), showing terminal spikes of sporangia.*

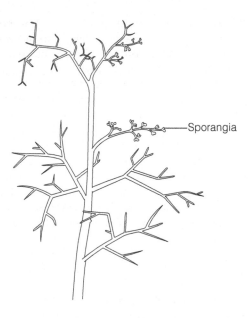

Fig. 3. Psilophyton *(Trimerophytopsida), showing lateral clumps of sporangia.*

**Later
developments**

Increasing numbers of fossil plants appear in rocks of the later Devonian era, 400–360 million years ago, suggesting that land plants diversified rapidly in this period (*Table 1*). These fossils include many short herbaceous species like those already described, and some shrubby ones with elaborate rhizome systems. Plants became taller, with monopodial branching and, by the end of the Devonian, there were many trees. These mainly belonged to groups with living members, the Lycopsida, Equisetopsida and Polypodiopsida (ferns). They, along with the first seed plants, covered the world with trees and reached their full diversity and abundance in the Carboniferous. Decay was poor owing to widespread swampy conditions and the quantity of lignin in the wood, which is resistant to decay. Many are preserved as fossils in the coal. They are thought to be instrumental in reducing the carbon dioxide in the atmosphere to a level similar to today and, in so doing, drying and cooling the climate over this period (Topics L6, Q5).

**Origins and
evolution**

Plants on land must be able to withstand large changes in temperature and humidity, wind and rain and have some means of withstanding desiccation. To grow upright they need a conducting system throughout the plant for water and nutrients, some structure in the ground to anchor and absorb water and a mechanically strong body. They also require inorganic nutrients from the soil which, in the era of the origins of land plants, would have been poor in the compounds vital for nutrient cycling (Topic E4). Many of the problems mentioned above are eased by small size.

Land plants are adapted to a low sodium environment and osmoregulate using potassium (K^+), sodium being toxic (Topic E4). Evolution from marine organisms would require major changes in ion transporters at the cell level and it is possible that they colonized the land via brackish or fresh water.

One further problem is reproduction. Land plants need reproductive structures that do not require water, which makes the sporophyte the generation more likely to succeed since spores are more resistant to desiccation. Sexual reproduction involves motile sperms which need a damp substrate to swim in; modern ferns and other spore-bearing plants still have this requirement for sexual reproduction. There is some evidence that the earliest plants mainly occurred in wet environments, perhaps forming swards about 10–20 cm high. Only the well-developed sporophytes of the late Devonian allowed plants to grow well above the damp soil surface.

Life cycle

Most of the fossil remains of these early vascular plants are of sporophyte plants. There appears to have been no dehiscence mechanism in the sporangia of *Cooksonia* and some others, though some plants had epidermal cells aligned in spirals, possibly associated with dehiscence.

Gametophytes from these plants are little known, but some plants that may have been gametophytes were upright with archegonia and antheridia borne in cup-like structures on the tops of the stems (*Fig. 4*). If so they must have resembled the 'inflorescences' of some mosses (Topic O3) in habit and function. Some of these early plants may have had sporophytes and gametophytes that looked alike, i.e. **isomorphic**, as in some algae. If most gametophytes were small and growing on the soil surface, as in living relatives of these plants, they are unlikely to have been preserved. Little is known about the gametophytes of these plants.

Homospory and heterospory

One of the most significant developments in vascular plants has been the evolution of heterospory from homospory. In living homosporous plants all spores are morphologically identical and they germinate and grow into a gametophyte bearing male and female gametes. Living heterosporous plants bear two different kinds of spore, one type that develops into a gametophyte bearing only male gametes, the other into a gametophyte bearing only female gametes. Male gametophytes normally derive from smaller spores than female gametophytes. The size of the gametophytes in living heterosporous plants is reduced, often to microscopic size. As far as is known, most of the fossil plants considered here were homosporous. A few Devonian Zosterophyllopsida bore two sizes of spores and could have been heterosporous and fossil evidence for two sizes of spores gets more frequent through the Devonian era.

Heterospory has evolved several times in vascular plant groups, both living and fossil. It is present in lycopsids and ferns, both of which have homosporous members too, and is universal among seed plants; indeed it was a vital precursor to the evolution of the seed (Topic Q1).

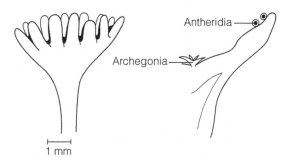

Fig. 4. *Tip of stem of a gametophyte,* Lyonophyton, *showing probable antheridia and archegonia.*

P2 CLUBMOSSES AND QUILLWORTS

Key Notes

Vegetative structure of Lycopsida

There are three groups of living clubmosses and quillworts, comprising 1000 herbaceous species distributed worldwide. Clubmosses are terrestrial or epiphytic, and the quillworts aquatic. The clubmosses have branched stems with microphyll leaves, roots and, in *Selaginella*, rhizophores; the quillworts have a basal corm and long microphylls.

Reproduction in Lycopsida

Sporangia are produced in leaf axils, often in separate strobili with small leaves. *Lycopodium* and its relatives are homosporous but *Selaginella* and *Isoetes* are heterosporous. Microspores may be very numerous in the sporangia but only four or a small number of megaspores are present in *Selaginella*, up to 100 in *Isoetes*.

Gametophyte of Lycopsida

Homosporous species have green or subterranean gametophytes, sometimes long-lived. Heterosporous species have reduced gametophytes contained within the sporangial wall. Female gametophytes may be dispersed after fertilization.

Fossil Lycopsida

Fossils go back to Devonian times and are similar to living members, particularly *Selaginella*. Heterosporous trees occur in Carboniferous rocks and are important constituents of coal.

Related topics

Early evolution of vascular plants
(P1)

Vegetative structure of Lycopsida

The lycopsids today consist of about 1000 species of small nonwoody plants known as **clubmosses** and **quillworts**, occurring throughout the world. Clubmosses are terrestrial and epiphytic and extend from tropical forests to the Arctic. One of the two groups of clubmosses, *Selaginella* species, can be abundant on the floor of tropical forests. The Isoetales (quillworts) are mainly aquatic plants. A few are grown as ornamentals.

The living clubmosses (*Fig. 1*) have a shoot system that branches, often dichotomously, but sometimes with a main stem and side branches. These stems have a central vascular system with no pith and are covered with **microphyll** leaves. Microphylls (literally 'small leaves') are generally small, a few millimeters in length, and characterized by a single vascular strand through the middle. They are thought to have derived from flattened spines, as in *Zosterophyllum* (Topic P1). There is no gap in the conducting system of the stem where the leaf branches off. By contrast, the leaves of most other vascular plants, known as **megaphylls**, are thought to have derived from a flattened and fused branching system that, in nearly all, has lost its stem buds and so has become **determinate** in its growth, i.e. it cannot continue growing. Megaphylls

(a)

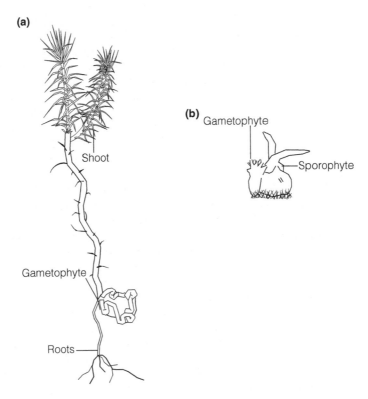

Fig. 1. *A clubmoss,* Lycopodium, *showing subterranean gametophyte and sporophyte growing from it.*

have a network of vascular strands (Topic D5) and there is a gap in the stem vascular system where the leaf branches off.

Rhizomes may be present and roots (Topic D2) branch from them. The roots branch dichotomously and regularly. In *Selaginella* the roots branch from a curious structure intermediate between stems and roots, called a **rhizophore**. These arise at angles in a stem and are unbranched, often aerial, but form roots when they touch the ground.

The quillworts have only a basal **corm**, a short (about 1 cm) swollen stem or rootstock, which can show limited secondary growth (Topic D4) and from which arises a rosette of remarkably long microphylls, often 10 cm or longer, and rhizoids, which, between them, completely conceal the corm. They closely resemble rosettes of aquatic flowering plants with which they often grow and do not resemble the clubmosses.

Reproduction in Lycopsida

One group of Lycopsida, including about 200 species of *Lycopodium* and its relatives, is **homosporous**; the other two groups, *Selaginella* (about 700 species), and *Isoetes* (about 70 species), are **heterosporous** (Topic P1). In all groups the sporangia are produced singly in the axils of leaves. In a few homosporous species there are fertile sections of the stem but in most there are separate **strobili** (singular strobilus). These consist of specialized scale-like microphylls with a different morphology from the vegetative leaves, often raised above the main stem. Nearly all *Selaginella* species have strobili and in *Isoetes* all leaves may bear sporangia.

The sporangium is normally 1–2 mm across, though larger in *Isoetes*. The sporangium consists of a wall, initially of several cell layers, though the inner layers break down as the spores mature. It dehisces along a line of thin-walled cells. In the heterosporous groups, strobili may contain a mixture of male sporangia and female sporangia, often with female at the base, or bear them in separate strobili. In the male sporangia, or **microsporangia**, there are numerous **microspores**, in *Isoetes* possibly as many as one million. In **megasporangia** of *Selaginella* there is usually one spore mother cell undergoing meiosis to produce four **megaspores**, though a few species may produce eight or more. In *Isoetes* more than 100 megaspores are produced in each sporangium.

Gametophyte of Lycopsida

In homosporous forms two types of gametophyte are known, both independent multicellular structures (*Fig. 1*). One type is cylindrical to 3 mm long and grows near the soil surface, becoming green and branched in the light, with sex organs, usually mixed, at the base of the branches. It lives for up to a year. The second type is subterranean or epiphytic and may live for 10 years then reaching 2 cm in length. It is oblong, disc-like or branched and antheridia are produced first in the center, archegonia towards the edge later. All are infected with fungi that occupy a defined place and are essential for nutrition of the gametophyte and act as mycorrhizae do in angiosperms (Topic M1).

In heterosporous groups the gametophyte is much reduced and develops entirely within the spore wall, beginning development before leaving the sporangium. In microspores the first cell division leads to one vegetative cell known as the **prothallus** cell, and a second cell that divides to form the antheridium. The antheridium consists of a sterile jacket, but this and the prothallus cell disintegrate at maturity. In *Selaginella* biflagellate sperms, usually 128 or 256, all contained within the spore, are produced; in *Isoetes* just four multiflagellate sperms are produced. By this time the microspore has dispersed. The spore wall ruptures to release the sperms. In the megaspore, a multicellular gametophyte grows and the spore wall ruptures. Archegonia similar to those of bryophytes (Topic O3) grow by the rupture. Dispersal of the female gametophyte occurs at different times in different species but in a few species it is retained until after fertilization when the embryo is growing.

Fossil Lycopsida

The clubmosses and quillworts are living remnants of a once much larger group with a rich fossil record. Fossil lycopsids are known from the early Devonian onwards, mostly resembling living homosporous clubmosses like *Lycopodium*, being herbaceous and most with dichotomous branching. Sporangia were sometimes stalked and borne on the leaves. By the Carboniferous, herbaceous heterosporous plants appear, and so closely resemble *Selaginella* that they are placed in the same genus.

The best known fossil lycopsids were trees, up to 30 m or even more, the tallest trees known from that time. They are common fossils, beautifully preserved from the Carboniferous period in Britain and North America, and are likely to have been the dominant trees in tropical and warm temperate climates. They can be ascribed to a genus *Lepidodendron* that had a long unbranched trunk with dichotomous branching at its tip (*Fig. 2*). Strobili were produced at the branch ends and these were heterosporous. Microsporangia and megaspo-

rangia were similar to those of living heterosporous lycopsids and the gametophytes were enclosed within the spore wall. In at least one species only one
megaspore was produced within the sporangium and the whole structure was
enclosed by a leaf-like outgrowth, to be dispersed together.

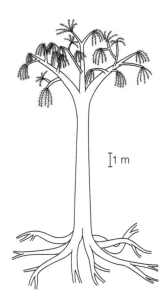

Fig. 2.   Reconstruction of fossil lycopsid Lepidodendron, *from Carboniferous rocks.*

P3 HORSETAILS

Key Notes

Vegetative structure of Equisetopsida	There are about 20 species of horsetails, worldwide in distribution. All are herbaceous but can reach 10 m. They have a jointed, ribbed stem with microphylls and whorls of branches and extensive rhizome and root systems. Silica is deposited giving them their roughness.
Reproduction in Equisetopsida	Strobili are produced at the ends of shoots, with sporangia at the branch tips. All are homosporous, the spores having elaters.
Gametophyte of Equisetopsida	These are green structures on the soil surface that may bear only antheridia or only archegonia, but archegonial ones produce antheridia as they age.
Fossil Equisetopsida	The earliest fossils come from the Devonian, and some Carboniferous fossils resemble *Equisetum*. Others are trees up to 20 m high, some heterosporous.
Related topics	Early evolution of vascular plants (P1) Ferns (P4)

Vegetative structure of Equisetopsida

There is only one living genus and about 20 species in the Equisetopsida, known as **horsetails**, *Equisetum*. They are herbaceous perennial plants, mostly up to 1 m in height, worldwide in distribution, but mostly in the north temperate where they are often common. South and Central American species are taller, reaching 10 m, and in these the shoots are perennial. In the smaller northern plants only the below-ground parts survive the winter.

The living horsetails have a characteristic rough jointed stem (*Fig. 1*). This is ribbed, with scale-like microphyll leaves at the joints where they readily come apart. Most species have whorls of side branches from the main stem. They have an extensive rhizome system, the rhizomes being similar to the above-ground stem with nodes from which roots arise. This allows horsetails to colonize large patches, and their ability to regrow from a fragment of rhizome means they can be pernicious weeds. The stems have extensive deposits of silica along their length which gives them their rough texture, and their alternative name, 'scouring rushes', refers to the use of some as scourers. Growth abnormalities, such as no internodes, no side branches or dichotomous branching, are frequent in horsetails and this suggests that their growth regulation is poor.

Molecular evidence has shown that the Equisetopsida form an early branch from the line leading to ferns and seed plants and some authors place them with ferns in the Pteropsida.

Reproduction in Equisetopsida

Sporangia are produced in strobili at the tip of either a vegetative shoot or a distinct, brownish fertile shoot (*Fig. 1*). The strobilus has short side branches each with an umbrella-shaped tip bearing 5–10 sporangia on the underside

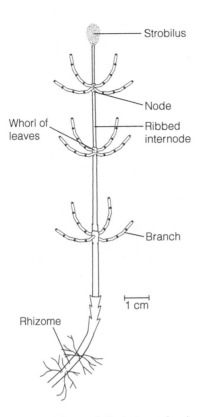

Fig. 1.   *A horsetail,* Equisetum, *showing vegetative shoot with terminal strobilus.*

(*Fig. 2*), similar in structure to those of Lycopsida. The **spores** are unusual in that they contain chloroplasts and, at maturity, have four **elaters** (*Fig. 2*). These are short band-like structures with a spoon-shaped tip that coil around the spore at high humidity and uncoil as they dry, aiding dispersal. All horsetails are homosporous.

Gametophyte of Equisetopsida

Although the spores are apparently all identical in horsetails, gametophytes show a partial division of the sexes. They are green structures on a damp soil surface, with a colorless base and a much branched upper part, though they remain less than 1 cm long. Gametophytes may produce antheridia or archegonia or both, any

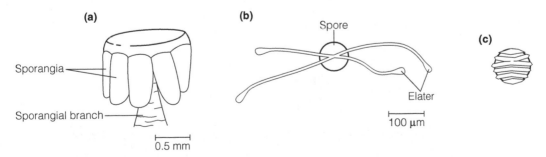

Fig. 2.   *(a) A branch of the strobilus of a horsetail showing sporangia, (b) a spore with attached elaters.*

archegonial plants usually producing antheridia as they get older. The determination of sex seems to be at least partly environmental, and is flexible. The antheridia have a two-layered jacket and produce multi-flagellate sperm; the archegonia are similar to those of bryophytes (Topic O3).

Fossil Equisetopsida

The fossil history of the group is rich, with the earliest fossils dating from the Devonian where rhizomatous herbaceous plants with forked leaves and divided sporangial branches are known. In the Carboniferous other fossils appear, varied in the structure of their strobili, including climbing plants and a fossil closely resembling *Equisetum* itself, so this may be an ancient genus.

The best known fossil equisetopsids were large trees forming an important constituent of coal, *Calamites* and its relatives. *Calamites* (Fig. 3) grew to 18 m or so in height but had many features similar to those of living horsetails. They had jointed ribbed stems, extensive branching and scale-like leaves, though these were often bigger than those of living horsetails. There was an underground rhizome system from which trunks grew and it is likely that these trees formed large patches in swampy ground. The stem had a large central pith, but secondary thickening was extensive outside this. Strobili were borne on the side branches rather than at the tips but varied in structure, some having leaf-like bracts by each sporangium. Most species were homosporous but, especially near the end of the Carboniferous and into the Permian, some had two different sizes of sporangium and were heterosporous in parallel with heterosporous lycopsids. Some species retained the megaspores in their sporangia on the parent plant to be dispersed as a unit.

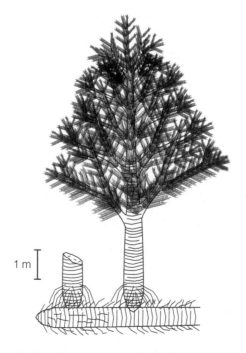

1 m

Fig. 3.   Reconstruction of Carboniferous member of the Equisetopsida, Calamites.

P4 FERNS

Key Notes

General characteristics	Ferns are a large group of plants occurring throughout the world. Most have rhizomes with roots and some have a trunk, though there is no secondary thickening. Leaves are variable but typically large and pinnate derived from a branch system. They are homosporous except for one small group.
Eusporangiate ferns	Three small groups, the Marattiales, the adderstongues, and the whisk-ferns, have a eusporangium similar to those of other spore-bearing plants. Marattiaceae have a rich fossil record and resemble typical tree ferns except in the sporangia. Adderstongues are small, do not resemble typical ferns and have usually two branches, one bearing the large sporangia. Whisk-ferns have no roots or leaves and resemble the earliest vascular plants.
Leptosporangiate ferns	These have a stem, usually a rhizome, and roots with no secondary thickening. Leaves are often large and pinnate but may be simple or one cell thick. They have a leptosporangium with explosive dehiscence, dispersing the spores. Sporangia are borne in sori on the underside of leaves or on reproductive leaves separate from the vegetative leaves.
Water ferns	Two small groups of ferns are aquatic, the Marsileaceae on mud and Salviniaceae that float. They are small plants, heterosporous and do not resemble typical ferns. Sporangia are on separate branches and do not dehisce. Megasporangia usually contain one spore.
The gametophyte	Typical leptosporangiate ferns have a prothallus. It is usually flat and green, about 10 mm across on damp soil, and bears antheridia and archegonia on its underside. It has rhizoids. Eusporangiate ferns have larger longer-lived prothalli, some being subterranean. The heterosporous ferns have much reduced gametophytes retained within the spore wall. Some reproduce vegetatively.
Ecology of ferns	Though often common they do not dominate except for bracken. They are mainly found in dense shade in woods or rock crevices and as pioneers of gaps in rainforest and as epiphytes. Bracken can cover moorland through vegetative growth.
Fossil ferns	Many fossils are known from mid-Devonian onwards, resembling modern eusporangiate ferns, and tree ferns are common in coal. Leptosporangiate ferns first appear in the Cretaceous. Heterospory is known from the Carboniferous onwards.
Ferns and man	Bracken was formerly used as bedding and kindling, and young shoots of ferns are eaten, although some, including bracken, are carcinogenic. The main use now is ornamental.
Related topics	Early evolution of vascular plants (P1) Horsetails (P3) Clubmosses and quillworts (P2) Early seed plants (Q1)

General characteristics

The ferns are a numerous and important group of vascular plants, with about 12 000 species and growing throughout the world. They are divided into four living groups, just one of which contains the great majority of ferns. They range from small epiphytes to tree ferns 20 m high to floating aquatics. Many ferns have a perennial rhizome, either underground or growing along a branch if epiphytic, with roots attached. Characteristically ferns have large **pinnate** leaves, i.e. with numerous leaflets, but some have simple leaves and there is a great range in size. The leaves are megaphylls (Topic P2). In most these have become determinate in their growth, without stem buds, so they cannot continue growing, although a few climbing ferns have stem buds in their leaves and these are indeterminate in their growth. The conducting system is typical of vascular plants (Topic D3), with the xylem cells being only tracheids in most, although cells resembling vessels are found in **Marattiales**, **bracken**, *Pteridium aquilinum*, and some of the **water ferns**, perhaps independently evolved from those of flowering plants but almost identical in structure. There is no secondary thickening and most ferns are herbaceous. The trunks of tree ferns are supported by a woody stem, usually 10–15 cm in diameter and widest at the top, and a mass of adventitious roots forming a 'skirt' around the base. All ferns except the water ferns are homosporous.

Eusporangiate ferns

Three small groups of ferns have sporangia similar to those of the lycopsids and horsetails (Topics P2, P3), known as **eusporangia**. In these, several initial cells form the sporangium and this develops a wall several cells thick. One or more of the cell layers disintegrates to form nutritive tissue for the developing spores. There is a line of weaker cells in the wall which splits to allow spores to be dispersed in the wind.

The **Marattiales** is a small living group closely resembling small typical tree ferns with large pinnate leaves and sporangia on the lower surface. The sporangia, however, are typical eusporangia, large with numerous spores and dehiscing by a slit at the tip. The group has a rich fossil record extending back to the Carboniferous. The stems have xylem cells resembling vessels as well as tracheids (Topic D3), and there are resin and tannin ducts that are a feature of many fossil ferns.

The **adderstongue** and moonwort group consists of small rather insignificant plants that do not resemble normal ferns but have one vegetative and one fertile leaf (occasionally more), perhaps the remains of a dichotomously branching stem (*Fig. 1*). The large eusporangia are borne in lines on the fertile leaf,

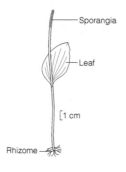

each dehiscing with a slit at the tip to produce 2000 or more spores. They are regarded as a remnant of an ancient group of ferns and one of their most peculiar features is the large number of chromosomes, presumably derived from high polyploidy. One species of adderstongue, *Ophioglossum reticulatum*, has approximately 1260 chromosomes, the largest number known for any living organism.

The **whisk-fern** group consists of two genera, *Psilotum* (*Fig. 2*) and *Tmesipteris*, of such simple vegetative construction that they resemble fossils of some of the first vascular land plants (Topic P1). Molecular and DNA evidence has shown that they are related to ferns, and some southern hemisphere ferns resemble them. They have branched stems, dichotomous in *Psilotum* and with flattened side branches in *Tmesipteris*, a branched rhizome but no leaves or roots. The sporangia are large and, in *Psilotum*, three are fused together at each point, associated with a small scale.

Leptosporangiate ferns

The fourth group, the typical ferns, comprises the great majority of living fern species (*Fig. 3*). They have unique **leptosporangia** (*Fig. 4*). The leptosporangium arises from a single initial cell and the mature sporangium always has a wall one cell thick, all other cells disintegrating as the spores mature. The number of spores is less than in a typical eusporangium, normally 64 or fewer. There is an incomplete ring of cells in the sporangial wall which have a thin cell wall on the outside and thickened inner walls. On drying out at maturity the thin sections of the cell wall are sucked in and the thickened walls drawn towards each

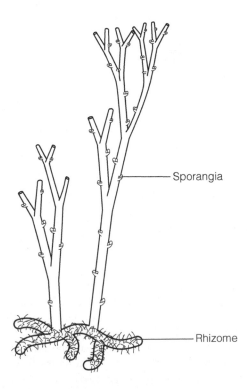

Sporangia

Rhizome

Fig. 2. A whisk-fern, Psilotum.

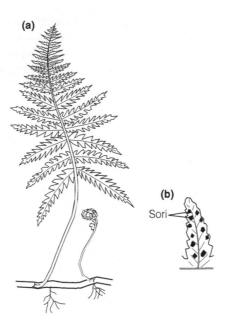

Fig. 3. A typical leptosporangiate fern, Phegopteris: *(a) leaf with developing leaf, (b) underside of leaflet showing sori.*

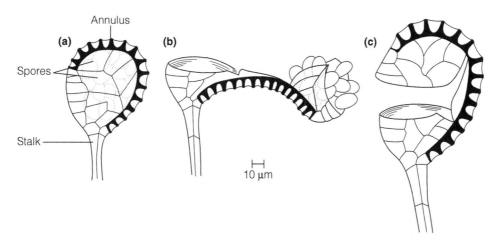

Fig. 4. A leptosporangium: (a) before dehiscence, (b) drying out, (c) after dehiscence.

other. This leads to a split in the part of the sporangial wall not covered by the ring of cells and the ring becoming inverted. Considerable tension builds up which eventually leads to a loss of cohesion in the water molecules in the cell; they then become gaseous leading to a sudden release of tension and an explosive return of the sporangium to its original position. Spores are ejected in the process. The ring of thick-walled cells is at one end of the sporangial wall in some ferns, in others it is obliquely or transversely positioned. A few ferns, such as the royal ferns, have sporangia intermediate between eusporangiate and leptosporangiate types.

The stem of typical ferns is usually an underground or epiphytic rhizome and most produce leaves only at its tip. A few have a branching rhizome, like bracken and, in tree ferns, the stem is a trunk. Many fern leaves are large and pinnate but others are simple, as in the British hartstongue and the tropical birdsnest fern (both *Asplenium*) and some tropical climbing ferns have branched leaves of indeterminate growth. In the filmy ferns, *Hymenophyllum*, the leaf is one cell thick, as in mosses. Simple leaves are regarded as derived from pinnate leaves. Fern leaves unfurl from a tight spiral giving the characteristic 'fiddle' heads (*Fig. 3*).

In many ferns all leaves are similar and these may bear sporangia, usually on the underside in groups known as **sori** (singular sorus) (*Fig. 3*) that, in some species are covered by a protective flap from the leaf, the **indusium**. Some ferns have separate fertile leaves of a different shape, usually with narrower segments and sometimes with the lamina much reduced.

Water ferns

The aquatic ferns do not resemble typical ferns and are the only **heterosporous** ferns. Members of one of the two families, the Marsileaceae, have a rhizome submerged or in mud and aerial leaves a few centimeters high resembling clover with two or four leaves or grass. The sporangia derive from one cell, so are leptosporangiate but have no specialized dispersal. They are borne on separate short stems in a hard oblong or spherical body a few millimeters long covered with rough hairs, the 'pills' of the pillwort (*Pilularia*; *Fig. 5*). They produce sori inside these, each sorus bearing both megasporangia and microsporangia. Only one spore matures in each megasporangium. Spores are released into water.

The two genera of floating water ferns are not attached to a substrate and both spread vegetatively in a similar way to other floating flowering plants. *Salvinia* (*Fig. 6*) has whorls of three simple leaves 3–10 mm long on each rhizome and no roots. *Azolla* has a tiny two-lobed leaf, with a cavity in the lower lobe containing a cyanobacterium that fixes nitrogen (Topic M2) and roots trailing into the water from the nodes. Megasporangia and microsporangia are produced separately on the undersides of the leaves and spores are released into the water. In *Azolla* there is one megaspore per sporangium, but *Salvinia* has more than one.

The gametophyte

In typical ferns the gametophyte is a prothallus that is most commonly a small green heart-shaped structure up to about 1 cm across that lives on the surface

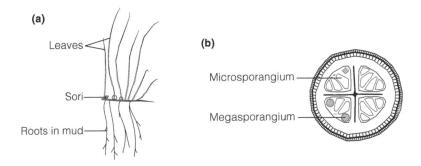

Fig. 5.   Pillwort, Pilularia: *(a) whole plant, (b) cross-section through sorus.*

Leaf

1 mm

Rhizome

Fig. 6.   The floating fern Salvinia (side view).

of damp soil (*Fig. 7*), sometimes in dense patches. Some are filamentous and a few are subterranean and associated with fungi. The body of the prothallus consists of undifferentiated parenchyma cells with a thickened central part bearing rhizoids that absorb water. Antheridia and archegonia are produced on the underside. The antheridium is a rounded jacket containing 16–32 motile sperm. The archegonia are produced mainly near the notch in most species, slightly later than the antheridia. Each archegonium has a neck of several cells surrounding two canal cells that degenerate at maturity, and an egg cell at their base, similar to those of bryophytes (Topic O3).

The eusporangiate ferns and royal ferns have larger and longer-lived prothalli than typical ferns and all have fungi associated with them acting as mycorrhizae (Topic M1); some are colorless and subterranean. The heterosporous water ferns have much reduced gametophytes that develop within the spore wall, as in heterosporous lycopsids (Topic P2). They develop in similar ways to *Selaginella*, although they are unrelated. In the floating group the microspores aggregate together and hooked hairs develop that may aid attachment to the megasporangia.

Some gametophytes can persist and spread vegetatively by gemmae similar to those of bryophytes (Topic O2). A striking example is the delicate Killarney

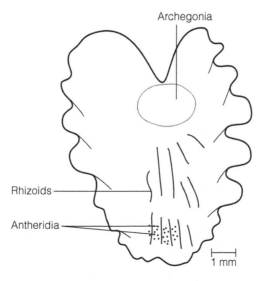

Archegonia

Rhizoids

Antheridia

1 mm

Fig. 7.   Typical fern prothallus from below.

fern, rare in the British Isles, that was long thought to have been made extinct by collectors in many places. This has recently been found solely in the gametophyte stage in many of its previous localities and several new ones. It only matures into a sporophyte in conditions of high and even rainfall, such as SW Ireland. Some other ferns may reproduce this way.

Ecology of ferns

Although often common, ferns rarely dominate the vegetation. They are mainly found in damp places since most are limited in their distribution by the need for damp ground for spore germination and gametophyte growth. The enormous numbers and wide distribution of spores makes them good colonizers of suitable sites and many are pioneers, quick to exploit gaps. They are frequently found in shady situations under dense tree canopy or in rock crevices. In the tropics many ferns are epiphytic and they can be among the commonest epiphytes in drier parts of rainforest; epiphytic ferns include the birdsnest fern which can grow to 3 m across. Several species, such as bracken, spread vegetatively with branching rhizomes. These can spread to cover large areas, usually where the soil has eroded or degraded as a result of cultivation or forest clearance, and often well away from damp ground. Bracken is one of the world's most common plants, sometimes covering large areas of moorland, and in places is considered a serious plant pest.

Very few animals eat ferns although there are a number of specialist insect feeders. Grazing animals may eat developing shoots. Some ferns contain high concentrations of insect molting hormone which may make them toxic.

Fossil ferns

Fossils of ferns are numerous from the mid-Devonian onwards. They were important in the coal measures of the Carboniferous. There were many tree ferns closely resembling the living Marattiales, and nonwoody species with various branching patterns. The fossils provide evidence, mainly from leaf form, that ferns probably evolved from the *Psilophyton* group (Topic P1). Sporangia of all the early ferns were eusporangiate. True leptosporangia, as seen in most modern ferns, probably did not appear until the Cretaceous (see *Table 1* in Topic P1). Living leptosporangiate ferns evolved in parallel with the flowering plants in recent epochs.

Several fossil ferns from the mid-Carboniferous and later were heterosporous. They are not directly related to the living heterosporous water ferns and heterospory has evolved several times within the ferns.

Ferns and man

Considering their abundance human use of ferns has been limited, although bracken was extensively used for animal bedding and as kindling in Europe. The young developing leaves of bracken and some other ferns have been harvested as food for centuries, sometimes much sought after as a delicacy and still eaten particularly in the Far East. Unfortunately some, including bracken, are carcinogenic and there is a relationship between regular eating of fern shoots and throat or esophageal cancer.

The main current use for ferns is as ornamentals. Their feathery leaves have long been admired and some rare species have been much sought after, making them rarer. Leaves are used in flower arrangements as background and orchids and other epiphytes are often grown on pieces of tree fern trunk or in a fibrous soil made from crushed fern leaves.

Q1 EARLY SEED PLANTS

Key Notes

Progymnospermopsida	These were trees of Devonian and Carboniferous times and had a combination of vegetative parts resembling those of conifers and sporangia resembling those of eusporangiate ferns.
Evolution of the seed	Seed plants are heterosporous and the female sporangia (ovules) develop only one spore and are retained on the parent plant. The gametophytes are reduced and enclosed within the spore wall. They have a protective integument, perhaps formed from sterile sporangia. Male spores are known as pollen grains. The seed derives from the fertilized ovule.
Pteridosperms	These were the first seed plants, appearing in the Devonian, with leaves that resembled ferns. Numerous fossils are known from late Devonian to Permian times. Some had one vascular bundle, some several and most had secondary thickening. The female gametophyte resembled that of heterosporous ferns and lycopsids.
Cordaitales	The other earliest seed plants were the Cordaitales which were among the largest trees of the Carboniferous with strap-like leaves and clustered reproductive structures. They resembled living conifers and are related to them.
Other fossil groups	Several other seed plant groups are known only from fossils and some of the five living groups have long fossil records. The fossil Bennettitales and Caytoniales were common from Triassic to Cretaceous and resembled the pteridosperms. The Bennettitales had flower-like reproductive structures, the Caytoniales a cupule surrounding the ovules.
Related topics	The seed (H1) Ferns (P4) Conifers (Q2)
	Cycads, ginkgo and Gnetopsida (Q3) Evolution of flowering plants (Q4)

Progymnospermopsida The Progymnospermopsida is an entirely fossil group found in mid Devonian to early Carboniferous rocks (see Topic P1, Table 1) in which they are often common. All were trees with a main trunk to 8 m high and lateral branches. The wood had marked secondary thickening, unlike any ferns, and closely resembled that of living conifers (Topics D4, Q2). Their leaves were densely packed on the lateral branches, the branch system resembling a fern frond. Some of the side branches terminated in sporangia, of the eusporangiate type (Topic P4), like contemporary fossil ferns. Homosporous and heterosporous forms are known.

The striking feature of the progymnosperms is the combination of sporangia typical of primitive ferns with vegetative structure typical of conifers. Whatever the precise origins of the seed plants, the progymnosperms provide a fascinating link between early spore-bearing plants and seed plants. There appear to have been three main branches among the Euphyllophytina from an ancestor similar to *Psilophyton*, one leading to the horsetails, one to the ferns and the third to the seed plants, along with the progymnosperms.

Evolution of the seed

The seed habit derives from a **heterosporous** condition similar to what is seen in some living lycopsids and ferns (Topics P2, P4) and fossil progymnosperms. The male spores and female spores are produced in separate sporangia and the gametophyte of both sexes is much reduced in size in all heterosporous plants, in most growing entirely within the spore wall. In many, the whole female sporangium is retained on the parent plant until the female gametophyte is fertilized. It is a sporangium that always matures just one spore, the three others produced by meiosis aborting, and is known as the **ovule**. It does not dehisce at all and the main body of the sporangium becomes a mass of parenchyma cells known as the **nucellus** (Topic H1). The male spores are usually (but not always) smaller than the female spores and the male gametophyte is limited to a few cells inside the spore wall. The male spores of seed plants are the **pollen** grains, the sporangium the **anther** (Topic G1). The seed derives from the female sporangium after fertilization.

In the seed the nucellus is surrounded, except for a small opening, by another structure, the **integument**, and it is this structure that distinguishes a seed from the female sporangium of other heterosporous plants. The origin of the integument is obscure, though the most likely possibility is that it is derived from the fusion of a number of sterile sporangia surrounding the fertile one. There are some intermediate stages represented among early fossil seed plants. When mature a rupture appears in the nucellus and the integument and, in all seed plants except the flowering plants and some Gnetopsida (Topic Q3), recognizable archegonia are produced on the gametophyte next to the rupture. In the flowering plants the female gametophyte is reduced to the embryo sac (Topic G2), commonly eight nuclei, within the spore wall, the synergid cells perhaps representing what remains of the archegonium. In all seed plants the seed is retained on the parent sporophyte until after a new embryo starts to grow and only then is it dispersed as a whole unit. It will be seen that a fertilized seed consists of parts of three different genetic constitutions: an outer integument and nucellus (sporangium) from the parent sporophyte, the haploid female gametophyte (or the endosperm in flowering plants; Topic H1) inside this and the new developing sporophyte inside this.

In all the **gymnosperms** (from Greek 'naked seed') the seed is exposed, usually on a modified leaf, but in the flowering plants, or **angiosperms** (Greek 'hidden seed') it is enclosed by the **ovary** that becomes the fruit at maturity. It is suggested that the ovary derives from a cup-like outgrowth from the sporophyte which is seen in some gymnosperms (Topic Q4).

Pteridosperms

Fossil seed plants appear first in Devonian rocks and they are abundantly represented as fossils from mid-Devonian onwards. They are highly variable in structure and it is possible that the seed habit evolved more than once.

The pteridosperms, or 'seed-ferns', Lyginopteridopsida, is the general name for these earliest seed plants. Many fossils of vegetative parts that closely resemble ferns from the mid-Carboniferous and later were heterosporous, and some retained the female gametophyte on the leaves until after fertilization with the embryo beginning to grow. These show intermediate stages in the evolution of the integument. There were trees and climbers, some large, and their stems were varied, some possessing a single vascular strand, others with many vascular strands but, unlike ferns, many had secondary thickening giving rise to tapered trunks. They were common plants of the Carboniferous with the tree lycopsids and equisetopsids. Their leaves were large and pinnate like those of ferns.

Like all seed plants they retained the female sporangia (each with one spore) on the leaves. In at least one of the earliest pteridosperm fossils the integument of the seed was a set of finger-like processes, suggesting its origin from a group of sterile sporangia. The female gametophyte was multicellular with archegonia similar to those of heterosporous ferns and lycopsids (Topics P2, P4). The pollen grains were probably trapped by a drop of fluid in a gap in the integument, as seen in living cycads (Topic Q3).

Fossil evidence suggests that their descendants include the cycads (Topic Q3) and the extinct Caytoniales and Bennettitales.

Cordaitales

The other common fossil seed plant group in the Carboniferous was the Cordaitales (or Cordaites), normally classified with the living conifers among the Pinopsida. They were massive trees to 30 m high with simple strap-shaped leaves, often quite large but with simple traces in the vascular system. Their vegetative parts strongly resembled those of some living conifers such as the monkey-puzzle, *Araucaria*. Male and female reproductive structures were separated onto different shoots, perhaps on the same tree. Shoots of both sexes had overlapping bracts (modified leaves) with sporangia at their tips (*Fig. 1*). Their relationship with the pteridosperms is obscure, but they more closely resemble the progymnosperms in habit and vegetative structure. It is possible that Cordaitales and conifers may have developed the seed habit independently from pteridosperms and other groups.

Other fossil groups

The seed plants gradually increased in prominence through the Carboniferous and Permian and by the end of the Permian they became dominant with the decline of tree lycopsids and equisetopsids. By this time cycads, ginkgos and a few true conifers had appeared (Topics Q2, Q3) along with some other groups that have since become extinct.

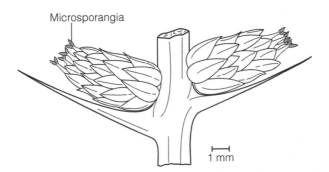

Microsporangia

1 mm

Fig. 1.   Cone-like male reproductive structure of the Cordaitales.

Of the extinct groups, the Bennettitales had unbranched or sparingly branched trunks and large pinnate, fern-like leaves like the pteridosperms. They resembled living cycads (Topic Q3) but differed in details of leaf anatomy. They appeared in the Permian and are abundantly represented in the fossil record from the Triassic to the Cretaceous. Their reproductive structures were borne on very short separate stems on the trunks, with the ovules towards the tip. The ovules were surrounded by modified leaves bearing the male sporangia (anthers), and these anthers were surrounded by bracts making the whole structure look remarkably like a flower (*Fig. 2*). It is likely that some, at least, were insect-pollinated, but this structure probably arose independently from the flowers of the angiosperms.

Another fossil group, the Caytoniales from Triassic and Jurassic rocks, may be ancestral to the flowering plants, since the ovules, and later the seeds, were surrounded by cup-like outgrowths, known as cupules (*Fig. 3*), resembling the ovary of flowering plants.

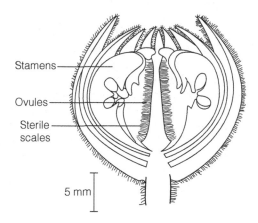

Fig. 2.   The hermaphrodite flower-like reproductive structure of fossil Bennettitales.

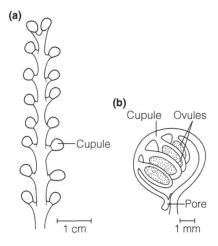

Fig. 3.   The female reproductive structure of fossil Caytoniales showing cupule surrounding the ovule.

Q2 CONIFERS

Key Notes

Ecology and human uses	The conifers comprise about 560 species of woody plants, mainly occurring in temperate and boreal regions. They form extensive forests and include the tallest, oldest and most massive of trees. They are important as sources of timber and wood pulp. Leaves decay slowly and accumulate on the ground forming an acid litter; periodic fires may be characteristic of conifer forests. Major radiation was in the Triassic.
Stems and roots	The stems resemble those of angiosperms except that the xylem has only tracheids and the phloem only one specialized type of cell, the sieve element. The tracheids have bordered pits in lines. The roots of most have ectomycorrhiza associated with them.
Leaves	The leaves are mostly needle-like or scale-like with a single central vein resembling microphylls but have a vascular trace in the stem. A few are broader and fossil leaves are forked. Resin canals are often present.
Male reproductive structures	Anthers are borne in pairs on the underside of scales forming male cones. The pollen grains usually have air bladders mainly involved with orientation at the micropyle for fertilization. All are wind-pollinated and pollen is produced in enormous quantity. The male gametophyte consists of four cells in most conifers contained within the pollen grain.
Female reproductive structures	Ovules are borne on ovule scales in female cones, or are solitary in yews. The female gametophyte initially has a free nuclear stage before forming cells and archegonia. A drop of fluid is secreted by the micropyle to trap pollen when it is mature.
Fertilization and the seed	Pollen tubes penetrate the archegonium and the largest gamete fertilizes the egg. Several embryos may be formed initially but one occupies the mature seed. It has several cotyledons. Seeds are released by the cone scales separating, and some are an important food source for animals.
Related topics	Meristems and primary tissue (D1) Early seed plants (Q1) Woody stems and secondary growth (D4)

Ecology and human uses

The Coniferales is the largest and most important seed plant class except for the flowering plants. Most conifers are trees, a few are shrubs; all are woody. They are found throughout the world but achieve their greatest abundance as the dominants of the boreal forests of the northern hemisphere (Topic L3) with high diversity mainly in western North America and east Asia. They cover extensive areas in temperate parts of the southern hemisphere too. They are much less common in the tropics with only a few tropical genera, mainly in the mountains. In total there are seven living families and about 560 species. They are the tallest and most massive of trees, at least three species reaching to well over 100 m, all taller than any angiosperm. They can live to a great age and are among the oldest known organisms, with the bristlecone pines of California at over 4000 years old and the huon pine of Tasmania possibly reaching 10 000 or more years. Most have specialist reproductive structures, the cones that give the class its name. A few, such as the yew, bear seeds individually. They are an ancient group, with fossil conifers first appearing in late Carboniferous rocks (Topic P1, see *Table 1*) and they were abundant in the Triassic.

Conifers are enormously important economically since their wood is widely used for furniture and other construction and they are one of the main sources of pulp for paper. They have been planted extensively throughout the temperate world, particularly on poor soils, and huge part-managed coniferous forests occur in Eurasia and North America. They are extensively used for ornament and many are resinous, with the resins widely used in gums and varnishes. Turpentine derives from pine resin.

The decay of many conifer leaves is slower than their production by the tree, leading to an accumulation of fallen leaves in a conifer forest that can acidify the ground and provide a fire hazard. Periodic fires are characteristic of many conifer woodlands and one of the main ways in which the accumulation of fallen leaves is removed.

Stems and roots

The stems of conifers are all woody and have xylem made of **tracheids** with no vessels (Topics D1, D4). There is extensive **secondary thickening** and the secondary tracheids form regular ranks. In temperate species tracheids that are formed in the spring are wider than later ones forming clear annual rings. The tracheids have lines of large circular pits with conspicuous borders in their cell walls. The xylem is interlaced with rays containing living parenchyma cells and sometimes **resin** cells (Topic D4). There are differences in xylem structure between different conifers that are well preserved in fossils, e.g. wood similar to that of living *Araucaria* (monkey-puzzle and its relatives) appeared first in the Carboniferous; pine-like wood did not appear until the Cretaceous. The phloem has a simpler structure than that of flowering plants with only one specialized type of cell, an elongated **sieve element** (Topic D1), with parenchyma cells rich in starch and mitochondria next to them. A pith is present in the young stages of stem growth.

Many conifers have a single straight stem with much smaller side branches, giving rise to the characteristic narrow pyramidal shape. The straightness of the stem has greatly helped foresters use the timber. Others, such as most pines, have a much more rounded crown with more even branching, and a few, such as the junipers, are shrubs with gnarled stems or several stems arising from the base.

The roots of conifers resemble those of angiosperms (except that the xylem has only tracheids). **Ectomycorrhizae** (Topic M1) are associated with all except the southern hemisphere families Podocarpaceae and Araucariaceae and the fungi aid in the decay of the leaf litter. Araucariaceae and Podocarpaceae have arbuscular mycorrhizae with a similar function, Podocarpaceae containing these within root nodules.

Leaves

The leaves of living conifers are all simple and most are shaped like **needles**, sometimes long, or **scales**, semicircular or flattened in cross-section. A few southern hemisphere conifers in the Araucariaceae and Podocarpaceae have larger, broader leaves, up to 5 cm wide; a few fossils show forked leaves. The great majority of conifers retain their leaves through the year, each leaf living for 2 years or more, up to 15 years in *Araucaria*. A few have deciduous leaves, such as the larches (*Larix*) and swamp cypress (*Taxodium*). Leaves are borne either directly on branches or on small pegs or scales and sometimes in bundles or whorls. Some species, e.g. pines, have two different forms of leaf growing together, needle-like and scale-like, and some cypresses have needle-like leaves in juvenile stages, with the scale leaves appearing later. In the pines, the coast redwood, *Sequoia*, and some others the small leaf-bearing lateral branches are shed with the leaves.

The leaf structure resembles that of flowering plants with a thick cuticle, palisade and spongy mesophyll, but all conifers have simple leaf venation, either with one central vein or a few parallel veins (Topic D5). In this they resemble **microphylls** (Topic P1), although the traces in the stem vascular tissue are not similar to those of clubmosses and other microphyll-bearing plants, and their origin remains anomalous. Most conifers have **resin** canals through the mesophyll, this resin consisting of acidic phenols, terpenes and other complex molecules. It can be present in large quantity and may protect the leaves from insect attack and make the leaves resistant to decay. A few conifers have other aromatic oils in their leaves, e.g. *Thuja*, giving characteristic scents, often species-specific, when the leaves are crushed.

Male reproductive structures

The **male sporangia**, or anthers, are borne on the underside of specialized fertile leaves in short strobili (**cones**) (*Fig. 1*). These leaves often have expanded tips to which the anthers are attached. There are two anthers per fertile leaf in many conifers but in some there are more. The cones themselves are produced in the axils of scale-leaves in pines, or at the tips of lateral shoots in other families.

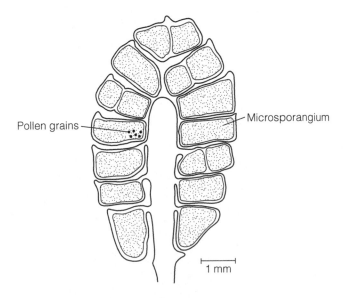

Fig. 1.   *Cross-section of a male cone of a pine,* Pinus.

The anthers have a thin wall, of one or a few cell layers. The outer cells have uneven thickening, in rings or a reticulate pattern, which are involved in dehiscence. They take about a year to develop to maturity and to release the pollen. The pollen grains resemble those of flowering plants except that pines and some other conifers have characteristic air bladders formed from an extension of the outer pollen wall giving a most distinctive appearance (*Fig. 2*). These bladders may aid with wind dispersal but are mainly involved with the orientation of the pollen as it fertilizes the ovules. All conifers are wind-pollinated and pollen can be produced in enormous quantity, frequently coming off in visible clouds when the cones are mature. By lakes they can form yellowish lines in the water or 'tide-lines' as water recedes.

The male gametophyte is much reduced and formed within the outer wall of the pollen grain. The largest male gametophytes are found in *Agathis*, the southern hemisphere white pines in the Araucariaceae that have a male prothallus with up to 40 cells. In pines there are four cells by the time the pollen is shed, two vegetative prothallus cells, a pollen tube nucleus and the generative cell. The generative cell gives rise to a sterile cell and two unequal-sized sperm cells (after another division). In the cypresses there are no vegetative cells, only the generative cell and tube nucleus. In all conifers the sperms have no flagella and are not motile.

Female reproductive structures

The female reproductive branch is the familiar pine or fir cone (*Fig. 3*). It has two ovules attached to each fertile scale leaf, but the cone differs markedly from the male in that each fertile scale has a bract underneath it, in some partially

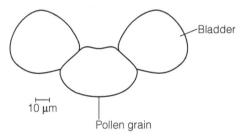

Fig. 2.   Pollen grain of a pine, showing bladders.

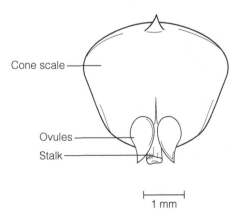

Fig. 3.   Surface view of a scale from the female cone of a pine, showing two ovules at its base.

fused with the ovule scale. This and evidence from fossil conifers suggests that the cone is a compound structure with each fertile scale derived from a whole shoot. The cone can take 2 years to mature. A few conifers, notably the **yew** and its relatives, Taxaceae, do not have cones and the ovule is solitary, borne at the tip of a minute shoot in the leaf axils. They are often placed in a separate class, Taxales.

The female sporangium (Topic Q1) has a single integument, and four spores are produced, though only one is functional. The female spore starts to divide to produce the gametophyte at an early stage, and in northern species there is usually a dormant period in the first winter. Eventually many free nuclei are produced and cell walls form once there are around 2000 nuclei. Between one and six archegonia are produced next to the **micropyle** (Topic G1) and, in these, the egg is surrounded by neck cells and a canal cell as in other vascular plants and bryophytes (Topic O3).

When the ovule is receptive the cone scales open apart slightly and a drop of sticky liquid is exuded from the micropyle. Pollen grains are caught in this drop, which is then resorbed, and after pollination the cone scales may close up again. The air bladders on the pollen grains serve to orient the pollen grain as it approaches the micropyle.

Fertilization and the seed

Once a pollen grain has reached the micropyle, the two male **gametes** are formed and a pollen tube similar to that of flowering plants grows actively through the nucellus to reach the archegonia. The male gametes are discharged into the egg along with the tube nucleus and the sterile nucleus. Fertilization is achieved by the fusion of the larger of the male gametes with the egg nucleus, the other three male nuclei degenerating.

If more than one archegonium has been fertilized several **embryos** may be formed, and each fertilized egg may divide to form several embryos, so competition can occur. Eventually one embryo outcompetes the others and absorbs the nutrients from the female gametophyte, filling the seed when it is mature. The embryo has several **cotyledons** (the food stores and first leaves; Topic H1). In most conifers the seeds are released when the cone scales separate at maturity. The seeds may be nutritious and can be a major food for some birds, such as crossbills, and small mammals. Some species have a resistant integument that allows the seed to lie dormant, and a few only germinate after a fire has passed over them. The **yew** has a fleshy outgrowth from the integument attractive to birds.

In comparison with most flowering plants the stages of reproduction are slow, taking about a year from female cones being receptive to the mature seed, but in pines fertilization does not occur until a year after the female cones become receptive and the cycle takes 2 years.

Q3 CYCADS, GINKGO AND GNETOPSIDA

Key Notes

Cycads	There are about 76 species of cycads of the tropics and subtropics. They have stout trunks with a pith and little secondary thickening, usually unbranched, and long pinnate leaves, interspersed with scale leaves. Fossils date back to the Permian and include slender branched trees.
Reproduction in cycads	All are dioecious, bearing either male or female fertile leaves, usually in cones. The sporangia are in pairs or groups. Pollen grains have three cells when released and produce two enormous flagellate sperm after they have reached the pollination drop by the ovules. The female gametophyte has many cells and archegonia. Fertilization can take 5 months from pollination.
Ginkgo	One species survives of an ancient group well represented as fossils. It is a tall ornamental tree with unique fan-shaped leaves and secondary thickening in the trunk.
Reproduction in ginkgo	Ginkgo is dioecious with male sporangia in catkin-like cones and ovules in pairs. The pollen has four nuclei on dispersal, and, at fertilization, produces two flagellate sperms. Fertilization takes 5 months.
Gnetopsida	Two genera of trees shrubs and climbers, *Gnetum* and *Ephedra*, and the unique desert plant *Welwitschia* make up the group. They have vessel-like xylem cells and phloem resembling that of angiosperms, but are probably not closely related to angiosperms.
Reproduction in Gnetopsida	They are dioecious or monoecious and the sporangia are borne in pairs or whorls. Pollen grains have two to five cells and produce two nonflagellate sperms when they reach the pollination drop. This drop may attract insects for pollination. Two eggs are fertilized, paralleling the double fertilization in angiosperms although one embryo aborts.
Related topics	Pollen and ovules (G2) Conifers (Q2) Early seed plants (Q1) Evolution of flowering plants (Q4)

Cycads

The cycads, Cycadopsida, are a group of rather palm-like plants that somewhat resemble the fossil pteridosperms (Topic Q1). Cladistic analysis indicates that they may be an early branch from the seed plant line. Their current distribution ranges from southern North America to Chile (about 36 species), South Africa (15 spp.) and eastern Asia to Australia (25 spp.) but fossils are much more widespread, dating back to the Permian. Between the Triassic and Cretaceous they were abundant and varied. All living cycads are woody slow-growing plants with large pinnate leaves in a cluster at the tip, occupying understorey positions in plant communities, though common in places. The trunks are thick, normally unbranched, and typically to about 2 m tall, though a few grow to 15 m and some have branched horizontal stems with leaf rosettes at ground level. They have a large central pith, restricted secondary thickening and a cortex. This form only appears as fossils in Tertiary deposits; before that many were slender and branched.

Scale leaves occur interspersed with the foliage leaves and persist to cover part of the trunk. Age can be inferred from the leaf scars and some can live to 1000 years or more. They have a deep tap root and surface roots often associating with N-fixing cyanobacteria (Topic M2). The starchy central pith is eaten in some places as sago (the main sago plants are palms) and some species are grown as ornamentals.

Reproduction in cycads

All cycads are dioecious and most produce reproductive structures on specialized leaves in terminal or lateral **cones** (*Fig. 1*). These cones can be enormous, weighing up to 40 kg. The anthers are borne on the underside of thick, scale-like leaves. The male gametophyte grows first within the wall of the pollen grain. The pollen is released when the gametophyte has three cells. The pollen grain bursts once it has reached a drop of fluid by the micropyle of an ovule, and the cells divide to give rise to two or more sperms (one genus, *Microcycas*, has up to 16 sperms). There is a short structure like a pollen tube through which the sperms swim. The sperms are like those of ferns in that they have numerous flagella and swim actively to fertilize the ovules, but they are about 1000 times the size of fern sperms.

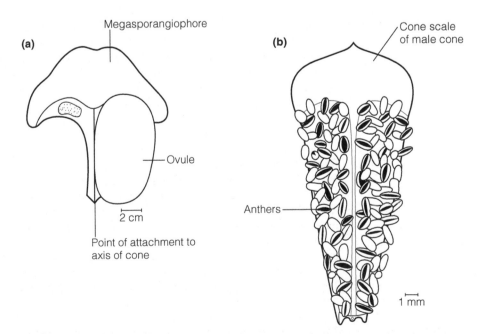

Fig. 1.   Cone scales of cycads: (a) female scale, (b) male scale with anthers.

The ovules vary from 6 mm to 6 cm in diameter and are borne in pairs, or as six or eight together, on large scale-leaves. The female gametophyte, developing entirely within the ovule wall, first has free nuclear divisions. Cell walls are formed when there are 1000 or more nuclei. Archegonia are formed by the micropyle with a single ventral canal cell above an egg that is the largest among land plants, up to 3 mm in diameter. A pollination drop is secreted by the micropyle containing sugars and amino acids and the pollen is trapped in this. Some cycads are pollinated by insects, mainly weevils, so the drop is an attractant as well as a pollen trap; others may be wind-pollinated. The whole process of fertilization in cycads takes about 5 months. Once fertilized, the ovule grows quickly, again with a period of free nuclear divisions first, absorbing the female gametophyte tissue. One, two or three cotyledons are formed and the seed germinates as soon as there are favorable conditions.

Ginkgo

The ginkgo or maidenhair tree, *Ginkgo biloba*, is the sole living member of the Ginkgoopsida and is a truly astonishing survivor. It is native of China but only survives in cultivation and has now been planted throughout the temperate world, particularly in towns where it is resistant to pollution and insect attack. The seeds are eaten in Japan and China. It resembles conifers in its vegetative structure and cycads in its reproduction but has several unique features. It is a tall elegant deciduous tree with a unique leaf shaped like a notched fan turning bright yellow in autumn. Similar leaves are found in Jurassic deposits from many sites and even the Triassic about 200 million years ago. The trunk has secondary thickening and the tracheids have bordered pits in a single row (Topic D4); in this it resembles conifers. It characteristically has long shoots with numerous leaves and short shoots bearing leaf whorls and the reproductive structures. The relationships of the Ginkgoopsida are obscure but fossils date back to the early Permian and the group was widespread and varied during the Triassic and Jurassic.

Reproduction in ginkgo

Ginkgo is dioecious. The male cones resemble small catkins and are made up of scale leaves, each of which bears two anthers on its underside. Each pollen grain contains four cells of the male gametophyte by the time it is shed. It is dispersed by the wind until caught in the pollination drop of a female plant, which then retracts the drop drawing the pollen into the ovule. There the pollen grain bursts, pollen tubes are produced and two more cell divisions of one of the four cells give rise to two sperms. These have many flagella, like those of cycads though the cells are smaller, and swim to the egg cell.

The ovules are borne in pairs on stalks from the leaf axils on short shoots (*Fig. 2*). The female gametophyte, like cycads, has a free nuclear stage and forms cell walls once several thousand nuclei are present. There are chloroplasts in the female gametophyte. Two archegonia are formed, each with a neck of four cells, a canal cell and an egg. By the time the pollination drop is retracted into the ovule there is a small chamber for the male gametophyte to grow. As in cycads, the time from pollination to fertilization can be about 5 months. After fertilization usually one of the pair of ovules grows to about 2 cm diameter and the outer part of the integument becomes fleshy and smells of rancid butter. The embryo, as in cycads, develops a free nuclear stage before cell walls develop and two cotyledons are produced.

Gnetopsida

The Gnetopsida consists of three rather disparate genera, *Gnetum*, *Ephedra* and *Welwitschia*. *Gnetum* comprises about 40 species of climbers and small trees in

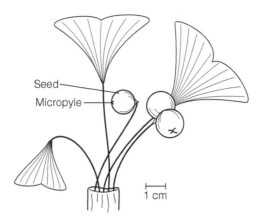

Fig. 2.   Shoot of a female ginkgo, showing developing seeds.

tropical rainforests of South America, Africa and south-east Asia. They closely resemble angiosperms except in their reproductive structures, and their leaves are almost identical to those of a dicotyledonous angiosperm. One species is cultivated for its edible seeds in Asia. *Ephedra* has 40 species of much-branched shrubs, with a few small trees and climbers, in arid warm temperate parts of Eurasia, north Africa and North and South America. They have photosynthetic stems with whorls of scale leaves. *Ephedra* provides the important drug ephedrin. *Welwitschia* has a single extraordinary species confined to the deserts of south-western Africa. It is unlike any other plant in that it produces a woody central crown at the top of a mainly underground stem, from which two leaves grow continuously, fraying at the ends and splitting. In mature plants, which can live for more than 1000 years, these can reach 3 m or more long and 1 m wide.

The Gnetopsida, unlike any other seed plants except angiosperms, have xylem with cells with such large pits that they can be described as vessels and phloem with cells associated with sieve elements resembling companion cells. On morphological grounds and by some molecular studies they appeared to be closely related to the angiosperms, but the most comprehensive molecular studies suggest that they are closer to conifers and other gymnosperms and that the similarities have arisen independently. Fossils date back only to the Cretaceous.

Reproduction in the Gnetopsida

Male and female organs are borne in separate cone-like structures (*Fig. 3*), usually on separate plants, although a few species are monoecious. The anthers are borne in pairs or small groups on stalks in the axils of scale-like leaves (bracts). Sterile ovules may occur in these male cones but fertile ones are in separate cones, again in pairs or whorls.

The pollen grains contain two (*Welwitschia*), three (*Gnetum*) or five cells (*Ephedra*) of the male gametophyte by the time they are shed. The female gametophyte has a free nuclear stage as in other gymnosperms. In *Ephedra* two archegonia are formed with many neck cells, but these are not differentiated in *Gnetum* or *Welwitschia*. The ovules, including sterile ones in the male cones, secrete a sugary pollination drop by the micropyle. This is attractive to insects, and pollen dispersed by insects or wind is caught in this fluid. There the pollen grains burst and one of the cells divides into two nonmotile sperms to fertilize

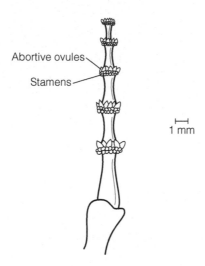

Fig. 3.   *Male cone of* Gnetum.

the egg. By the time the pollen reaches the egg in *Ephedra* (and perhaps others) the egg has divided too and each sperm fertilizes an egg, but one of the eggs aborts. The interval between pollination and fertilization is only about 1 day, in marked contrast to cycads and ginkgo. The double fertilization of Gnetopsida has parallels with that in angiosperms and may suggest how that originated (Topics G2, Q4).

In the developing embryo only *Ephedra* has free nuclear divisions, the other two genera developing cells immediately. Many embryos may be formed initially but only one with two cotyledons is present in the ripe seed. Mature seeds may have a fleshy integument or surrounding bracts or develop wings for dispersal by wind.

Q4 EVOLUTION OF FLOWERING PLANTS

Key Notes

Origin of the flower

The fertile parts of a flower are a further reduction from those of other seed plants. The anther is simple and the male gametophyte consists of a pollen tube nucleus and two sperm nuclei. The ovule is enclosed by a carpel and the female gametophyte is reduced to an embryo sac. Fertilization of the endosperm is unique to angiosperms.

The earliest flowering plants

The earliest known fossils come from the early Cretaceous and resemble pollen grains of living Chloranthaceae. DNA evidence suggests that a shrub, *Amborella*, and water lilies diverged first in angiosperm evolution. It is likely that the primitive flowers were mainly small but variable in size with few but indefinite numbers of fertile parts, some hermaphrodite, others unisexual. A range of large and small flowers appeared soon after.

Early evolution

The sepals derive from a whorl of leaves around the fertile parts. Some petals derived from sterile stamens, others from a second whorl of sepals. There were two main lines of evolution, one to large flowers and specialist insect pollination, mainly by beetles, the other to small flowers in inflorescences pollinated by wind. Subsequent developments led to smaller flowers with floral parts in fives or threes.

Later evolution

In the late Cretaceous/early Tertiary, flowering plants radiated to dominate the world, in conjunction with specialist flower-feeding insects. Innovations included fused petals, carpels inserted below the petals and bilateral symmetry. Some had inflorescences. Wind-pollination evolved many times. Most modern plant families appeared.

Specialization in flowers

A few plants are specialized to one or a few pollinator species. The orchids may attract just one species of bee and figs can only be visited by one group of wasps with which they have a close interdependent relationship. Generalization in pollination is much more common.

Evolution of vegetative structure

Flowering plants are enormously variable but the earliest were probably nonwoody shrubs from which developed trees and herbaceous plants. Vessels appeared early but a few families have none. Dicot leaves are mainly net-veined, many monocot leaves parallel-veined, but there are exceptions and some variable families.

Other aspects of evolution

Primitive fruits and seeds are likely to have been small. Many fruit and seed types have evolved in parallel in different families and one trend has been towards fusion of fruits. Pollen has developed increased number of apertures. Self-incompatibility is likely to be primitive and a stimulus for rapid evolution. C4 carbon fixation and CAM metabolism have evolved from C3 several times each.

Classification	The distinctions between plant families are based mainly on flower form but there are often associated fruit and vegetative characters. Some large families vary little; some, particularly those with primitive features, vary more. The two main divisions, monocots and dicots, are confirmed by molecular evidence, except for those families that branched off early in evolution, now known as primitive dicots.
Three families	As examples of features on which the classification is based, the daisies, grasses and orchids are all large important families each with many unifying features.
Related topics	The flower (G1) General features of plant evolution Fruits (H2) (Q5) Cycads, ginkgo and Gnetopsida (Q3)

Origin of the flower

The flower of an angiosperm (Topic G1) represents a further reduction and specialization of the heterosporous form discussed in relation to the other seed plants. The anther is simpler than that of the Bennettitales or Gnetopsida (Topics Q1, Q3); the 'fertile leaf' is, in most, a simple stalk, or **filament**. The male gametophyte is always reduced to a total of three cells (Topic G2). The origin of the carpel surrounding the ovules is obscure but it may be a modified leaf, or a cupule as in the Caytoniales (Topic Q1). In nearly all angiosperms the ovary completely encloses the ovules. The carpel also has a stigma and usually a style joining this to the ovary. The carpel may have evolved in response to insect visitation of the flowers, perhaps as a protection for the vulnerable and nutritious ovule. Insects could transfer pollen to other flowers and **insect pollination** is thought to be the primitive pollination type in angiosperms.

The female gametophyte that develops within the ovule, known as the **embryo sac**, is much reduced compared with other seed plants (Topic G2). No archegonia are produced and, in many, nuclei are formed without cell walls. From pollination to fertilization in angiosperms takes a few hours in most plants and, in contrast to all other seed plants, the **double fertilization** leads to growth of the endosperm as the seed's food store. This means that the food store only grows in fertilized seeds, potentially saving resources. In all other seed plants the female gametophyte provides the food store.

The earliest flowering plants

The earliest evidence of flowering plants is fossil pollen grains and fragmentary flowers and fruits from lower **Cretaceous** rocks, about 110 million years old (some 130 million-year-old fragments may be angiosperms), suggesting that flowering plants must have evolved before that, perhaps in the **Jurassic**, about 150 million years ago. Molecular evidence suggests an earlier origin, in the Triassic or possibly as early as the late Carboniferous, though this seems unlikely with no intervening fossils at all. All evidence suggests that they evolved long after the other seed plant groups (see Topic P1, *Table 1*).

The pollen and form of the flower of the earliest fossils are similar to those of living members of the **Chloranthaceae**, a small family from tropical America, east Asia and New Zealand. The flowers of these plants are simple and tiny, up to 3 mm long, some genera bearing hermaphrodite flowers, others unisexual, borne in inflorescences (*Fig. 1*). Each flower contains one or three stamens

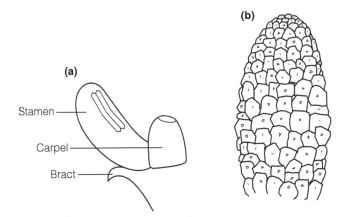

Fig. 1. Flowers of living members of the Chloranthaceae: (a) hermaphrodite flower of Sarcandra, (b) inflorescence of male flowers of Hedyosmum (from Endress, P., 1994, Diversity and Evolutionary Biology of Tropical Flowers, Cambridge University Press).

and/or one carpel and some have a bract (small modified leaf) under them. They have no petals or sepals. Hermaphrodite species are pollinated by unspecialized insects; unisexual species by the wind. Several other living plant families are represented as fossils in slightly younger rocks, with a wide range of flower structure from large-flowered magnolias to catkin-bearing trees such as the planes. In some of these the flower structure is not rigidly set, and they have variable numbers of stamens and carpels surrounded by a **perianth**, i.e. modified leaves or bracts not clearly sepals or petals.

Analysis of molecular evidence gives a slightly different picture. A small shrub, *Amborella*, from New Caledonia that bears unisexual flowers, and the **water lilies**, which are herbaceous aquatics bearing large hermaphrodite flowers, diverged from the other angiosperms earlier than any other group. Both of these have numerous stamens and an indefinite number of perianth segments. The Chloranthaceae and its relatives are, on this evidence, closer to other families.

Whatever the most primitive plant is, it is likely that the earliest flowers had certain features since these are shared by all the plants mentioned above and are consistent with the fossil record. There was a **variable number** of stamens and carpels, each separate and inserted above the perianth where this was present. **Unisexual** and **hermaphrodite** forms appeared early, and some had a perianth or bracts underneath the fertile parts but no well-defined petals or sepals. They are likely to have been visited and pollinated by unspecialized beetles, flies and wasps, though wind-pollination probably arose early on.

Early evolution

Adaptive radiation of flowering plants, i.e. their spread and diversification, occurred through the Cretaceous in parallel with other seed plant groups and some living families appeared. Outer whorls of the flower became differentiated early on, the outermost layer, referred to as sepals even if there is only one whorl, probably evolving from bracts. When there are two different whorls, the inner whorl is referred to as the petals, but some probably evolved from a second whorl of sepals, others from stamens losing their anthers and becoming organs purely for attracting insects. Some water lilies have intermediate organs, stamens with flattened coloured filaments, between the fertile stamens and larger entirely sterile petals. Cultivated 'double' flowers such as roses have some stamens replaced by petals (Topic N4). One main early line of evolution was towards a fixed number in each whorl, often five in the dicots and three in the monocots, making the flowers radially symmetrical (**actinomorphic**).

There was specialization for pollination, by beetles along one line of evolution including the magnolias, and by wind along another leading to the catkin-bearing trees. The insect-pollinated group had hermaphrodite flowers of, at first, increasing size and complexity, though later insect-pollinated flowers are smaller. By the end of the Cretaceous many families with actinomorphic insect-pollinated flowers had appeared, such as buttercups, pinks and heathers. The wind-pollinated group retained small unisexual flowers, with few parts borne in inflorescences, usually with male and female on the same plant. This **monoecious** breeding system (Topic G3) includes less than 5% of living species but several of these are dominants such as the oaks and beeches, birches, hazels and planes.

Later evolution of flowers

At the end of the Cretaceous and beginning of the **Tertiary**, between 75 and 50 million years ago, there was a second and larger adaptive radiation which saw the increasing dominance of the angiosperms, the appearance of the majority of modern plant families and the associated decline of the gymnosperms. This was closely associated with the adaptive radiation of specialist insect pollinators and leaf feeders, the butterflies and moths, long-tongued flies and bees. Innovations included the fusion of the lower part of the petals and/or the sepals into a tube and the fusion of the carpels. Bilaterally symmetrical, or **zygomorphic**, flowers appeared, usually facing sideways rather than upwards and with the lower petal(s) forming a lip as a landing platform for insects, e.g. members of the thyme family (*Fig. 2*). These were adaptations for specialist insect pollination and at least in part serve to exclude other visitors. Adaptations for pollination by birds or bats appeared later, mainly derived from specialist insect-pollinated flowers. The main trends are outlined in *Table 1*.

Aggregation of flowers into **inflorescences** is an adaptive trend in many plant families. It is taken to extreme specialization in the daisy family in which there

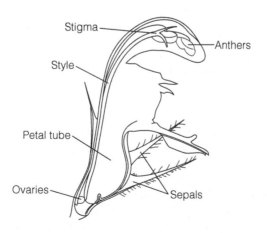

Table 1.   *Evolutionary trends in flowers during the end Cretaceous – early Tertiary radiation*

Radial symmetry → bilateral symmetry
Separate petals → fused petals
Large numbers of floral parts → fixed small numbers of parts
Large flowers → small flowers
Carpels inserted above the petals → carpels inserted below the petals
Carpels free → carpels fused
Flowers with pollen as food reward → flowers with nectar as food reward
Pollination by unspecialized insects → pollination by specialized bees; butterflies/moths; long-tongued flies; birds or bats
In insect-pollinated species, hermaphrodite flowers → unisexual flowers
Insect-pollinated species → wind-pollinated species

Many trends occurred in parallel in different families; some also happened in reverse.

are often two different forms of flower in each inflorescence, the whole aggregation, sometimes of more than 100 flowers, resembling a single flower (*Fig. 3*).

Specialization for wind-pollination has nearly always involved a reduction in size and numbers of floral parts. Some families are nearly all wind-pollinated, e.g. grasses, but wind-pollinated species occur in numerous mainly insect-pollinated families and have evolved many times. Evolution in the reverse direction, from wind- to insect-pollination has occurred rarely.

Specialization in flowers

A few flowers show extremes of specialization in their pollination. In many orchids each flower needs a single visit for the successful pollination of hundreds of seeds so they can become highly selective in the species that they use. Some tropical species have a long spur growing out of the back of the flower with nectar at its base, which only moths with extremely long thin tongues can reach. The European bee orchid group, and a few other groups of orchids in other parts of the world, have flowers that resemble the female of one bee species both in looks and in scent, each orchid species resembling a different bee; they are visited and successfully pollinated by the males of that bee species.

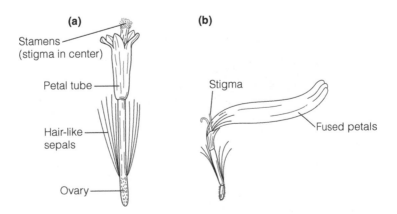

(a)
Stamens
(stigma in center)
Petal tube
Hair-like sepals
Ovary

(b)
Stigma
Fused petals

Fig. 3.   Florets from the composite inflorescence of a daisy: (a) hermaphrodite disc floret;

The figs, a large and important group of tropical trees and 'stranglers' (Topic L1) have flowers in an inverted inflorescence with flowers opening into a central cavity, surrounded by a solid mass (*Fig. 4*). The inflorescence can only be penetrated by small wasps through a hole at the tip. The wasps lay their eggs, grow and mate in the inflorescences before collecting pollen and dispersing to another fig. Plant and wasp are totally interdependent. There is a similar interdependence between yuccas and moths, but otherwise such specialization is rare.

More generalist flowers are much commoner than the specialists, particularly among the dominant vegetation, which are usually pollinated either by wind or by a range of insects and sometimes birds as well. Pollinating insects (and other animals) can vary manyfold in numbers from year to year and place to place, so specialization to one or a few species can be risky.

Evolution of vegetative structure

The first angiosperms were almost certainly small nonwoody **shrubs** and herbs, like *Amborella*, water lilies and Chloranthaceae today. They may have occupied gaps and early successional stages among dominant gymnosperms. The first evidence of woody angiosperms appears at the end of the Cretaceous, after which a great range of vegetative form evolved from tall long-lived trees (although none grows as tall or lives as long as certain conifers) to short-lived plants. They may float in fresh water without attachment to any soil or spread by underground rhizomes to cover a large area. Their leaves range from minute duckweeds 1 mm or so in diameter to some palm leaves that can exceed 15 m in length, or they may be reduced to spines or absent. Most woody plants are dicots and have secondary thickening (Topic D4). Palms, bamboos and some other monocots have numerous separate vascular bundles and most only start to produce a trunk when the bud has grown broad at ground level; their trunk often tapers only slightly or not at all. In a few monocot trees the vascular bundles line up and they have some secondary thickening. Evidence for the direction of evolution in these plants is thin, and vegetative form is flexible, many families containing a wide variety of form and leaf shape.

Xylem vessels evolved early on, but a few angiosperms including *Amborella* and the Winteraceae, a family with flowers retaining many primitive features,

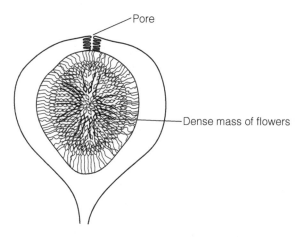

Fig. 4.   Inverted inflorescence of a fig.

have none and in water lilies the xylem cells are somewhat intermediate between tracheids and vessels.

The earliest angiosperm leaves were net-veined and small, like weedy dicot leaves today, and most dicots are net-veined. Among monocots there is great variation, although many have spear-shaped leaves with parallel veins. Some monocot families, especially the arum family, have enormous variation in leaf shape, with net venation like the dicots and some even developing holes in the leaf, like the 'swiss-cheese plant', *Monstera*, popular as a house plant.

Other aspects of evolution

Fruits do not follow a clear evolutionary path. The earliest fossil fruits are small with small seeds, similar to plants of a weedy nature today. Fleshy fruits must have appeared early and it is thought that a fleshy outgrowth from the seed, an **aril**, may be primitive. Trends in the evolution of fruits have gone in many directions, towards larger or smaller fruits or seeds. There has been **parallel evolution** of fruit types in many plant families, i.e. evolution along similar paths in unrelated groups. Frequently this has involved a reduction in size and in number of seeds per fruit and a fusion of fruits together into a single carpel. Some of the most successful families such as the grass and daisy families have one-seeded fruits, but another large family, the orchids, has many hundreds or thousands of tiny seeds per pod.

Pollen preserves well as fossils and the earliest pollen was mainly small, 10–20 μm in diameter, though one larger type, up to 50 μm, is known. They had no or just a single furrow. Later pollen had three furrows and by mid-Cretaceous multiple apertures. Apertures and furrows are known to speed pollen germination and examples of these and other types are known today (Topic G2).

Evidence from existing members of primitive dicot families suggests that self incompatibility is primitive among angiosperms, probably with a gametophytic recognition system (Topic G4). This will undoubtedly have stimulated the speed of evolution and may have been one of the main selection pressures favoring the enclosure of the ovules within an ovary in which the recognition occurs.

The primitive form of carbon fixation is undoubtedly C3 (Topic F3). It seems that C4 plants evolved some time in the mid to late Tertiary. It remains a feature of certain angiosperm families (most notably grasses), in each of which it has probably evolved independently. CAM plants have also evolved several times but it occurs among ferns and the Gnetopsida (*Welwitschia*) as well as succulent angiosperms and it probably evolved in the late Cretaceous. Both these types of carbon fixation probably appeared in response to declining carbon dioxide levels in the atmosphere and increasing aridity.

Classification

Classification of flowering plants is mainly based on the form of the flower as this is regarded as a **conservative** character (one that does not change fast in evolution – often a suite of connected characters). Frequently it is correlated with the structure of fruits and vegetative form. Many plant families are defined by the shape and orientation of the flowers, and some large families have remarkably uniform flowers and growth form. Some families, particularly those with primitive features, are more varied but classified together because of a suite of characters, e.g. the buttercup family. Some of the relationships between families are obscure.

Flowering plants have long been classified as two main divisions based on morphological characters: **dicots**, or dicotyledons, comprising about 70% of angiosperms, and **monocots**, or monocotyledons, the other 30%. The groupings are largely supported by molecular evidence of relationship except for the group of families that retain primitive features, comprising about 5% of angiosperm species including water lilies, magnolias and Chloranthaceae. These form a disparate group of more or less related families that diverged from the main angiosperm line earlier than the monocot/dicot split. They are now known as **primitive dicots** because of their earlier classification within the dicots.

Three families of plants

To illustrate the unifying features that are used to place plants in different families examples of three large and important families are given.

Daisy family (Asteraceae or Compositae)
The daisy family has over 20 000 species, all with specialized inflorescences of tiny flowers (*Fig. 3*) each maturing one seed. They are unspecialized in their pollination but many are extremely attractive to bees, butterflies and other flower visitors. Fruits are dry and small and many are effectively dispersed by the wind using a cottony sail deriving from persistent modified sepals. They are mostly herbaceous with very few small trees. They are associated particularly with dry regions and many have a deep tap root, but they flourish everywhere.

Grass family (Poaceae or Gramineae)
The grass family has about 9000 species and dominates large areas of the world. They are entirely wind-pollinated except for one or two secondarily adapted to insect pollination. The fruit is one-seeded and dry, dispersed by wind or sticking to animals. All are herbaceous except for the bamboo group that have a most unusual woody form. Growth is not from the stem tip but from nodes on the stem allowing particular tolerance of grazing and fire (Topic L3). They appear in the fossil record in the early Tertiary associated with large grazing mammals.

Orchid family (Orchidaceae)
The orchid family is the most species-rich of all with over 25 000 species. Their flowers have three sepals and three petals, one looking different from the others forming a lip. Their pollen is dispersed in aggregations (pollinia), and many species are adapted to one or a few insect species, usually bees. Many provide unusual food rewards such as oil or scent, or no food reward, deceiving their pollinators. The fruit is a pod with huge numbers of microscopic seeds deriving from one or few pollinations. All are herbaceous with specialist mycorrhizal infection in their roots (Topic M1) and many are epiphytic (Topic L1).

Q5 GENERAL FEATURES OF PLANT EVOLUTION

Key Notes

Adaptive radiation	Plants have not evolved in synchrony with animals. There have been no mass extinctions but the four major plant radiations have occurred in periods of continental upheaval and elevated CO_2 levels. Angiosperms are well adapted to the current low CO_2 levels.
Plant geography	Globally the world is divided into several phytogeographic regions, these regions reflecting the current distribution of the continents and their drift over the past 140 million years or so. There are connections between the regions and they differ from animal geographic regions, especially in the southern temperate that retain remnants of the flora of Gondwana.
Isolating mechanisms	Plants become isolated geographically or ecologically and can diverge. In any one place physiological barriers, barriers to pollination or chromosomal changes, especially polyploidy, can lead to isolation. In any one place, a stable climate over a long period leads to more species formation especially in diverse environments.
Polyploidy	If the chromosomes divide but the cell does not, it becomes tetraploid. If this happens in the reproductive cells, the new plant will be tetraploid and may become reproductively isolated from its parent diploid species. This has happened many times after hybridization and it can restore fertility. A majority of flowering plants are polyploid and it occurs in all other land plant groups.
Patterns of speciation	Plants are more diverse in the tropics and some particular places than in the north temperate region. Trees and wind-pollinated plants have fewer species than herbaceous or insect-pollinated plants. The most specialized insect-pollinated plants, orchids, have a large number of species, many of them rare.
Related topics	Ecology of flowering and pollination (G5) Polymorphisms and population genetics (L5) Plant communities (L3) Evolution of flowering plants (Q4)

Adaptive radiation

Since the Silurian era there has been continuous change in the earth's **climate**, CO_2 **concentration** of the atmosphere and the **configuration** of the continents and a few periods when there have been major upheavals, e.g. when continents have collided or split apart. The evolution of plants is connected closely with these general changes, though details of cause and effect are still uncertain. Plant evolution has not operated in synchrony with animal evolution and some changes do not conform easily to the eras that were named for discontinuities in the animal fossil record. One major difference is that there have been no **mass extinctions** among plants. This is likely to be because plants are inherently more resistant to rapid upheavals since they have simple nutrient requirements, long lives and potentially numerous dormant seeds and wide dispersal.

There have been four major adaptive radiations of land plant groups: the first land plants; the ferns and their allies; the gymnosperms; and the angiosperms. Each radiation appears to have been stimulated by major continental changes (coalescence or break up) and especially by the presence of large land masses at low latitudes and by a wide range of topography. The radiations are also associated with times of increased carbon dioxide concentration and a warmer, wetter climate. The three factors go together since continental changes can stimulate volcanic activity and this can increase CO_2 concentration which itself changes the climate (Topic L6). Ocean currents will also change, exaggerating or reducing climate change. Increasing CO_2 is known experimentally to enhance growth and affect plant reproduction leading to shorter lifespans, and alleviate stresses that may come from other sources, e.g. toxic substances. Warmer moister environments over more of the globe allow a greater area to be colonized.

Over millions of years after an upheaval, plants reduce the quantity of atmospheric CO_2, because some plant material does not fully decay and is laid down as peat or undecayed wood that eventually fossilizes as coal or oil. This leads to a drier, cooler environment with greater differentiation between equator and poles. In the late Devonian, when the ferns and earliest seed plants were increasing and actively evolving, the CO_2 levels were nearly 10 times what they are today, to decline to present-day levels by the end of the Carboniferous. Further increases by late Triassic and again in mid-Cretaceous times to about four times today's level were followed by adaptive radiations in gymnosperms and angiosperms respectively. Each radiation has coincided with a steady loss of the previously dominant groups, although living relatives remain. Currently CO_2 concentration is low, though rising with industry, and the climate is generally arid and showing marked contrast from equator to poles. Angiosperms are better suited to this than gymnosperms, themselves better than ferns, a factor contributing to the general dominance of angiosperms except in boreal environments. C4 and CAM metabolism are the newest adaptations to low CO_2 concentration.

Plant geography

In the last 140 million years or so, the way in which the continents have drifted, consolidated and separated again has affected which angiosperm families are distributed where and how much they have diversified. Most angiosperms cannot disperse from one continent to another and there are differences in the floras of the major continents at the genus and family level. The major **phytogeographic regions** of the world are given in *Fig. 1*. When the first unequivocal flowering plants were appearing about 140 million years ago the single supercontinent of **Pangea** was splitting into **Laurasia** in the north and **Gondwana** in the south. The Gondwanan continent split later, Africa and India separating from the rest by 120 million years ago in the Cretaceous; South America by the early Tertiary, 60 million years ago; and Antarctica, Australasia and New Zealand not finally separating until about 40 million years ago. The northern continent remained one continuous land mass until more recently and then was joined by India, then Africa and finally South America.

The result has been that the whole **north temperate** area is regarded as a single region, the **boreal**. In the tropics there are three regions: **South America**

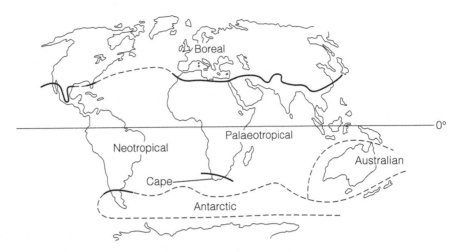

Fig. 1. Phytogeographic regions of the world.

with central America; tropical and subtropical **Africa** with tropical **Asia** forming the palaeotropical region; and **Australia**. There are strong affinities between all the tropical floras, and New Guinea and north-eastern Australia have some similarities to Asia, suggesting that there has been some migration between the regions. The southern temperate regions, now small in area, have a remnant flora of Gondwana. The **Cape** region of South Africa and the Australian regions have rather different Gondwanan floras of drier Mediterranean-climate and semi-desert floras that have evolved separately, the Cape being a separate and exceptionally diverse small region. The **Antarctic** region, comprising southern South America and New Zealand, reflects their earlier connection via Antarctica and these areas retain a flora of what was once widespread across cool-temperate Gondwana. Minor phytogeographic regions are formed by some **oceanic islands**, particularly in the Pacific, and a small Gondwanan fragment, New Caledonia, has >70% of its flora endemic, including what is probably the most primitive living plant, *Amborella*.

The phytogeographic regions have some similarity to animal geographical regions but are less well defined, owing to the fact that flowering plants evolved before modern vertebrate groups and migrated across the world earlier, maintaining affinities across regions. They differ mainly in the southern temperate region.

Isolating mechanisms

If a plant population becomes isolated from others and does not interbreed with them, new varieties and eventually species may form. Most commonly plants become isolated through being physically separated. If seeds are transported to an island by freak weather or via a bird, any plants that grow may be isolated. This **geographical isolation** may involve a plant on an island of suitable habitat such as a mountain peak surrounded by lowland. A population can become at least partially isolated **ecologically** within the distribution of the parent species if it colonizes a habitat with different soil conditions or dominant vegetation type, requiring markedly different adaptations. In these, and perhaps other circumstances, a **physiological** reproductive barrier may develop in one location, perhaps associated with a self-incompatibility system (Topic G4) or flowers may open at different times of year and not overlap in their flowering seasons. Plants may become isolated by becoming adapted to different pollinating insects, especially if the flowers are highly specialized to one or a small group of species, e.g. in the orchids. Plants may be isolated by chromosome number through **polyploidy** since this is flexible and variable in plants and those with different numbers may be incapable of reproducing together.

Species need time to develop their differences. If a varied environment remains stable for a long period, species become more specialized and the species diversity increases. The most diverse regions, then, will be those that are most diverse topographically and ecologically and that have remained stable for a long period.

In the long term the drifting of the continents has meant that plants once connected have been isolated for millions of years. Some continents have drifted to a different climatic zone or joined a different continent in new ecological conditions. The result has been the formation of new species, and in time genera and families with new adaptive radiations in isolation.

Polyploidy

If a plant's chromosome number doubles at some point in the reproductive cycle, around or during meiosis or just after fertilization, the basic number for the plant can be doubled. This has occurred frequently and can happen spontaneously within a species, when it is known as **autopolyploidy**. Some individuals of the species will then be **tetraploid** and have four (or more) sets of chromosomes alongside the diploids with two. Morphologically the diploid and tetraploid plants are often identical or nearly so but they may be completely isolated from each other reproductively and have different distributions. Some authors regard these as separate species but most, for practical reasons, regard them as chromosomal varieties. Some are not completely isolated from each other and can reproduce successfully across the ploidy levels, though there may be abnormalities at meiosis leading to partial sterility.

Following hybridization between two species, polyploidy can restore fertility. If a hybrid is partially or completely sterile, this is normally due to the inability of the chromosomes of the two parents to pair correctly at meiosis (Topic C6) because they are too different. If the chromosomes double without cell division, then at meiosis there will be an exact copy of each chromosome, so pairing at meiosis will be possible and fertility restored. This has happened numerous times in plants, many tetraploids being the product of a hybridization followed by a doubling of chromosomes. This is known as **allopolyploidy**. The new polyploid is likely to be reproductively isolated from both parents immediately, so, in effect, forms a new species in a single generation. Polyploids are frequently self-compatible even if the parents have a self-incompatibility mechanism (Topic G4) so one plant is often viable as the founder of a new species.

A small majority of flowering plant species are probably derived by polyploidy at some stage in their ancestry since they have a high number of chromosomes (50 or more). It occurs in numerous angiosperm families and is common among all other land plants too, the highest chromosome number known (over 1000) being found in an adderstongue fern (Topic P4).

Patterns of speciation

The number of species is highly variable between families and between genera and these can be related to climate, mode of growth and reproduction. There are more species in the tropics and in certain places like the Cape region than in the north temperate. Although there are many reasons for this, one of them must be the period of isolation. In the last million years the boreal region has had great climatic changes through **glaciations**, probably extinguishing many species so that only adaptable plants with good dispersal powers remain and many species are widespread. The southern continents have not been subject to the major glaciations and climatic shifts of the northern hemisphere and here there is exceptional diversity within small areas, especially where there is great topographical diversity as in the Cape region. This will promote γ diversity (Topic L3), and in the most species-rich regions each species has a more restricted distribution than in the less diverse regions.

In general, trees are less diverse than herbaceous plants and wind-pollinated plants less diverse than insect-pollinated. Trees and wind-pollinated plants disperse their pollen greater distances and more generally than herbaceous plants (most insect-pollinated trees can be pollinated by a range of insects) connecting populations over a large area. Among insect-pollinated families the most specialized to particular insects are the most species-rich. The extreme example is the orchids, in which each species has specialized to one or a few insect species as pollinators. This has allowed numerous species to form (Topic Q4), most of which are rare members of the plant communities in which they grow. Many are capable of hybridization with related species producing fertile offspring, but they do not normally hybridize because the insect species remain constant.

FURTHER READING

General reading

Bell, P.R. and Hemsley, A.R. (2000) *Green Plants: Their Origin and Diversity*, 2nd Edn. Cambridge University Press, Cambridge.

Esau, K. (1977) *Anatomy of Seed Plants*. J. Wiley & Sons, New York.

Hopkins, W.G. (2004) *Introduction to Plant Physiology*, 3rd Edn. J. Wiley & Sons, New York.

Mauseth, J.D. (1998) *Botany: An Introduction to Plant Biology*, 2nd Edn. Saunders College Publishing, Philadelphia, Pa.

Raven, P.H., Evert, R.F. and Eichorn, S.E. (1999) *Biology of Plants*, 6th Edn. W.H. Freeman, New York.

Stern, K.R. (2003) *Introductory Plant Biology*, 9th Edn. McGraw Hill, Boston, Mass.

Taiz, L. and Zeiger, E. (2002) *Plant Physiology*, 3rd Edn. Sinauer Associates, Sunderland, Mass.

Westhoff, P., Jeske, H., Jurgens, G., Kloppstech, K. and Link, G. (1998) *Molecular Plant Development – from Gene to Plant*. Oxford University Press, Oxford.

More specialist reading

Section B

Anderson, M. and Roberts, J. (eds) (1998) *Arabidopsis*. Academic Press/CRC Press, Boca Raton, Fla.

Blalock, E.M. (ed.) (2003) *A Beginner's Guide to Microarrays*. Kluwer Academic Publishers, Boston, Mass.

Bowman, J. (ed.) (1994) *Arabidopsis: An Atlas of Morphology and Development*. Springer-Verlag, New York.

Cann, A.J. (2003) *Maths from Scratch for Biologists*. J. Wiley & Sons, Chichester.

Dennis, C. and Surridge, C. (2000) *A. thaliana* genome. *Nature* **408**, 791.

Fowler, J., Cohen, L. and Jarvis, P. (1998) *Practical Statistics for Field Biology*, 2nd Edn. J. Wiley & Sons, Chichester.

Gresshoff, P.M. (ed.) (1994) *Plant Genome Analysis*. CRC Press, Boca Raton, Fla.

Koncz, C., Chua, N-H. and Schell, J. (eds) (1992) *Methods in Arabidopsis Research*. World Scientific, Singapore.

Krebs, C.J. (1999) *Ecological Methodology*, 2nd Edn. Benjamin Cummings, Menlo Park, Calif.

Kuzoff, R.K. and Gasser, C.S. (2000) Recent progress in reconstructing angiosperm phylogeny. *Trends Plant Sci.* **5**, 330–336.

Miglani, G.S. (1998) *Dictionary of Plant Genetics and Molecular Biology*. Food Products Press, New York.

Ohlrogge, J. and Benning, C. (2000) Unravelling plant metabolism by EST analysis. *Curr. Opin. Plant Biol.* **3**, 224–228.

Pyke, K. and Lopez-Juez, E. (1999) Cellular differentiation and leaf morphogenesis in *Arabidopsis*. *Crit. Rev. Plant Sci.* **18**, 527–546.

Sokal, R.R. and Rohlf, F.J. (1995) *Biometry*, 3rd Edn. W.H.Freeman & Co., New York.

Westhoff, P., Jeske, H., Jurgens, G., Kloppstech, K. and Link, G. (1998) *Molecular Plant Development – from Gene to Plant*. Oxford University Press, Oxford.

Wilson, Z.A. (ed.) (2000) *Arabidopsis: A Practical Approach*. Oxford University Press, Oxford.

Zupan, J. and Zambryski, P. (1997) The *Agrobacterium* DNA transfer complex. *Crit. Rev. Plant Sci.* **16**, 279–295.

Section C

Brett, C.T. and Waldron, K.W. (1996) *Physiology and Biochemistry of Plant Cell Walls*. Chapman & Hall, London.

Emons, A.M.C. and Mulder, B.M. (2000) How the deposition of cellulose microfibrils builds cell wall architecture. *Trends Plant Sci.* **5**, 35–40.

Gunning, B.E.S. and Steer, M.W. (1995) *Plant Cell Biology: Structure and Function*. Jones and Bartlett, Boston.

Rose, J.K.C. (ed.) (2003) *The Plant Cell Wall, Annual Plant Reviews*. Blackwell, Oxford.

Section D

Bowman, J.L. and Eshed, Y. (2000) Formation and maintenance of the shoot apical meristem. *Trends Plant Sci.* **5**, 110–115.

Laux, T. and Schoof, H. (1997) Maintaining the shoot meristem – the role of CLAVATA1. *Trends Plant Sci.* **2**, 325–327.

Sinha, N. (1999) Leaf development in angiosperms. *Annu. Rev. Plant Physiol. Plant Mol. Biol.* **50**, 419–446.

van der Schoot, C. and Rinne, P. (1999) Networks for shoot design. *Trends Plant Sci.* **4**, 31–37.

Section E

Assman, S.M. (2003) OPEN STOMATA1 opens the door to ABA signaling in *Arabidopsis* guard cells. *Trends Plant Sci.* **8**, 151–153.

Blatt, M.R. (2000) Ca^{2+} signalling and control of guard-cell volume in stomatal movements. *Curr. Opin. Plant Biol.* **3**, 196–204.

Bray, E.A. (1997) Plant responses to water deficit. *Trends Plant Sci.* **2**, 48–54.

Crawford, N.M. and Glass, A.D.M. (1998) Molecular and physiological aspects of nitrate uptake in plants. *Trends Plant Sci.* **3**, 389–395.

Flowers, T.J. and Yeo, A.R. (1992) *Solute Transport in Plants*. Blackie, London.

Fox, T.C. and Guerinot, M.L. (1998) Molecular biology of cation transport in plants. *Annu. Rev. Plant Physiol. Plant Mol. Biol.* **49**, 669–696.

Gilroy, S. and Jones, D.L. (2000) Through form to function: root hair development and nutrient uptake. *Trends Plant Sci.* **5**, 56–60.

Kozela, C. and Regan, S. (2003) How plants make tubes. *Trends Plant Sci.* **8**, 159–164.

Maathuis, F.J.M. and Sanders, D. (1999) Plasma membrane transport in context – making sense out of complexity. *Curr. Opin. Plant Biol.* **2**, 236–243.

Marschner, H. (1995) *Mineral Nutrition of Higher Plants*, 2nd Edn. Academic Press, London.

Maurel, C. (1997) Aquaporins and water permeability of plant membranes. *Annu. Rev. Plant Physiol. Plant Mol. Biol.* **48**, 399–429.

Zimmermann, U., Wagner, H.J., Schneider, H., Rokitta, M., Haase, A. and Bentrup, F.W. (2000) Water ascent in plants: the ongoing debate. *Trends Plant Sci.* **5**, 145–146.

Section F

Boudet, A.M. (1998) A new view of lignification. *Trends Plant Sci.* **3**, 67–71.

Boudet, A.M., Kajita S. and Grima-Pettenati, J. (2003) Lignins and lignocelluloses: a better control of synthesis and improved uses. *Trends Plant Sci.* **8**, 576–581.

Cushman, J.C. and Bohnert, H.J. (1999) Crassulacean acid metabolism: molecular genetics. *Annu. Rev. Plant Physiol. Plant Mol. Biol.* **50**, 305–332.

Hankamer, B., Barber, J. and Boekema, E.J. (1997) Structure and membrane organization of photosystem II in green plants. *Annu. Rev. Plant Physiol. Plant Mol. Biol.* **48**, 641–671.

Hill, S.A. (1998) Carbohydrate metabolism in plants. *Trends Plant Sci.* **3**, 370–371.

Kromer, S. (1995) Respiration during photosynthesis. *Annu. Rev. Plant Physiol. Plant Mol. Biol.* **46**, 45–70.

Lea, P.J. and Leegood, R.C. (eds) (1999) *Plant Biochemistry and Molecular Biology*, 2nd Edn. John Wiley & Sons, Chichester.

Nimmo, H.G. (2000) The regulation of phosphoenolpyruvate carboxylase in CAM plants. *Trends Plant Sci.* **5**, 75–80.

Osmond, B., Badger, M., Maxwell, K., Bjorkman, O. and Leegood, R. (1997) Too many photos: photorespiration, photoinhibition and photooxidation. *Trends Plant Sci.* **2**, 119–121.

Patrick, J.W. (1997) Phloem unloading: sieve element unloading and post-sieve element transport. *Annu. Rev. Plant Physiol. Plant Mol. Biol.* **48**, 191–222.

Raghavendra, A.S. (ed.) (1998) *Photosynthesis: A Comprehensive Treatise.* Cambridge University Press, Cambridge.

Wink, M. (ed.) (1999) *Biochemistry of Plant Secondary Metabolism.* CRC Press, Boca Raton, Fla.

Winter, H. and Huber, S.C. (2000) Regulation of sucrose metabolism in higher plants: localization and regulation of activity of key enzymes. *Crit. Rev. Plant Sci.* **19**, 31–67.

Section G

Bentley, B. and Elias, T. (eds) (1983) *The Biology of Nectaries.* Columbia University Press, New York.

Bittencourt, N.S., Gibbs, P.E. and Semir, J. (2003) Histological study of post-pollination events in *Spathodea campanulata* Beav. (Bignoniaceae), a species with late-acting self-incompatibility. *Am. J. Bot.* **91**, 827–834.

Blackmore, S. and Barnes, S.H. (1991) *Pollen and Spores: Patterns of Diversification.* Academic Press, London.

Blackmore, S. and Ferguson, I.K. (1986) *Pollen and Spores: Form and Function.* Academic Press, London.

Brugiere, N., Cui, Y. and Rothstein, S.J. (2000) Molecular mechanisms of self-recognition in *Brassica* self-incompatibility. *Trends Plant Sci.* **5**, 432–438.

Burd, M. (1994) Bateman's principle and plant reproduction: the role of pollen limitation in fruit and seed set. *Bot. Rev.,* **60**, 89–137.

Charlesworth, D. (1995) Multi-allelic self-incompatibility polymorphisms in plants. *Bioessays* **17**, 31–38.

Cheung, A.Y. (1996) Pollen–pistil interactions during pollen-tube growth. *Trends Plant Sci.* **1**, 45–51.

Franklin-Tong, N. and Franklin, F.C.H. (2003) Gametophytic self-incompatibility inhibits pollen tube growth using different mechanisms. *Trends Plant Sci.* **8**, 598–605.

Ganders, F.R. (1979) The biology of heterostyly. *N.Z. J. Bot.* **17**, 607–635.

Igic, B. and Kohn, J.R. (2001) Evolutionary relationships among self-incompatibility RNases. *Proc. Natl Acad. Sci. USA* **98**, 13167–13171.

Kay, Q.O.N., Daoud, H.S. and Stirton, C.H. (1981) Pigment distribution, light reflection and cell structure in petals. *Bot. J. Linn. Soc.* **83**, 57–84.

Kevan, P.G. (1978) Floral coloration, its colorimetric analysis and significance in anthecology. In: *Pollination of Flowers by Insects* (ed. A.J. Richards). Academic Press, London, pp. 51–78.

O'Neill, S. (1997) Pollination regulation of flower development. *Annu. Rev. Plant Physiol. Plant Mol. Biol.* **48**, 547–574.

Proctor, M., Yeo, P. and Lack, A. (1996) *The Natural History of Pollination.* HarperCollins, London.

Richards, A.J. (1997) *Plant Breeding Systems*, 2nd Edn. Chapman & Hall, London.

Sage, T.L., Strumas, F., Cole, W.W. and Barrett, S.C.H. (1999) Differential ovule development following self- and cross-pollination: the basis of self-sterility in *Narcissus triandrus* (Amaryllidaceae). *Am. J. Bot.* **86**, 855–870.

Seavey, S.R. and Bawa, K.S. (1986) Late-acting self-incompatibility in angiosperms. *Bot. Rev.* **52**, 195–219.

Sims, T.L. (1993) Genetic-regulation of self-incompatibility. *Crit. Rev. Plant Sci.* **12**, 129–167.

Taylor, L.P. and Hepler, P.K. (1997) Pollen germination and tube growth. *Annu. Rev. Plant Physiol. Plant Mol. Biol.* **48**, 461–491.

Section H

Fenner, M. and Thompson, K. (2005) *The Ecology of Seeds.* Cambridge University Press, Cambridge.

Holdsworth, M., Kurup, S. and McKibbin, R. (1999) Molecular and genetic mechanisms regulating the transition from embryo development to germination. *Trends Plant Sci.* **4**, 275–280.

Lang, G.A. (ed.) (1996) *Plant Dormancy: Physiology, Biochemistry, and Molecular Biology.* CAB International, Wallingford.

Lovett Doust, J. and Lovett Doust, L. (eds) (1988) *Plant Reproductive Ecology.* Oxford University Press, Oxford.

Mauseth, J.D. (1988) *Plant Anatomy.* Benjamin/Cummings, Menlo Park, Calif.

Mol, J., Grotewold, E. and Koes, R. (1998) How genes paint flowers and seeds. *Trends Plant Sci.* **3**, 212–217.

Nathan, R. and Muller-Landau, H.C. (2000) Spatial patterns of seed dispersal, their determinants and consequences for recruitment. *Trends Ecol. Evol.* **15**, 278–285.

Section I

Ballare, C.L. (1999) Keeping up with the neighbours: phytochrome sensing and other signalling mechanisms. *Trends Plant Sci.* **4**, 97–102.

Cushman, J.C. and Bohnert, H.J. (2000) Genomic approaches to plant stress tolerance. *Curr. Opin. Plant Biol.* **3**, 117–124.

Furuya, M. (1993) Phytochromes – their molecular species, gene families, and functions. *Annu. Rev. Plant Physiol. Plant Mol. Biol.* **44**, 617–645.

Glenn, E.P., Brown, J.J. and Blumwald, E. (1999) Salt tolerance and crop potential of halophytes. *Crit. Rev. Plant Sci.* **18**, 227–255.

Hart, J.W. (1990) *Plant Tropisms and Other Growth Movements.* Unwin Hyman, London.

Jackson, M. (1997) Hormones from roots as signals for the shoots of stressed plants. *Trends Plant Sci.* **2**, 22–28.

Khurana, J.P., Kochhar, A. and Tyagi, A.K. (1998) Photosensory perception and signal transduction in higher plants – molecular genetic analysis. *Crit. Rev. Plant Sci.* **17**, 465–539.

Palme, K. and Galweiler, L. (1999) PIN-pointing the molecular basis of auxin transport. *Curr. Opin. Plant Biol.* **2**, 375–381.

Reeves, P.H. and Coupland, G. (2000) Response of plant development to environment: control of flowering by daylength and temperature. *Curr. Opin. Plant Biol.* **3**, 37–42.

Viestra, R.D. (2003) The ubiquitin/26S proteasome pathway, the complex chapter in the life of many plant proteins. *Trends Plant Sci.* **8**, 135–142.

Wang, H.Y. and Deng, X.W. (2003) Dissecting the phytochrome A-dependent signalling network in higher plants. *Trends Plant Sci.* **8**, 172–178.

Section J

Abeles, F.B., Morgan, P.W., Saltveit, M.E. Jr, (1992) *Ethylene in Plant Biology.* Academic Press, San Diego, Calif.

Barak, S., Tobin, E.M. and Andronis, C. (2000) All in good time: the *Arabidopsis* circadian clock. *Trends Plant Sci.* **5**, 517–522.

Bennett, M.J., Roberts, J. and Palme, K. (2000) Moving on up: auxin-induced K⁺ channel expression regulates gravitropism. *Trends Plant Sci.* **5**, 85–86.

Bethke, P.C. and Jones, R.L. (1998) Gibberellin signaling. *Curr. Opin. Plant Biol.* **1**, 440–446.

Bowman, J.L. and Eshed, Y. (2000) Formation and maintenance of the shoot apical meristem. *Trends Plant Sci.* **5**, 110–115.

Chang, C. and Shockey, J.A. (1999) The ethylene-response pathway: signal perception to gene regulation. *Curr. Opin. Plant Biol.* **2**, 352–358.

Chen, M., Chory, J. and Fankhauser, C. (2004) Light signal transduction in higher plants. *Annu. Rev. Genet.* **38**, 87–117.

Clouse, S.D. and Sasse, J.M. (1998) Brassinosteroids: essential regulators of plant growth and development. *Annu. Rev. Plant Physiol. Plant Mol. Biol.* **49**, 427–451.

Creelman, R.A. and Mullet, J.E. (1997) Biosynthesis and action of jasmonates in plants. *Annu. Rev. Plant Physiol. Plant Mol. Biol.* **48**, 355–381.

D'Agostino, I.B. and Kieber, J.J. (1999) Molecular mechanisms of cytokinin action. *Curr. Opin. Plant Biol.* **2**, 359–364.

de Vries, S.C. (1998) Making embryos in plants. *Trends Plant Sci.* **3**, 451–452.

Dolan, L. (1997) SCARECROW: specifying asymmetric cell divisions throughout development. *Trends Plant Sci.* **2**, 1–2.

Guilfoyle, T.J. (1998) Aux/IAA proteins and auxin signal transduction. *Trends Plant Sci.* **3**, 205–207.

Howell, S.H. (1998) *Molecular Genetics of Plant Development.* Cambridge University Press, Cambridge.

Kakimoto, T. (1998) Cytokinin signaling. *Curr. Opin. Plant Biol.* **1**, 399–403.

Kieber, J.J. (1997) The ethylene response pathway in arabidopsis. *Annu. Rev. Plant Physiol. Plant Mol. Biol.* **48**, 277–296.

Kuhlemeier, C. and Reinhardt, D. (2001) Auxin and phyllotaxis. *Trends Plant Sci.* **6**, 187–189.

Laux, T. and Schoof, H. (1997) Maintaining the shoot meristem – the role of CLAVATA1. *Trends Plant Sci.* **2**, 325–327.

Lin, C.T. and Shaltin, D. (2003) Cryptochrome structure and signal transduction. *Annu. Rev. Plant Biol.* **54**, 469–496

Lovegrove, A. and Hooley, R. (2000) Gibberellin and abscisic acid signalling in aleurone. *Trends Plant Sci.* **5**, 102–110.

Malamy, J.E. and Benfey, P.N. (1997) Down and out in Arabidopsis: the formation of lateral roots. *Trends Plant Sci.* **2**, 390–396.

Perrot-Rechenmann, C. and Hagen, G. (eds) (2002) *Auxin Molecular Biology.* Kluwer Academic Publishers, Dordrecht.

Sinha, N. (1999) Leaf development in angiosperms. *Annu. Rev. Plant Physiol. Plant Mol. Biol.* **50**, 419–446.

Stevenson, J.M., Pepera, I.Y., Heilmann, I., Persson, S. and Boss, W.F. (2000) Inositol signaling and plant growth. *Trends Plant Sci.* **5**, 252–258.

Taylor, L.P. and Hepler, P.K. (1997) Pollen germination and tube growth. *Annu. Rev. Plant Physiol. Plant Mol. Biol.* **48**, 461–491.

van der Schoot, C. and Rinne, P. (1999) Networks for shoot design. *Trends Plant Sci.* **4**, 31–37.

Viestra, R.D. (2003) The ubiquitin/26S proteasome pathway, the complex chapter in the life of many plant proteins. *Trends Plant Sci.* **8**, 135–142

Section K

Allard, R.W. (1999) *Principles of Plant Breeding*, 2nd Edn. J. Wiley & Sons, New York.

Chrispeels, M.J. and Sadava, D.E. (2003) *Plants, Genes, and Crop Biotechnology*. Jones and Bartlett, Boston, Mass.

de Maagd, R.A., Bosch, D. and Stiekema, W. (1999) *Bacillus thuringiensis* toxin-mediated insect resistance in plants. *Trends Plant Sci.* **4**, 9–13.

Dornenburg, H. and Knorr, D. (1996) Generation of colors and flavors in plant cell and tissue cultures. *Crit. Rev. Plant Sci.* **15**, 141–168.

Evans D.E., Coleman J.O.D. and Kearns A. (2003) *Plant Cell Culture: the Basics*. Bios Scientific Publishers, London.

Flavell, R.B. (2000) Plant biotechnology – moral dilemmas. *Curr. Opin. Plant Biol.* **3**, 143–146.

Hanley, Z., Slabas, T. and Elborough, K.M. (2000) The use of plant biotechnology for the production of biodegradable plastics. *Trends Plant Sci.* **5**, 45–46.

Hansen, G. and Wright, M.S. (1999) Recent advances in the transformation of plants. *Trends Plant Sci.* **4**, 226–231.

Holmberg, N. and Bulow, L. (1998) Improving stress tolerance in plants by gene transfer. *Trends Plant Sci.* **3**, 61–66.

Kappeli, O. and Auberson, L. (1998) How safe is safe enough in plant genetic engineering? *Trends Plant Sci.* **3**, 276–281.

Komari, T., Hiei, Y., Ishida, Y., Kumashiro, T. and Kubo, T. (1998) Advances in cereal gene transfer. *Curr. Opin. Plant Biol.* **1**, 161–165.

Smeekens, S. (1997) Engineering plant metabolism. *Trends Plant Sci.* **2**, 286–288.

Zupan, J. and Zambryski, P. (1997) The Agrobacterium DNA transfer complex. *Crit. Rev. Plant Sci.* **16**, 279–295.

Section L

Allen, L.H. (1992) Free-air CO_2 enrichment field experiments – an historical overview. *Crit. Rev. Plant Sci.* **11**, 121–134.

Brown, A.H.D., Clegg, M.T., Kahler, A.L. and Weir, B.S. (1990) *Plant Population Genetics, Breeding and Genetic Resources*. Sinauer, Sunderland, Mass.

Cheplick, G.P. (ed.) (1998) *Population Biology of Grasses*. Cambridge University Press, Cambridge.

Cox, C.B. and Moore, P.D. (1985) *Biogeography: An Ecological and Evolutionary Approach*, 4th Edn. Blackwell, Oxford.

Crawley, M.J. (ed.) (1997) *Plant Ecology*, 2nd Edn. Blackwell, Oxford

Drake, B.G., Gonzalez-Meler, M.A. and Long, S.P. (1997) More efficient plants: a consequence of rising atmospheric CO_2? *Annu. Rev. Plant Physiol. Plant Mol. Biol.* **48**, 609–639.

Gibson, D.J. (2002) *Methods in Comparative Plant Population Ecology*. Oxford University Press, Oxford.

Grime, J.P., Hodgson, J.G. and Hunt, R.P. (1988) *Comparative Plant Ecology*. Unwin Hyman, London.

Grubb, P.J. (1977) The maintenance of species richness in plant communities: the importance of the regeneration niche. *Biol. Rev.* **52**, 107–145.

Harper, J.L. (1977) *Population Biology of Plants*. Academic Press, London.

Hubbell, S.P. (2001) *The Unified Neutral Theory of Biodiversity and Biogeography*. Princeton University Press, Princeton, N.J.

Hubbell, S.P., Foster, R.B., O'Brien, S.T. *et al.* (1999) Light-gap disturbances, recruitment limitation, and tree diversity in forest gaps. *Science* **283**, 554–557.

Malhi, Y. and Grace, J. (2000) Tropical forests and atmospheric carbon dioxide. *Trends Ecol. Evol.* **15**, 332–337.

Richards, P.W. (1996) *The Tropical Rain Forest*, 2nd Edn. Cambridge University Press, Cambridge.

Silvertown, J., Franco, M. and Harper, J.L. (1997) *Plant Life Histories: Ecology, Phylogeny and Evolution*. Cambridge University Press, Cambridge.

Silvertown, J.W. and Charlesworth, D. (2001) *Introduction to Plant Population Biology*, 4th Edn. Blackwell, Oxford.

Tian, H., Hall, C.A.S. and Qi, Y. (1998) Modelling primary productivity of the terrestrial biosphere in changing environments: toward a dynamic biosphere model. *Crit. Rev. Plant Sci.* **17**, 541–557.

Section M

Bais, H.P., Park, S.-W., Weir, T.L., Calloway, R.M. and Vivanco J.M. (2004) How plants communicate using the underground information superhighway. *Trends Plant Sci.* **9**, 26–32

Bladergroen, M.R. and Spaink, H.P. (1998) Genes and signal molecules involved in the rhizobia-Leguminoseae symbiosis. *Curr. Opin. Plant Biol.* **1**, 353–359.

Bohme, H. (1998) Regulation of nitrogen fixation in heterocyst-forming cyanobacteria. *Trends Plant Sci.* **3**, 346–351.

Bohn, H.F. and Federle, W. (2004) Insect aquaplaning: *Nepenthes* pitcher plants capture prey with the peristome, a fully wettable water-lubricated anisotropic surface. *Proc. Natl Acad. Sci. USA* **101**, 14138–14143.

Brown, V.K. and Gange, A.C. (1990) Insect herbivory below ground. *Adv. Ecol. Res.* **20**, 1–58.

Cohn, J., Day, R.B. and Stacey, G. (1998) Legume nodule organogenesis. *Trends Plant Sci.* **3**, 105–110.

Dempsey, D.A., Shah, J. and Klessig, D.F. (1999) Salicylic acid and disease resistance in plants. *Crit. Rev. Plant Sci.* **18**, 547–575.

DeWit, P.J.G.M. (1997) Pathogen avirulence and plant resistance: a key role for recognition. *Trends Plant Sci.* **2**, 452–458.

Dicke, M., Agrawal, A.A. and Bruin, J. (2003) Plants talk, but are they deaf? *Trends Plant Sci.* **8**, 403–405.

Gibson, C.C. and Watkinson, A.R. (1992) The role of the hemiparasitic annual *Rhinanthus minor* in determining grassland community structure. *Oecologia* **89**, 62–68.

Harrison, M.J. (1998) Development of the arbuscular mycorrhizal symbiosis. *Curr. Opin. Plant Biol.* **1**, 360–365.

Harrison, M.J. (1999) Molecular and cellular aspects of the arbuscular mycorrhizal symbiosis. *Annu. Rev. Plant Physiol. Plant Mol. Biol.* **50**, 361–389.

Juniper, B.E., Robins, R.J. and Joel, D.M. (1989) *The Carnivorous Plants*. Academic Press, London.

Karban, R. and Baldwin, I.T. (1997) *Induced Responses to Herbivory*. University of Chicago Press, Chicago, Ill.

Mendgen, K. and Hahn, M. (2002) Plant infection and the establishment of fungal biotrophy. *Trends Plant Sci.* **7**, 352–356.

Palo, R.T. and Robbins, C.T. (eds) (1991) *Plant Defenses against Mammalian Herbivory*. CRC Press, Boca Raton, Fla.

Press, M.C. and Graves, J.D. (eds) (1995) *Parasitic Plants*. Chapman & Hall, London.

Press, M.C., Scholes, J.D. and Barker, M.G. (eds) (1999) *Plant Physiological Ecology*. Blackwell Science, Oxford. Chapters by Read, D.J. (mycorrhizae); Press, M.C. *et al.* (parasitic plants) and Crawley, M.J. (herbivory).

Saikkonen, K., Wali, P., Helander, M. and Faeth, S.H. (2004) Evolution of endophyte-plant symbioses. *Trends Plant Sci.* **9**, 275–280.

Slack, A. (1979) *Carnivorous Plants*. Alphabooks, Dorset.

Smith, S.E. and Read, D.J. (1996) *Mycorrhizal Symbiosis*. Academic Press, London.

Strauss, S.Y. and Agrawal, A.A. (1999) The ecology and evolution of plant tolerance to herbivory. *Trends Ecol. Evol.* **14**, 179–185

Udvardi, M.K. and Day, D.A. (1997) Metabolite transport across symbiotic membranes of legume nodules. *Annu. Rev. Plant Physiol. Plant Mol. Biol.* **48**, 493–523.

van der Heijden, M.G.A., Klironomos, J.N. *et al.* (1998) Mycorrhizal fungal diversity determines plant biodiversity, ecosystem variability and productivity. *Nature* **396**, 69–72.

Section N

Baker, H.G. (1970) *Plants and Civilization*, 2nd Edn. Macmillan Press, London.

Balick, M.J. and Cox, P.A. (1996) *Plants, People, and Culture: The Science of Ethnobotany*. Scientific American Library, New York.

Chrispeels, M.J. and Sadava, D.E. (1994) *Plants, Genes and Agriculture*. Jones and Bartlett, Boston, Mass.

Cobbett, C.S. (2000) Phytochelatin biosynthesis and function in heavy-metal detoxification. *Curr. Opin. Plant Biol.* **3**, 211–216.

De Cleene, M. and Lejeune, M.C. (1999–2003) *Compendium of Symbolic and Ritual Plants in Europe*, vols. 1 and 2. Mens en Kultuur, Ghent.

Duke, J.A. (1985) *CRC Handbook of Medicinal Herbs*. CRC Press, Boca Raton, Fla.

Hobhouse, H. (1992) *The Seeds of Change*. Papermac, London.

Lewington, A. (1990) *Plants for People*. Natural History Museum, London.

Mabey, R. (1996) *Flora Britannica*. Sinclair Stevenson, London.

Meagher, R.B. (2000) Phytoremediation of toxic elemental and organic pollutants. *Curr. Opin. Plant Biol.* **3**, 153–162.

Salt, D.E., Smith, R.D. and Raskin, I. (1998) Phytoremediation. *Annu. Rev. Plant Physiol. Plant Mol. Biol.* **49**, 643–668.

Simmonds, N.W. (ed.) (1976) *Evolution of Crop Plants*. Longman, London.

Sections O and P

Bold, H.C., Alexopoulos, C.J. and Delevoryas, T. (1980) *Morphology of Plants and Fungi*. Harper & Row, New York.

Bold, H.C. and Wynne, M.J. (1985) *Introduction to the Algae*, 2nd Edn. Prentice-Hall, Upper Saddle River, N.J.

Cove, D.J., Knight, C.D. and Lamporter, T. (1997) Mosses as model systems. *Trends Plant Sci.* **2**, 99–105.

Gifford, E.M. and Foster, A.S. (1989) *Morphology and Evolution of Vascular Plants*. W.H. Freeman, San Francisco, Calif.

Kenrick, P. and Crane, P.R. (1997) The origin and early evolution of plants on land. *Nature* **389**, 33–39.

Niklas, K.J. (1997) *The Evolutionary Biology of Plants*. University of Chicago Press, Chicago, Ill.

Qiu, Y.-L. and Palmer, J.D. (1999) Phylogeny of early land plants: insights from genes and genomes. *Trends Plant Sci.* **4**, 26–30.

Richardson, D.H.S. (1981) *The Biology of Mosses*. Blackwell, Oxford.

Watson, E.V. (1971) *The Structure and Life of Bryophytes*, 3rd Edn. Hutchinson, London.

Wellman, C.H., Osterloff, P.L. and Mohiuddin, W. (2003) Fragments of the earliest land plants. *Nature* **425**, 282–284.

Willis, K.J. and McElwain, J.C. (2002) *The Evolution of Plants*. Oxford University Press, Oxford

Section Q

Barkman, T.J. *et al.* (2000) Independent and combined analysis of sequences from all three genomic compartments converge on the root of flowering plant phylogeny. *Proc. Natl Acad. Sci. USA* **97**, 13166–13171.

Briggs, D. and Walters, S.M. (1997) *Plant Variation and Evolution*, 3rd Edn. Cambridge University Press, Cambridge.

Comes, H.P. and Kadereit, J.W. (1998) The effect of Quaternary climatic changes on plant distribution and evolution. *Trends Plant Sci.* **3**, 432–438.

Dodd, M.E., Silvertown, J.W. and Chase, M.W. (1999) Phylogenetic analysis of trait evolution and species diversity variation among angiosperm families. *Evolution* **53**, 732–744.

Endress, P.K. (1994) *Diversity and Evolutionary Biology of Tropical Flowers*. Cambridge University Press, Cambridge.

Friedman, W.E. (1992) Evidence of a pre-angiosperm origin of endosperm: implications for the evolution of flowering plants. *Science* **255**, 336–339.

Friis, E.M., Chaloner, W.G. and Crane, P.R. (eds) (1987) *The Origins of Angiosperms and their Biological Consequences*. Cambridge University Press, Cambridge.

Furness, C.A. and Rudall, P.J. (2004) Pollen aperture evolution – a crucial factor for eudicot success? *Trends Plant Sci.* **9**, 154–158.

Kuzoff, R.K. and Gasser, C.S. (2000) Recent progress in reconstructing angiosperm phylogeny. *Trends Plant Sci.* **5**, 330–336.

Leitch, I.J. and Bennett, M.D. (1997) Polyploidy in angiosperms. *Trends Plant Sci.* **2**, 470–476.

Lloyd, D.G. and Barrett, S.C.H. (eds) (1996) *Floral Biology: Studies on Floral Evolution in Animal-Pollinated Plants*. Chapman & Hall, London.

Norstog, K. (1987) Cycads and the origin of insect pollination. *Am. Sci.* **75**, 270–279.

Owens J.N., Takaso, T. and Runcons, C.J. (1998) Pollination in conifers. *Trends Plant Sci.* **3**, 479–485.

Proctor, M., Yeo, P. and Lack, A. (1996) *The Natural History of Pollination*. HarperCollins, London.

Soltis, D.E. and Soltis, P.S. (1999) Polyploidy: recurrent formation and genome evolution. *Trends Ecol. Evol.* **14**, 348–352.

Soltis, D.E. *et al.* (2004) Genome-scale data, angiosperm relationships, and 'ending congruence': a cautionary tale in phylogenetics. *Trends Plant Sci.* **9**, 477–483.

Willis, K.J. and McElwain, J.C. (2002) *The Evolution of Plants*. Oxford University Press, Oxford.

INDEX

Bold type is used to indicate the main entry where there are several.

ABC model of flower
    development, 184
Abscisic acid (ABA), 127, 157,
    174–5, 178
    seed dormancy and, 103,
        136, 174, 178
    stomata and, 67, 178
Abscission, 127, 156–7, 172
Absorption spectrum, 79, **80**
Acacia, 58, 228
    bullshorn, 234
Accessory pigment, 80
Acer (maple), 133, 256
Acetyl coenzyme A (CoA), 94,
    98–9, 260
Acid growth theory, 179
Actin, 21
Actinomorphy(ic), 320
Actinomycete, 229
Adderstongue, 296–7, 328
Adenosine triphosphate, *see*
    ATP
Adventitious roots, 49
Aeciospore, 236
Aeollanthus biformifolius, 266
Aerenchyma, 43, 66, 158, 161,
    171
Agamospermy(-ous), 114–15
Agamous, 184
Agar, 270
Agathis, 310
Agave, 91, 210, 253, 257
Aglaophyton, 285
Agrobacterium tumefaciens, 13,
    173, 192–7
Alcohol, 262
Alder (*Alnus*), 229
Aldolase, 94
Aldrovanda vesiculosa, 246–7
Alectra, 243
Aleurone layer, 139–40
Alfalfa (*Medicago*), 229
Algae, 2–3, 204, **269–72**, 287
    blue-green, *see* cyanobacteria
    brown, 2–4, **270–1**
    green, 2–4, **270–2**, 274

red, 2–4, **270–1**
Alginate, 271
Alkaloid, 103, 159, 233, 251,
    259–60
Allium, embryo, 127
Allspice, 252
Alnus, 229
Alternation of generations, 2,
    271, 274
Aluminum, 78, 159, 160, 204
Alyxia rubricaulis, 266
Amborella, 319, 322, 327
Amelioration, 158, 160
Amino acid biosynthesis,
    98–103, 230
    in nectar, 108
    non-protein, 159
    1–aminocyclopropane -
        carboxylic acid (ACC), 172
α-amylase, 93, 140
Amyloplast, 22, 150–2
Amylose, 93
Anabaena, 228
Andreaea, 281
Aneuploidy, 9
Angiosperm
    (Angiospermopsida),
    classification, 1, 16, 304, 326
Aniseed, 252
Annual rings, 53–4, 255, 308
Annual plant, *see* ephemeral
Ant(s), as seed dispersers, 133,
    234
Antenna complex, 81
Anther, 304, 309–10, 313,
    315–18 (*see also* stamen)
Antheridium(a), of bryophytes,
    278–9
    of ferns, 300
    of horsetails, 293–4
    of lycopsids, 290
    of seed plants, 304
    of vascular plants, 287
Anthocerophyta, *see* hornwort
Anthocyanin, 103, 106–7, 130
Anthoxanthin, 106

Antibody production, 196
Antioxidants, in nectar, 108
Antipodal cell, 110
Antiporter, 71
Antirrhinum majus, 5, 7, 184
Antisense, 180
Apetala, 184
Apiaceae, *see* umbellifer
Apical cell, 165
Apical dominance, 169
Apoplast, 21, 23, 63
Apoplastic transport, 73–4
Apple, 130, 252
Aquaporin, 60
Arabidopsis thaliana
    (arabidopsis), **5–7**, 52, 177,
    184, 266
    flower, 106
    fruit, 129
    genome project, 7
    mutants, 14, 175
    seedling, 128
Araucaria, Araucariaceae, 305,
    308–10
Archaea, 1
Archegonium(a), of
    bryophytes, 278–9
    of ferns, 300
    of flowering plants, 318
    of horsetails, 293
    of lycopsids, 290
    of seed plants, 304, 311, 314
    of vascular plants, 287
Architecture, plant, 52
Aril, 126, 133, 323
Arum family (Araceae), 113,
    201, 323
Asclepiad, *see* milkweed
Asexual reproduction, 17, 112,
    114–15, 269–70, 274
Asparagine synthetase, 100
Asplenium, 299
Asteraceae, *see* composite
ATP, ATP synthesis, 27, 68–70,
    82–7, 98, 92, 95–6
Atropa belladonna, 103

Autoclave, in cell culture, 188
Aux/IAA proteins, 179
Auxin, and abscission, 157
  and cell division, 38, 168
  and elongation, **168**, 178
  and flowering, 115
  cytokinin ratio, 169, 189
  mechanism of action, 171,
    178–9
  reporter gene, 151
  tropisms and, 150–1
Avocado, 250
Azolla, 228–9, 299

Bacillus thuringiensis, 196
Bacteria, 226, 230, 232, **239–40**
  (*see also* nitrogen fixation)
Balsam, 133
Bamboo, 200, 210–11, 322, 324
Banana, 130, 252
Banksia, 136
Banyan, sacred, 263
Baobab, 263
Barberry, 236
Bark, 54
Barley, 139–40, 162, 172, 250,
  252
Basal cell, 165
Basidiomycete, 235–6 (*see also*
  mycorrhiza)
Basil, 252
Bats, as flower visitors, 122,
  320–1
Bean, 126, 128–9, 154, 228–9,
  233, 250
Beech, 256, 320
Bees, as flower visitors, 108,
  122–3, 252, 320–1, 324
Beetles, as flower visitors,
  319–20
  as herbivores, 231
Bennettitales, 284, 305–6, 318
Betalain, 106
Betula, Betulaceae, *see* birch
Biennial, 210–11
Binary vector system, 194
Bindweed family, *see*
  Convolvulaceae
Bioinformatics, 8, 14
Biological species concept, 17
Biome, 203, 206, 222
Bio-mining, 265
Bioremediation, 265–6

Birch, 118, 133, 320
Birds, as flower visitors, 108,
  122, 320–1
  as fruit/seed dispersers,
    133–4
Birdsfoot trefoil (*Lotus*), 229
Birdsnest, *see Monotropa*, fern,
  birdsnest, and orchid,
  birdsnest
Blackwood, African, 256
Bladderwort, 246–7
Blight, bacterial, 239–40
  potato, 240, 251
Blue light responses, 143, 146
Boerhavia (Nyctaginaceae), 210
Bombacaceae, 258
Bonsai, 264
Boswellia, 262
Bracken, 296, 299, 301
Bract, 316
Bramble (*Rubus*), 114
Brassica crops, 250–1
Brassica juncea, 266–7
Brassicaceae, *see* cabbage
  family
Brassin/ brassinosteroid, 175
Brazil nut, 126
Breadfruit, 252
Breeding systems, 112–15
Bromeliad (Bromeliaceae), 201
Broomrape (*Orobanche*,
  Orobanchaceae), 242–3
Bryophyta, *see* moss
Bryophyte, 141, 201, 204, 223,
  244, **273–81**, 285, 294, 300,
  311
Bud dormancy, 174
Bulbil, 113, 115
Bulk flow, 60
Bundle sheath, 58, 88–9
Bur, 133–4
Burseraceae, 262
Butchers' broom, 58
Buttercup family, 320, 323
Butterflies, as flower visitors,
  108, 122–3, 320–1, 324
  as herbivores, 231–2,
Butterwort, 246
Buxbaumia, 277
Buzz-pollination, 107

C3/C4 plants, *see*
  photosynthesis

Ca^{2+} ATPase, 180–1
Cabbage, 250–1
Cabbage family, 118–19, 129,
  224, 226, 232
Cactus, 66, 91, 234
Cadastral genes, 184
Caffeine, 262
Calamites, 294
Calcicole/calcifuge, 78
*Calcium, 67, 70, 78, 151,
  180–182*, 204
Calcium carbonate, 219
Callose, 111, 118, 237
Calluna vulgaris, see heather
Callus, 43, 166, 169, 189, 195
Calmodulin, 181
Calvin cycle, 82, 84, 140
Calyptra, 281
CAM, *see* crassulacean acid
  metabolism
Cambium, 42, 51, 53
Camellia sinensis, 262
Campion, 113
Canavalia ensiformis, 233
Cannabis, 257, 262
Canola, *see* rape, oilseed
Cape region (South Africa),
  208, 327–9
Capillary action, 61
Capsella bursa-pastoris, embryo,
  126
Capsicum, 252
Capsule, of bryophytes, 4, 274,
  279–81
Caraway, 252
Carbohydrate metabolism,
  92–6
Carbon cycle, 219–22
Carbon dioxide (CO$_2$)
  concentration, 139, 219–22,
  285–6, 325–6
Carbon-fixation reaction, 82–4
Carboxylation, 84
Cardamom, 252
Cardiac glycoside/
  cardenolide, 103, 259–60
Carex, 224
Carnivorous plants, 245–7
Carotenoid, 80, 103, 106, 109,
  130
Carpel, **105–8**, 112, 116, 129
  evolution of, 318–20
Carrot family, *see* umbellifer

Casparian strip, 46–7, 66
Cassava, 233–4, 250–1
Castor-oil, 126
Casuarina, 227
Catharanthus roseus, 103, 259
Cavitation, 60
Caytoniales, 284, 305–6, 318
cDNA library, 10–11
Ceanothus, 229
Ceiba pentandra, 258
Cell, cycle/division, **36–9**, 164, 175
  structure, **21–3**
  wall, 2, 21, 24–5, 150, 158, 165
Cellulose, 2, 24–5, **101–2**, 109, 232, 255
Central oscillator, 147
Century plant, *see Agave*
Cephalotaceae (*Cephalotus*), 245–6
Cerastium glomeratum, 114
Chara, Charales, Charophyceae, 150, 272
Chelating agents, 265
Chemotropism, 149
Chenopodiaceae, 224, 226
Cherry, 130
Chimera, 264
Chloranthaceae, 318–9, 322, 324
Chlorenchyma, 41, 43
Chlorophyll, 2, **79–81**, 270
Chlorophyta, *see* algae, green
Chloroplast, 22, 27–8, 99
  development, 28
  genes, 18
Chlorosis, 78
Chocolate, 262
Chromatid, 35
Chromatin, 33
Chromoplast, 27, 130
Chromosome, 33, 35
  sex, 115
Chrysanthemum, 233
Chrysolina beetle, 234
Cinnamon, 252
Circadian rhythm, 67, 146, **147–8**, 154
Citric acid cycle, **94–5**, 99
Citrus fruit, 130, 252
Classification of plants, 3, 16–18, 323
Claviceps purpurea, 238

Clematis, 201
Climacteric, 171
Climate change, 219–22
Climax community, 207
Climber, 200–1, 234, 305, 314–15
Clonal spread, 210, 214, 217
Cloning, 264
  gene, 10–12
Clove, 252
Clover, 228–9
  white, 232
Clubmoss, *see* lycopsid
Coal, 219–20, 326
Coca, Cocaine, 103, 233, 260, 262
Cocoa, 252, 262
Coconut, 122, 125–6, 262
Cocoyam, 251
Codeine, 103, 260
Codon, 34
Co-enzyme A (CoA), 94
Coffee (*Coffea*), 103, 232, 262
Cohesion-tension, 59–64
Coleoptile, 127, 140, 150
Coleus, 154
Collenchyma, 43, 257
Commercial applications, 190–1
Commiphora, 262
Companion cell, 44
Composite (Asteraceae, daisy family), 114, 118, 122, 130, 320, 323–4
  fruit, 122
  pollen, 110
  seeds, 132
Compound leaf, 56
Cone, 305, 308–11, 313–14
Conifer (Coniferophyta, Pinopsida), 2, 136–7, 200, 206, 224, 227, 241, 256–7, 284, 303–5, **307–11**, 314–15, 322
Conocephalum, 274–5
Convolvulaceae (bindweed family), 118, 243
Cooksonia, 283–5, 287
Coppice, 257
Coral reef, 270
Corchorus, 257
Cordaitales (Cordaites), 3, 284, 305

Cork, 42, 53–5
Corm, 289
Corn, *see* maize
Corpus, 52, 183
Cortex, 50
Cortical microtubules, 26
Cotton, 43, 133, 233, 257–8
Cotyledon, **126–8**, 139, 311, 314, 316
Cowpea, 243
Cowslip (*Primula veris*), 119
Crassulaceae, 91
Crassulacean acid metabolism (CAM), **88–91**, 323, 326
Creosote bush, 210
Crista, 27
Cross breeding, 9, 186
Crossbill, 311
Crown gall disease, 192
Crustacea, 246
Cryptochrome, 146
Cryptothallus, 277
Cucumber family, 110, 130
Cultivar, 263
Cupule, of Caytoniales, 306
Cuscuta, 243
Cuticle, 66, 158–9, 236
Cutin, 66, 125
Cuttings, 264
Cyanobacteria, 2, 228–30, 275, 277, 299, 313
Cyanogenesis, 232
Cycad (Cycadopsida), 3, 199, 200, 229, 284, 305–6, **312–14**
Cyclic electon flow, 84
  photophosphorylation, 84
Cyclin and cyclin dependent protein kinase, 38
Cyperus papyrus, 256
Cypress, 309–10
  swamp, 309
Cytisus, 264
Cytochrome *bf* complex, 33
Cytogenetics, 8–9
Cytokinesis, 36
Cytokinin, **173–174**, 178
  and cell division, 38, 169
Cytoplasm and cytosol, 2, 21
Cytoskeleton, 21

Dacrydium franklinii, 210
Daffodil, 118

Daisy family, *see* composite
Daisy, florets, 321
Dandelion (*Taraxacum*), 114
Dasheen, 251
Dead-nettle, 320
Defense 158–9, 236
Deficiens mutant, 184
Delphinium nelsonii, larkspur, 123
Dermal cells, 43
Dermatogen, 165
Desert, 200, 203–4, 222, 254, 276
Desmotubule, 26
Diacylglycerol (DAG), 182
Diatom, 269–70
Dicotyledon (dicot), 3, 127, 200, 284, 315, 320, 322–4
Dicot, primitive, 3, 106, 120, 323–4
Differentiation, 165–6
Digitalis purpurea, 103, 259–60
Digitoxin, 103
Dihydroxyacetone phosphate, 94
Dimorphotheca pluvialis, seeds, 132
Dinitrogenase, 230
Dinoflagellate, 269–70
Dioecious (dioecy), 106, **112–13**, 115, 122, 252, 279, 314
Dionaea muscipula, 246–7
Diversity, 139, **207–8**, 227, 328–9
DNA, 10–11, 18–19, **34–5**, 127
fingerprint, 18
junk, 19
particle gun, 194–5
polymerase, 10, 35, 193–5
variation, 216–17
Dodder, 243
Dodo, 136
Dominance, 9, 207–8
Dormancy, seed, 134–7, 250
Double flowers, 263, 320
Droseraceae, 246
Drought, C4 as adaptation to, 90
Dry-rot, 256
Duckweed, 322
Duplication, gene, 9

Ebony, 54

Ecotype, 215
Eel-grass (*Zostera*), 204
Eichhornia crassipes, 267
Elaeis guineensis, 262
Elaioplast, 27
Elaiosome, 133
Elater, 280, 293
Electophoresis, 17–18
Electrochemical gradient, 69–70
Electron, flow, 84
transport, 83, 95
Elephant, 234
Elicitor, 159, 237
Elm, 214
Elongation, 150
zone, 45
Embolism, 59, 62
Embryo, 126, 139, 174, 311
culture, 188
sac, 108, **110–11**, 118, 304, 318
Embryogenesis, 165
Embryoid, 190
Endodermis, 46, 63, 74, 162, 242
Endomembrane system, 22
Endophyte, fungal, 237
Endoplasmic reticulum, 21–2, 29, **30**, 100, 150, 182
Endoreduplication, 35, 39
Endosperm, 111, **125–6**, 139–40, 318
Endosymbiont theory, 27
Enterolobium, 136
Ephedra, 3, 314–16
Ephemeral plant, 210–13, 234, 242
Epidermis, 41, 50, 57
Epigeal germination, 127–8, 139
Epinasty, 153, 171
Epiphyll, 276
Epiphyte, 200–1, 274, 276, 288, 290, 296, 301
Equisetopsida, *Equisetum*, 3, 284, 286, 305(*see also* horsetail)
Ergot, 238
Ericaceae, *see* heather
Ethnobotany, 259
Ethylene, 157, 171–2, 176–8, 180–1
Etiolation, 145

Etioplast, 27, 140
Eucalyptus, 136–7, 257
Eudicotyledon, 3(*see also* dicotyledon)
Euglena, 269–70
Euphorbia, 233
Euphorbiaceae, 66
Euphyllophytina, 3, 285, 304
Eusporangium(a), -ate, 296, 298, 300–3
Evocation, of floral meristem, 183–4
Exine, 109
Exon, 35
Expansin, 179
Exponential, 212
Expressed sequence tag (EST), 11
Extensor, 154
Extinction, mass, 326

Fagus sylvatica, 256
Fatty acid, 100
Fe protein, 230
Fern, 2–4, 16, 199–201, 263, 284–7, **295–301**, 303–4, 326
birdsnest, 299, 301
filmy, 299
gametophyte, 244
hartstongue, 299
Killarney, 300–1
polypody, 233
royal, 298, 300
seed, *see* pteridosperms
water, 296, 299–300
Ferrodoxin, 83
Fertilization, double, 111, 316, 318
Festuca arundinacea, 238
Fiber, 41, 43, 257–8
Fig (*Ficus*), 129, 201, 263, 322
Filicopsida, *see* ferns
Fire, as ecological factor, 136, 205, 308, 324
Fissidens, 275
Fix genes, 230
Flavonoid, 103, 106
Flax (linseed), 250, 257
Flexor, 154
Flies, as flower visitors, 122, 319–21
Floral development, 7, **183–4**
Floret, disc and ray, 132

Flower, description of, 105–8
  origin, 318
Forget-me-not, 110
Founder effect, 218
Foxglove, 259–60
Frankincense, 262
Frost, limiting plant
    distribution, 203
Fructose, 102, 108
Fructose 1,6-bisphosphate,
    94
Fruit, 121–2, 125, **129–30**
  dispersal, 131–4, 231, 250
  evolution of, 323
  ripening, 130, 171, 179
Frullania, 274
Fucus, 5, 271
Funaria hygrometrica, 5
Functional genomics, 12
Fungi, 235–8 (*see also*
    mycorrhiza)
  as pathogens, 208, 225–6

Gall, 232
Gametophyte, **3–4**, 125, 223,
    270–2, 287, 290, 304
  of angiosperm, 318
  of bryophyte, 274–7
  of conifer, 310
  of fern, 299–301
  of horsetail, 293–4
Gaps in vegetation, 138–41
Gelidium, 270
Gemmae, 274–5, 300
Gene, development, 176–82
  flow, 216–18
  homeotic, 183–4
  re-arrangement, 9
  reduplication, 35
Genetic drift, 218
  engineering, modification,
    19, 186, 191, 194–6, 253,
    265–6
  map, 8–9
Genomics, 8
Germination, 131–2, 135–41,
    208, 212
Gibberellic acid (gibberellin,
    GA), 103, 139–40, 172–3
Gibberellic acid response
    complex/element
    (GARC/E), 140, 178
Ginger, 252

Ginkgo, *Ginkgo biloba*,
    Ginkgoopsida, 3, 263, 284,
    305, **314–15**
Glaciation, 205, 207, 329
Global warming, 219–22
Glucose, 93, 102, 108
β-glucuronidase (GUS), 150–1
Glutamine synthase/glutamate
    synthase, 77, 98
Glutathione, 77
Glyceraldehyde 3-phosphate,
    85, 94
Glycerolipid, 100
Glycine, 229
Glycolysis, 92, 94
Glycophyte, 161–2
Glycoprotein, in
    incompatibility, 119
Glycosylation, 31
Gnetopsida, 3, 17, 284, 304,
    314–16, 318, 323
Gnetum, 3, 314–16
Golgi apparatus, 21–2, 26, 29,
    31, 47, 102
Gondwana, 326–7
Goosefoot family, 224, 226
Gossypium, 257–8
Graft, 252, 264
Gramineae, *see* grass
Grana, 28, 81
Grape, 252
Grapefruit, 252
Grass, 110, 118–22, 126–7, 130,
    132, 139–40, 200, 233, 250,
    321, **323–4**
Grasshopper, 231
Grassland, 200–1, 207, 210, 233
Gravireceptor, 149–50
Gravitropism, 149–50
Ground tissue, 41–3, 50
Groundnut, 243
Growth and development,
    163–6
Growth substances, *see*
    hormone
GS-GOGAT, *see* glutamine
    synthase
Guanacaste (*Enterolobium*), 136
Guard cell, 57–8, 66–7
Gum, 308
Gymnosperm, 304, 320, 326 (*see
    also* conifer, cycad, ginkgo,
    Gnetopsida)

Habituation, 190
Halophyte, 20, 78, 161–2
Hartig net, 224–5
Hatch/Slack pathway, 89
Haumaniastrum katangense, 266
Haustorium, 242
Hay fever, 110, 122
Hazel, 320
Heartwood, 53, 55, 255
Heather, 243, 320
Heather family group, 224
Heathland, 200, 224
Hedera helix, 173
Hedyosmum, 319
Helianthus, 267
Heliconius butterfly, 232
Hemp, 257
  manila, 257
Henna, 262
Hepatophyta, *see* liverwort
Herbaceous stem, 50–2
Herbicide, herbicide tolerance,
    192, **195–6**
Herbivore, 139, 141, 158–9, 206,
    208, 213, **231–4**
Herbs, culinary, 252
Hermaphrodite, 105, 112–13,
    116, 121, 124, 318–21
Heterochrony, 183–4
Heterokont, 240, 251
Heterospory(ous), 287, 289,
    294, 299–304, 318
Heterotrophic plant, 200–1
Heterozygous, 217
Hevea brasiliensis, 103, 261
Histone, 34
Homeotic genes, 184
Homospory, 287, 289, 296, 303
Homozygous, 217
Honey, 108
Hordeum vulgare, 140
Hormone, 115, 167–75
  action, 176–80
Hornwort, 3–4, 273–81 (*see also*
    bryophyte)
Horsetail, 3–4, 223, 244, **292–4**,
    295, 304
Horticulture, 17, 263–4
Hurricanes, in plant
    distribution, 205
Hyacinth, water, 267
Hyaline cells, 275
Hybrid vigor, 124, 217

Hybridization, 17, 217, 250, 252, 263, 328–9
Hydroid cells, 275–6, 281
Hydrophyte, 65–6
Hydrostatic pressure, 45, 60
Hydrotropism, 149
Hymenophyllum, 299
Hyperaccumulator, 265
Hypersensitive response, 237
Hypha, 223–7, 235
Hypocotyl, 127
Hypodermis, 58
Hypogeal germination, 127–8, 139
Hyponasty, 153

Iberis intermedia, 266
Imbibition, in germination, 139
Inbreeding depression, 119, 124
Indian pipe, *see Monotropa*
Indole-3-acetic acid (IAA), 168, 170 (*see also* auxin)
Indole-3-butyric acid (IBA), 168
Induction, floral, 183
Indusium, 299
Infection thread, 229
Inflorescence, 129–30, 183–4
Inositol triphosphate (IP$_3$), 182
Insect color sensitivity, 106
Insertional mutagenesis, 8, 12
Integument, 108, 125–6, 130, 304–5, 316
Intercalary meristem, 42
Intercellular space, 43
Interception, 73
Interfascicular cambium, 51–2
Interphase, 36
Intine, 109
Intracellular messenger, 176, 180–2
Invertase, 96
Ion channel, 68, 71
Iron, 204,
Isatis tinctoria, 262
Iso-electric focussing, 11
Isoetes, Isoetales, 288–90 (*see also* lycopsids)
Isolating mechanisms, 327–8
Isoprene, 103, 173
Isozyme, 17, 19, 216
Iteroparous, 210–11
Ivy, 173

Jay, as seed disperser, 133
Juniper, 129, 308
Jute, 257

Kapok, 258
Karyogamy, 236
Kelp, 271
Kinetin, *see* cytokinin
Krantz anatomy, 58, **88–9**
Krebs Cycle, 94

Laburno-cytisus, 264
Laburnum, 264
Lamiaceae, 252, 320
Lamina, 56
Laminar flow cabinet, 188
Laminaria, 271
Lamium, 320
Larch (*Larix*), 309
Larkspur, 123
Larrea tridentata, 210
Lateral meristem, 42
Lateral roots, 47–8
Latex, 103, 261–2
Laurasia, 326
Laverbread, 270
Lawsonia inermis, 262
Leaves, 16, 56–58 (*see also* megaphyll, microphyll)
Lectin, 229
Leghemoglobin, 230, 250
Legume, 129, 133, 226, 228–30, 232–3
Lentil, *Lens culinaris*, 229, 243
Lepidodendron, 290–1
Leptosporangium, 297–9, 301
Leucoplast, 27
LHY, 147
Liana, 201
Lichen, 201, 276–7
Lignin, lignification, 24, 26, 43, 53, 103, 125, 232, 255
Lilac, mountain (*Ceanothus*), 229
Lime (tree), 256
Lime, liming, 78
Linseed (*Linum*, flax), 126, 257
Lipid, 29, 98
bilayer, 28–9
Liverwort, 2, 18, 273–81 (*see also* bryophyte)
Lolium perenne, 213
Long-day plant, 143

Longevity, seed, 135, 137
Loranthaceae, *see* mistletoe
Lotus, 229
Lotus, sacred, 135
Lycophytina, 3, 285
Lycopodium, 289–90
Lycopsid (Lycopsida, clubmoss, quillwort), 3–4, 244, 284–7, **288–90**, 294–5, 300, 304–5
Lyginopteridopsida, *see* pteridosperm
Lyonophyton, 287
Lysergic acid amide (LSD) and ergot disease, 238

Macrocystis, 271
Magnesium, 78
Magnolia, 319–20, 324
Magnoliopsida, 3
Maidenhair tree, *see* ginkgo
Maize, 47, 66, 91, 185–6, 221, 243, 250, 253, 266
Mallee (*Eucalyptus*), 136
Mallow family (Malvaceae), 133, 257
Mammals, as herbivores, 231–4
Manganese, 83
Mango, 252
Mangrove, 135, 205
Maple, *see Acer*
Marattiales, 296, 301
Marchantia, 279
Marchantiophyta, *see* liverwort
Marigold, 106
Marsileaceae, 299
Mass flow, 60, 73
Mast-fruiting, 139
Matrix space, 28
Media, for cell and tissue culture, 188
Medicago, 229
Medicinal plants, 259–60
Mediterranean climate, 206, 210, 327
Megaphyll, 288–9, 296
Meiosis, 3, 35, **39**, 109–10, 304, 328
Melon, water, 122, 129
Membranes, 29–32, 68–71
Mentha palustris, 103
Meristem, 37, **41–2**, 45, 47, 50, 201, 210, 233

Mesophyll, 57–8, 66, 88–90
Mesophyte, 65–6
Metals, heavy, 205
Methane, CH_4, 220
Microbodies, 22
Microcycas, 313
Microfibril, 24–5
Microphyll, 288–9, 292, 309
Micropyle, 108, 111, 311, 313–15
Microtubules, 21
Middle lamella, 24
Midrib, 56
Milkweed family, 110, 232
Millet, 243, 251
Mimosa pudica, 155
Mint, 252
Mistletoe, 201, 242
Mitochondrion (-a), 18, 21, 27, 87, 100
Mitosis, 3, 38
Mnium, 276
Model organisms, 5–7
Models, in plant ecology, 19
MoFe protein, 230
Molecular clock, 19
Molecular farming, 196
Molecular techniques, 9–15
Mollusc, 141, 231–2
Monarch butterfly, 232
Monkey-puzzle, 305, 308
Monocarpic, 210–11
Monocotyledon (monocot), 3, 48, 53, 106, 127, 200, 284, 320, 322–4
Monoecious (-y), 106, 112–16, 279, 315, 320
Monohybrid cross, 9
Monosaccharide, 101
Monotropa, Monotropaceae (birdsnest, Indian pipe), 224, 243
Monstera, 323
Moonwort, 296
Morphine, 103, 233, 260
Moss, 2–4, 16, 273–81, 287, 299 (*see also* bryophyte)
Moth, cactus, 234
    yucca, 322
Moths, as herbivores, 231–2
    as pollinators, 122, 320–2
Mouse-ear, 114
Mucilage, 47
Musa textiles, 257

Mutagenesis, 6, 12, 186
Mutants, 9, 12–13
Mutation, 18
MYBs, 140
Mycelium, 235
Mycoplasma, 239–40
Mycorrhiza, 48, 73, 132, 139, 141, **223–7**, 243–4, 257, 277, 290, 300, 308, 324
Myosotis, 110
Myrica gale, 229
Myrrh, 262

NADP-reductase, 82–3
Naphthalene acetic acid (NAA), 169
1 -N-naphthylphthalamic acid (NPA), 169, 171
Narcissus triandrus, 118
Nastic responses, 153–5
Natural selection, 216–18
Necrosis, 78
Nectar, 107, **108**, 122, 252
Needles, of conifers, 57–8, 308
Neighborhood, genetic, 216
Nelumbo (sacred lotus), 135
Nepenthaceae (*Nepenthes*), 245–6
Nereocystis, 271
Nettle, 257
Nicotiana tabacum, *see* tobacco
Nicotine, 233, 260, 262
Nif genes, 230
Nightshade, 107
    deadly, 103
Nitrate, 98
    transporter, 70
Nitrogen, 204, 225–6
    assimilation, 76–7
    fixation, 204, 226, **228–30**, 275, 313
    oxides, 161, 220
Nod factor, *nod* genes, 229–30
Nodulation, 229–30
Non-cyclic electron flow, 83
NPH-1, 150
Nucellus, 108, 125–6, 304, 311
Nucleolus, 33
Nucleus, nuclear envelope, 21–2, 29, **32–5**
Nutmeg, 252
Nutrient, 72–4
    functions, 75–8

transport, uptake, 68–74
Nyctinasty, 153–5
Nylon, 258

Oak, 256, 262, 320
    cork, 256–7
    European (*Quercus robur*), 232
Oat, 165
Octant stage, 165
Oil, 219–20, 326
    essential, storage, 100–1
    as floral food reward, 108, 324
Oligosaccharide, 175
Olive, 250, 263
Onion, 127
Ophioglossum, *see* adderstongue
Opiate, 259, 262
Orchid, 107, 110, 122, 125, 132, 201, 225–6, 243–4, 259, 263, 301, 321, 323, **324**, 328–9
    bee, 321
    birdsnest, 244
    coralroot, 244
Organ, **41–4**
Organochlorines, 253
Orobanche, *see* broomrape
Oryza, *see* rice
Osmosis, osmotic potential, 60–1, 162
Ovary, 107–8, 117–20, 129–30, 304, 321, 323
Ovule, 107–8, 111, 116–20, 123, 125, 129, 304, 306, 310–11, 318, 323
    of cycad, 314
    of ginkgo, 314
    of Gnetopsida, 315–16
Oxaloacetate, 94
Oxidative phosphorylation, 95
Ozone, 158, 161

Palisade mesophyll, 57, 161
Palm, 53, 199, 253, 313, 322
    date, 257
    oil, 262
    royal, 210
Pangea, 326
Papaver somniferum, 259
Papaveraceae, *see* poppy
Paper, 256, 308

Papyrus, 256
Parasitic plants, 132, 201, **241–4**
Parenchyma, 42, 130
Parsimony, 18
Parsley, 252
Parsnip, 233
Parthenocarpy, 170
Passion-flower (Passifloraceae),
    232
Patch clamping, 70
Pathogens, 75, 235–8
Pea, 129, 228–9, 252 (*see also*
    legume)
Pear, 130, 252
Peat, 204–5, 219, 222, 276–7,
    326
Pectin, pectic polysaccharide,
    24, 109, 119
Pepper, black, 252
Perforation plate, 63
Perianth, 319
Peribacteroid membrane, 230
Peristome, 281
Periwinkle, 103, 259
Peroxisome, 22, 87
Petal, **105–7**, 129–30, 263, 324
    origin, 319–20
Petiole, 56
Phaeophyta, *see* algae, brown
Phegopteris, 298
Phenolics, 103, 125, 233, 237,
    309
Phenological escape, 158, 160
Phenoxyacetic acid, 168
Phenylethylamine, 262
Phloem, 41–4, 52, 276, 308, 315
    secondary, 53–4
    transport, 72, 74, 92–5, 97
Phosphoenolpyruvate
    carboxylase (PEPCASE),
    89
Phospholipase C, 182
Phosphorescence, 269
Phosphorus, 77–8, 225–6, 253
Photolysis, 82
Photomorphogenesis, 143–6
Photoperiodism, 146–8
Photophosphorylation, 83–4
Photoreceptor, 143–5
Photorespiration, 82, 86, 221
Photosynthesis, 79–91, 140–1,
    220–1, 225–6
    C3, 88

C4, 58, **87–91**, 221, 323, 326
Photosystems I and II, 81–2
Phototropism, 146, 149–50
Phragmites, 267
Phragmosome, 36
Phyllotaxy, 52
Phylogeny, plant, 18–19, 216
Physcomitrella patens, 5
Phytoalexin, 237
Phytochrome, 138, 144–6, 155,
    184
    genes for, 18
Phytoecdysone, 232–3
Phytogeographic regions,
    326–7
Phytohormone, *see* hormone
Phytomere, 164
Phytophthora infestans, 240, 251
Pig, 234
Pigment, accessory, 80, 106
Pillwort, *Pilularia*, 299
Pine, Pinaceae, 58, 224, 233,
    256, 308–11
    bristlecone, 210, 308
    huon, 212, 308
    scots, 212
    white, 310
Pineapple, 129, 257
Pinguicula, 246
Pink family, 320
Pinopsida, *see* conifer
Pinus, see pine
Pioneer plants, 135, 137, 200,
    207, 301
Pistillata, 184
Pisum, 229
Pitcher-plant, 245–6
Pith, 50–1, 308
Pits, in xylem, 43, 63, 308, 314
Plane, 319–20
Plankton, 2, 269–70
Plant breeding, 17, 185–6, 253
Plants as symbols, 263
Plasma membrane, 21, 29, **32**
    proton pump, 29, 32, 67,
    68–70
Plasmodesmata, 21, 23–34, 26,
    96, 166
Plasmolysis, 60
Plastid, 27, 140, 173
Plastocyanin, 83
Plastoquinone, 83
Ploidy, 35 (*see also* polyploid)

Plum blossom (in China), 263
Poaceae, *see* grass
Podocarpaceae, 308–9
Polar transport, 170–1
Pollen, 107–8, **109–10**, 116–20,
    121–4, 304, 310, 314–15
    evolution, 323
    flow, 19, 216
    tube, 108–9, 111, 116–20, 123
Pollination, 106, 108, 112–15,
    121–4, 231, 310, 319–22,
    324
Pollinia, 107, 110, 122, 324
Pollution, 277
Polyamine, 175
Polycarpic, 210–11
Polygalacturonase, 157, 180
Polymerase chain reaction
    (PCR), 10, 12, 18
Polymorphism, 216
Polyphenol, 271
Polyploid(y), 17, 113–15, 126,
    217, 250, 277, **328**
Polypodiopsida, *see* fern
Polypody fern, *Polypodium
    vulgare*, 233
Polysaccharide, 23–4, 98
Polytrichum, 275–7
Poplar, 267
Poppy, family, 120
    opium, 259
Population dynamics, 212
Porphyra, 270
Positional cloning, 9
Potassium, 67, 286
Potato, 240, 250–1, 253
Prayer plant, 154
Pre-prophase band, 36
Pressure flow model, 95, 97
Pressure potential, 60
Primary pump, 68–71
Primary root, 48
Primates, as herbivores, 234
Primordium, 50
Primrose family, Primulaceae,
    118–19
Primula veris, see cowslip
Principal components analysis,
    19
Procambium, 42
Progymnosperm (-opsida), 3,
    284, 303–5
Prokaryotes, 1(*see also* bacteria)

Propagation, 187
Propagule (synthetic seed), 190
Protandry (-ous), 114
Protein kinase/C, 182
Proteome, proteomics, 8, 12
Prothallus, 290, 299–300
    of conifers, 310
Protists (Protista Protoctista),
    see algae
Protoderm, 42
Protogyny (-ous), 114
Proton ATPase, *see* plasma
    membrane proton pump
Protonema, 275
Protoplast, 188
Protoxylem, 43
Psilophyton, 285–6, 301, 304
Psilotum, Psilotales, 17–18, 297
Pteridium aquilinum, see
    bracken
Pteridosperm (-opsida), 284,
    304–5, 306, 312
Pteropsida, *see* fern
Puccinia graminis, 235–6
Pulse, 250
Pulvinus, 154
Putrescine, 175
Pyrenoid, 275
Pyrethroid, 232–3
Pyrophosphatase, 70
Pyruvate, 92, 94, 100

Quadrat, 19
Quantitative trait loci, 10
Quercus, see oak
Quiescent center, 164
Quillwort, *see* lycopsid
Quinine, 260

Radicle, 126
Rafflesia (Rafflesiaceae), 242–3
Random amplified
    polymorphic DNA
    (RAPD), 18
Rape, oil-seed (Canola), 126,
    185–6, 250–1
Ray, vascular, 55
Ray cell, 44
Reaction center, 80–2
Receptacle, 105, 108, 129–30
Receptors, 30, 177
Redwood, coast, 309
Reed, 267

Reporter gene construct, 150–1
Resin, 58, 108, 255, 262, 296,
    308–9
Resonance energy transfer
    (RET), 80–1
Respiration, 92–7
Restriction enzyme, 10, 18, 193
Reverse transcriptase, 10
RFLP, 18
Rhizobium, 228–9 (*see also*
    nitrogen fixation)
Rhizodermis, 43
Rhizofiltration, 266
Rhizoid, 223, 275, 285
Rhizome, 21, 114, 199, 283, 285,
    289, 292, 296–9
Rhizophore, 289
Rhizosphere, 47
Rhodophyta, *see* algae, red
Rhynia, 285
Rhyniopsida, 3, 283–5
Ribonuclease (RNase), 119–20
Ribosome, 21, 27, **30**, 119
Ribulose 5 -phosphate kinase,
    85
Ribulose bisphosphate/
    carboxylase/oxygenase,
    82–8
Riccia, 280
Rice, 5, 7, 66, 185, 228, 250, 253
RNA, 34–5, 119, 127
    polymerase, 35
Root, **45–9**, 114, 126, 150
    cap, 47, 126–7
    hair, **45–6**, 47, 63
    pressure, 61
    stock, 264
Rose, 263
Rose family (Rosaceae), 114
Rosette plant, 52, 172
Roystonea, 210
Rubber, 233, 261–2
Rubiaceae, 232
Rubus, see bramble
Rust fungus, 235–6
Rye, 253
Rye-grass, 213

S alleles, 117–20
S-adenosyl methionine (SAM),
    172
Sage, 252
Sago, 313

Salicylic acid, 159, 237
Salinity, 158, 161–2
Salix, 200, 227, 267
Salt gland, 162
Saltmarsh, 205
Salvinia, 299–300
Samara, 133
Sand dunes, 207, 210
Sandalwood, 242
Saponin, 236
Saprophyte, 224, 226, **243–4**
Sapwood, 53, 55
Sarcandra, 319
Sargassum, Sargasso sea, 271
Sarraceniaceae, *Sarracenia*,
    245–6
SAUR, 151, 180
Savannah, 200, 205–6, 222
Saxifraga cernua, saxifrage,
    115
Scales, bulb, 58
Scale-leaves, of conifers, 308–9
    of cycads, 313
Scent, of flowers, 324
Scion, 264
Sclereids, sclerenchyma, 41, 43,
    130, 257
Scr mutant, 150
Scrophulariaceae, 243
Scutellum, 127
Seaweed, 267, 270–2
Secondary compounds
    (products), 29, 98, 102, 191,
    232–4, 252, 260–2
Secondary coupled transporter,
    68–9, **71**
Secondary thickening , 53–5
Sedge, 224
Seed, 121, **125–8**
    bank, 135, **137**
    coat, 136
    dispersal (flow), 19, 131–4,
        137, 216–17, 222, 231
    dormancy, **135–7**, 210–11
    germination, 172
    origin of, 304
    set, 122
    storage reserves, 139
Seed-fern, *see* pteridosperm
Seismonasty, 153, **155**
Selaginella, 288–90, 300
Selection breeding, 186
Self-compatibility, 113, 328

Self-fertilization, 17, 112–14, 123–4, 210, 216–17, 250
Self-incompatibility, 110, 112–14, **116–20**, 123, 216–17, 323, 327–8
Self-thinning, 213–14
Semelparous, 210–11
Senescence, 170, 173
Sensitive plant, 155
Sepal, **105–6**, 129–30, 133, 319, 324
Sequoia, 309
She-oak, 227
Shikimic acid pathway, 103
Shikonin, 191
Shoot anatomy, 41–3
Short-day plant, 143–4
Sideroxylon, 136
Sieve element, of phloem, 41, 44, 308
Silene dioica (campion), 113
Silica, 232, 234
Sink tissue, 92, 95
Sisal, 257
Sleep movement, 148
Sloth, 234
Smut fungi, 235–6
Soft rot, 239
Solute potential, 60
Somatic embryo, 190
Somatic mutation, 252–3
Sorghum, 243, 251
Sorus(i), 298–9
Source tissue, 92, 95
Southern blotting, 18
Soy(a) bean (*Glycine*), 229, 252
Speciation in flowering plants, 328–9
Species definition in flowering plants, 17
Sperm, *see* each plant group
Spermatophytina, 3
Spermidine, spermine, 175
Sphacelia typhina, 238
Sphagnum, 204, 275–7, 281
Sphenopsida, *see* Equisetopsida
Spice, 252
Spiders, 246
Spinach, 172
Spirogyra, 272
Splachnaceae, 281
Spongy mesophyll, 57, 66

Sporangium, *see* eusporangium, leptosporangium and each plant group
Spore, 3, 274, 280–1, 287, 290–1, 293–4, 296–9, 304–5
Sporophyte, definition, 3–4 (*see also* each plant group)
Sporopollenin, 109
Spruce, 256
Squamosa, 184
S-state mechanism, in photosynthesis, 83
St Johnswort, 234
Stamen, **105–7**, 109, 112–14, 263, 306, 318–21
Starch, 92–4, 98
  sheath, 150
Statistics, 19
Statocyte, statolith, 150–2
Stele, 45–6
Steppe, 203–4
Stigma, 107–8, 109–11, 117–20, 123, 321
Stinging hairs, 234
Stolon, 199, 210
Stomata, 57–8, 61–2, **65–7**, 89
Stomatal aperture, 148, 174
Stonewort, 272
Strangler, 200–1, 322
Streptanthus polygaloides, 266
Streptomycete, 239
Stress response, 158–62, 174
Striga hermonthica (witchweed), strigol, 242
Strobilus(i), 289–90, 292–3, 309 (*see also* cone)
Stroma, 27, 81
Structural genetics, 8
Strychnine, 233
Style, 107–8, 110–11, 117, 123, 129
Suaeda maritime, 162
Suberin, 47, 53–4, 66
Subspecies, 17, 215
Succession, 141, 207
Succulence, 65–6
Sucker, 264
Sucrose, 92–6, 98, 101
Sugar beet, 162
Sugar cane, 243, 252
Sulfur, 77
  dioxide, 161, 277
Sundew, 246

Sunflower, 243, 250, 267
Suspension culture, 188–91
Suspensor, of embryo, 126–7
Swede, 251
Sweet gale (bog myrtle, *Myrica*), 229
Swiss-cheese plant, 323
Symplast, 21, 23, 63
Symplastic transport, 73–4
Symport, 71
Synaptonemal complex, 39
Synergid cell, 110
Synteny, 5
Systemic acquired resistance, 237
Systems biology, 15

Tambalacoque (*Sideroxylon*), 136
Tannin, 103, 159, 232–3, 255, 262, 296
Tapetum, of anther, 107, 109
Taraxacum, 114
TATA box, 53
Taxaceae, Taxales, *Taxus* (yew), 125, 129, 259–60 308, 311
Taxodium, 309
Taxol, 259–60
T-DNA, 8, 13, 193–4
Tea, 262
Teliospore, 236
Tendrils, 58
Teosinte, 185
Termite, 256
Terpene, 159, 233, 260, 309
Terpenoids, 103
Testa, 125, 130
Theobroma cacao, 262
Thermal cycler, in gene cloning, 10
Thermonasty, 153
Thigmonasty, 153
Thigmotropism, 149
Thlaspi, 266
Thuja, 309
Thylakoid, 28, 81, 83
Thyme family, 113, 252
Ti plasmid, 173, **192–5**
Tilia cordata, 256
Tissue, 41–4
Tissue culture, 166, 186, **187–91**, 260
Tmesipteris, 297

Tobacco, 243, 262
Tomato, 130, 179, 221, 233, 243, 252
Tonoplast, 29, **32**, 182
Topiary, 264
Totipotency, 166
Toxic gases, 158, 161
Toxic ions, 78, 158–60
Tracheid, 43, 63, 308, 314, 323
Tracheophyta, 3
Transamination, 100
Transcellular transport, 63
Transcriptome, 12
Transfer cell, 74
Transmembrane transport, 72
Transpiration, 59, 61
Transposon tagging, 7, 13–14
Triacylglycerol, 100–1
Trichome, 57
Trifolium, see clover
Trimerophytopsida, 3, 286
Triple response, 171
Triterpenes, 236
Triticum see wheat
Tropism, 149–52
Tryptophan, 262
Tsunami, 205
Tubulin, 21
Tundra, 205, 222
Tunica, 183
Turgor, 24, 60, 67, 165
  movements, 153
Turmeric, 252
Turnip, 251
Turpentine, 308

Ulmus, 214
Ultraviolet (UV), 106, 146
Umbellifer (Apiaceae carrot family), 113, 233, 252

Unisexual flower, 106, 112, 114, 318–21
Uredospore, 236
Uridine Diphosphoglucose (UDP Glucose), 25
Urtica, Urticaceae, 234, 257
Utricularia, 246–7

Vacuole 2, 21, 89–90, 106, 160 (*see also* tonoplast)
Vanilla, 252
Varnish, 308
Vavilov Centers, 251
Vectors, of disease, 240
  in plant genetic engineering, 193–4
Vein, 56
Venus fly-trap, 246–7
Vernalization, 136, 172
Vessel, of xylem, 43, 63, 296, 315, 322–3
Vicia faba, 154
Vinblastine, vincristine, 103, 259–60
Vir (Virulence) Region, 193–4
Virus, 240
Viscaceae, *see* mistletoe
Viscose, 256
Vision insect, 106–7
Volvox, 271–2

Wasps, as pollinators, 319
  fig, 322
Water hyacinth *see Eichhornia crassipes*
Water lily, 319, 322–4
Waterlogging, 158, 161
Water-relations in plants, 59–67, 276, 286–7
Waterwheel plant, 246–7
Wax, on plant surfaces, 66

Weevil, as
Welwitschia,
Wheat, 126,
  rust, 235
Whisk-fern, s
Willow, 115,
Wilt, bacterial,
Winteraceae, 3
Witches broom
Witchweed, *see*
Woad, 262
Woodland, 200,
Wrack, 271

Xanthophyll, *see* p
  accessory
Xerophyte, 64, **67–**
Xylem, 41–4, 52, 73,
  276, 296, 315,
  of conifers, 308
  parenchyma, 72
  secondary, 53–4
  water flow, 61–2

Yam, 250–1
Yeast, 252
Yew, *see* Taxales
Yucca, 322

Zea mays, see maize
Zea perennis, 253
Zeatin, 174
Zeaxanthin, 146, 175
Zostera, 110, 204
Zosterophyllum,
  Zosterophyllopsida, 3, 28
  287–8
Z-scheme, 83
Zygomorphy(ic), 320
Zygomycetes, 224
Zygotic embryo, 190

Lightning Source UK Ltd.
Milton Keynes UK
UKOW07f0514251016
286077UK00008B/220/P

Propagation, 187
Propagule (synthetic seed), 190
Protandry (-ous), 114
Protein kinase/C, 182
Proteome, proteomics, 8, 12
Prothallus, 290, 299–300
    of conifers, 310
Protists (Protista Protoctista),
    see algae
Protoderm, 42
Protogyny (-ous), 114
Proton ATPase, see plasma
    membrane proton pump
Protonema, 275
Protoplast, 188
Protoxylem, 43
Psilophyton, 285–6, 301, 304
Psilotum, Psilotales, 17–18, 297
Pteridium aquilinum, see
    bracken
Pteridosperm (-opsida), 284,
    304–5, 306, 312
Pteropsida, see fern
Puccinia graminis, 235–6
Pulse, 250
Pulvinus, 154
Putrescine, 175
Pyrenoid, 275
Pyrethroid, 232–3
Pyrophosphatase, 70
Pyruvate, 92, 94, 100

Quadrat, 19
Quantitative trait loci, 10
Quercus, see oak
Quiescent center, 164
Quillwort, see lycopsid
Quinine, 260

Radicle, 126
Rafflesia (Rafflesiaceae), 242–3
Random amplified
    polymorphic DNA
    (RAPD), 18
Rape, oil-seed (Canola), 126,
    185–6, 250–1
Ray, vascular, 55
Ray cell, 44
Reaction center, 80–2
Receptacle, 105, 108, 129–30
Receptors, 30, 177
Redwood, coast, 309
Reed, 267

Reporter gene construct, 150–1
Resin, 58, 108, 255, 262, 296,
    308–9
Resonance energy transfer
    (RET), 80–1
Respiration, 92–7
Restriction enzyme, 10, 18, 193
Reverse transcriptase, 10
RFLP, 18
Rhizobium, 228–9 (see also
    nitrogen fixation)
Rhizodermis, 43
Rhizofiltration, 266
Rhizoid, 223, 275, 285
Rhizome, 21, 114, 199, 283, 285,
    289, 292, 296–9
Rhizophore, 289
Rhizosphere, 47
Rhodophyta, see algae, red
Rhynia, 285
Rhyniopsida, 3, 283–5
Ribonuclease (RNase), 119–20
Ribosome, 21, 27, 30, 119
Ribulose 5 -phosphate kinase,
    85
Ribulose bisphosphate/
    carboxylase/oxygenase,
    82–8
Riccia, 280
Rice, 5, 7, 66, 185, 228, 250, 253
RNA, 34–5, 119, 127
    polymerase, 35
Root, 45–9, 114, 126, 150
    cap, 47, 126–7
    hair, 45–6, 47, 63
    pressure, 61
    stock, 264
Rose, 263
Rose family (Rosaceae), 114
Rosette plant, 52, 172
Roystonea, 210
Rubber, 233, 261–2
Rubiaceae, 232
Rubus, see bramble
Rust fungus, 235–6
Ryc, 253
Rye-grass, 213

S alleles, 117–20
S-adenosyl methionine (SAM),
    172
Sage, 252
Sago, 313

Salicylic acid, 159, 237
Salinity, 158, 161–2
Salix, 200, 227, 267
Salt gland, 162
Saltmarsh, 205
Salvinia, 299–300
Samara, 133
Sand dunes, 207, 210
Sandalwood, 242
Saponin, 236
Saprophyte, 224, 226, 243–4
Sapwood, 53, 55
Sarcandra, 319
Sargassum, Sargasso sea, 271
Sarraceniaceae, Sarracenia,
    245–6
SAUR, 151, 180
Savannah, 200, 205–6, 222
Saxifraga cernua, saxifrage,
    115
Scales, bulb, 58
Scale-leaves, of conifers, 308–9
    of cycads, 313
Scent, of flowers, 324
Scion, 264
Sclereids, sclerenchyma, 41, 43,
    130, 257
Scr mutant, 150
Scrophulariaceae, 243
Scutellum, 127
Seaweed, 267, 270–2
Secondary compounds
    (products), 29, 98, 102, 191,
    232–4, 252, 260–2
Secondary coupled transporter,
    68–9, 71
Secondary thickening , 53–5
Sedge, 224
Seed, 121, 125–8
    bank, 135, 137
    coat, 136
    dispersal (flow), 19, 131–4,
        137, 216–17, 222, 231
    dormancy, 135–7, 210–11
    germination, 172
    origin of, 304
    set, 122
    storage reserves, 139
Seed-fern, see pteridosperm
Seismonasty, 153, 155
Selaginella, 288–90, 300
Selection breeding, 186
Self-compatibility, 113, 328

Self-fertilization, 17, 112–14, 123–4, 210, 216–17, 250
Self-incompatibility, 110, 112–14, **116–20**, 123, 216–17, 323, 327–8
Self-thinning, 213–14
Semelparous, 210–11
Senescence, 170, 173
Sensitive plant, 155
Sepal, **105–6**, 129–30, 133, 319, 324
Sequoia, 309
She-oak, 227
Shikimic acid pathway, 103
Shikonin, 191
Shoot anatomy, 41–3
Short-day plant, 143–4
Sideroxylon, 136
Sieve element, of phloem, 41, 44, 308
Silene dioica (campion), 113
Silica, 232, 234
Sink tissue, 92, 95
Sisal, 257
Sleep movement, 148
Sloth, 234
Smut fungi, 235–6
Soft rot, 239
Solute potential, 60
Somatic embryo, 190
Somatic mutation, 252–3
Sorghum, 243, 251
Sorus(i), 298–9
Source tissue, 92, 95
Southern blotting, 18
Soy(a) bean (*Glycine*), 229, 252
Speciation in flowering plants, 328–9
Species definition in flowering plants, 17
Sperm, *see* each plant group
Spermatophytina, 3
Spermidine, spermine, 175
Sphacelia typhina, 238
Sphagnum, 204, 275–7, 281
Sphenopsida, *see* Equisetopsida
Spice, 252
Spiders, 246
Spinach, 172
Spirogyra, 272
Splachnaceae, 281
Spongy mesophyll, 57, 66

Sporangium, *see* eusporangium, leptosporangium and each plant group
Spore, 3, 274, 280–1, 287, 290–1, 293–4, 296–9, 304–5
Sporophyte, definition, 3–4 (*see also* each plant group)
Sporopollenin, 109
Spruce, 256
Squamosa, 184
S-state mechanism, in photosynthesis, 83
St Johnswort, 234
Stamen, **105–7**, 109, 112–14, 263, 306, 318–21
Starch, 92–4, 98
sheath, 150
Statistics, 19
Statocyte, statolith, 150–2
Stele, 45–6
Steppe, 203–4
Stigma, 107–8, 109–11, 117–20, 123, 321
Stinging hairs, 234
Stolon, 199, 210
Stomata, 57–8, 61–2, **65–7**, 89
Stomatal aperture, 148, 174
Stonewort, 272
Strangler, 200–1, 322
Streptanthus polygaloides, 266
Streptomycete, 239
Stress response, 158–62, 174
Striga hermonthica (witchweed), strigol, 242
Strobilus(i), 289–90, 292–3, 309 (*see also* cone)
Stroma, 27, 81
Structural genetics, 8
Strychnine, 233
Style, 107–8, 110–11, 117, 123, 129
Suaeda maritime, 162
Suberin, 47, 53–4, 66
Subspecies, 17, 215
Succession, 141, 207
Succulence, 65–6
Sucker, 264
Sucrose, 92–6, 98, 101
Sugar beet, 162
Sugar cane, 243, 252
Sulfur, 77
dioxide, 161, 277
Sundew, 246

Sunflower, 243, 250, 267
Suspension culture, 188–91
Suspensor, of embryo, 126–7
Swede, 251
Sweet gale (bog myrtle, *Myrica*), 229
Swiss-cheese plant, 323
Symplast, 21, 23, 63
Symplastic transport, 73–4
Symport, 71
Synaptonemal complex, 39
Synergid cell, 110
Synteny, 5
Systemic acquired resistance, 237
Systems biology, 15

Tambalacoque (*Sideroxylon*), 136
Tannin, 103, 159, 232–3, 255, 262, 296
Tapetum, of anther, 107, 109
Taraxacum, 114
TATA box, 53
Taxaceae, Taxales, *Taxus* (yew), 125, 129, 259–60 308, 311
Taxodium, 309
Taxol, 259–60
T-DNA, 8, 13, 193–4
Tea, 262
Teliospore, 236
Tendrils, 58
Teosinte, 185
Termite, 256
Terpene, 159, 233, 260, 309
Terpenoids, 103
Testa, 125, 130
Theobroma cacao, 262
Thermal cycler, in gene cloning, 10
Thermonasty, 153
Thigmonasty, 153
Thigmotropism, 149
Thlaspi, 266
Thuja, 309
Thylakoid, 28, 81, 83
Thyme family, 113, 252
Ti plasmid, 173, **192–5**
Tilia cordata, 256
Tissue, 41–4
Tissue culture, 166, 186, **187–91**, 260
Tmesipteris, 297

Tobacco, 243, 262
Tomato, 130, 179, 221, 233, 243, 252
Tonoplast, 29, **32**, 182
Topiary, 264
Totipotency, 166
Toxic gases, 158, 161
Toxic ions, 78, 158–60
Tracheid, 43, 63, 308, 314, 323
Tracheophyta, 3
Transamination, 100
Transcellular transport, 63
Transcriptome, 12
Transfer cell, 74
Transmembrane transport, 72
Transpiration, 59, 61
Transposon tagging, 7, 13–14
Triacylglycerol, 100–1
Trichome, 57
Trifolium, see clover
Trimerophytopsida, 3, 286
Triple response, 171
Triterpenes, 236
Triticum see wheat
Tropism, 149–52
Tryptophan, 262
Tsunami, 205
Tubulin, 21
Tundra, 205, 222
Tunica, 183
Turgor, 24, 60, 67, 165
  movements, 153
Turmeric, 252
Turnip, 251
Turpentine, 308

Ulmus, 214
Ultraviolet (UV), 106, 146
Umbellifer (Apiaceae carrot family), 113, 233, 252

Unisexual flower, 106, 112, 114, 318–21
Uredospore, 236
Uridine Diphosphoglucose (UDP Glucose), 25
Urtica, Urticaceae, 234, 257
Utricularia, 246–7

Vacuole 2, 21, 89–90, 106, 160
  (*see also* tonoplast)
Vanilla, 252
Varnish, 308
Vavilov Centers, 251
Vectors, of disease, 240
  in plant genetic engineering, 193–4
Vein, 56
Venus fly-trap, 246–7
Vernalization, 136, 172
Vessel, of xylem, 43, 63, 296, 315, 322–3
Vicia faba, 154
Vinblastine, vincristine, 103, 259–60
Vir (Virulence) Region, 193–4
Virus, 240
Viscaceae, *see* mistletoe
Viscose, 256
Vision insect, 106–7
Volvox, 271–2

Wasps, as pollinators, 319
  fig, 322
Water hyacinth *see Eichhornia crassipes*
Water lily, 319, 322–4
Waterlogging, 158, 161
Water-relations in plants, 59–67, 276, 286–7
Waterwheel plant, 246–7
Wax, on plant surfaces, 66

Weevil, as pollinators, 314
Welwitschia, 3, 314–16, 323
Wheat, 126, 130, 250
  rust, 235
Whisk-fern, *see* Psilotales
Willow, 115, 133, 200, 227, 267
Wilt, bacterial, 239
Winteraceae, 322
Witches broom, 237
Witchweed, *see Striga*
Woad, 262
Woodland, 200, 210–11
Wrack, 271

Xanthophyll, *see* pigment, accessory
Xerophyte, 64, **67–8**, 78
Xylem, 41–4, 52, 73, 242, 255, 276, 296, 315
  of conifers, 308
  parenchyma, 72
  secondary, 53–4
  water flow, 61–2

Yam, 250–1
Yeast, 252
Yew, *see* Taxales
Yucca, 322

Zea mays, see maize
Zea perennis, 253
Zeatin, 174
Zeaxanthin, 146, 175
Zostera, 110, 204
Zosterophyllum, Zosterophyllopsida, 3, 285, 287–8
Z-scheme, 83
Zygomorphy(ic), 320
Zygomycetes, 224
Zygotic embryo, 190